AF324007

Advances in Ring Theory

Proceedings of the 4th China–Japan–Korea
International Conference   24–28 June 2004

J. L. Chen
Southeast University, China

N. Q. Ding
Nanjing University, China

H. Marubayashi
Naruto University of Education, Japan

editors

World Scientific

NEW JERSEY · LONDON · SINGAPORE · BEIJING · SHANGHAI · HONG KONG · TAIPEI · CHENNAI

Published by

World Scientific Publishing Co. Pte. Ltd.

5 Toh Tuck Link, Singapore 596224

USA office: 27 Warren Street, Suite 401-402, Hackensack, NJ 07601

UK office: 57 Shelton Street, Covent Garden, London WC2H 9HE

British Library Cataloguing-in-Publication Data
A catalogue record for this book is available from the British Library.

ADVANCES IN RING THEORY
Proceedings of the 4th China–Japan–Korea International Symposium

ISBN 981-256-425-X

Printed in Singapore by World Scientific Printers (S) Pte Ltd

PREFACE

This volume of the Proceedings of the Fourth China-Japan-Korea International Symposium on Ring Theory took place from June 24 to June 28, 2004 in the beautiful city of Nanjing, well known for being China's capital of six ancient dynasties and having a rich cultural heritage. The conference is held once every four years on a rotating basis. The first conference was held in 1991 in Guilin, China. In 1995 the second conference took place in Okayama, Japan. At the second conference, Korea was included and the new member hosted the conference of 1999 in Kyongju, Korea.

The purpose of this conference is to contribute to the development of ring theory and to strengthen the ties of friendship between ring theorists of China, Japan and Korea. Over 100 mathematicians from 11 different countries and regions attended the conference, including some well-known ring theorists in the world.

This volume contains survey articles delivered by invited speakers as well as research articles. These peer-refereed articles provide the latest developments and trends in ring theory, module theory, representation theory and theory of Hopf algebras. The survey articles are very useful for researchers to get the overviews on various areas and also for younger researchers looking for fields to investigate, while the research articles contribute to the development of special areas in mathematics.

Providing a wide variety of the theory, this volume should be valuable to graduate students as well as to specialists in ring theory.

Our thanks go to all participants and contributors who invested a lot of their time to make the conference a great success. Our thanks also go to the referees who provided us, in a very short time, their reports in spite of their busy schedule.

Financial supports from National Natural Science Foundation and Ministry of Education of China, Southeast University and Nanjing University are greatly appreciated.

We would like to thank a number of staff and students of Southeast University and Nanjing University who spent a great deal of their time on various arrangements for the conference.

We also appreciate Professor Yiqiang Zhou and Professor Xiaosheng's advice and assistance, which greatly contributed to the success of the conference. Our thanks also go to the staff of Liuyuan Hotel, Southeast University, for their efficient arrangement of facilities and accommodations and for providing an enjoyable environment for the conference.

Finally we would like to announce that the Fifth Japan-Korea-China International Ring Theory Symposium is going to be held in Tokyo, in the summer of 2007. Detailed information on the symposium can be found in the following home-page: http://ring.cec.yamanashi.ac.jp/ ring/cjk2007/

Editors
Jianlong Chen (Southeast University, Nanjing, China)
Nanqing Ding (Nanjing University, Nanjing, China)
Hidetoshi Marubayashi (Naruto University of Education, Naruto, Japan)

May, 2005

CONTENTS

ON HARADA RINGS OF A COMPONENT TYPE

YOSHITOMO BABA

*Department Mathematics, Osaka Kyoiku University,
Kashiwara, Osaka, 582-8582 Japan
E-mail: ybaba@cc.osaka-kyoiku.ac.jp*

We define Harada rings of a component type. The class of them contains indecomposable serial rings. We consider the structure of them and show that they have weakly symmetric self-dualities.

1. Introduction.

We say that a module M is *non-small* if M is not a smll submodule of its injective hull $E(M)$. In [8] M. Harada studied a perfect ring satisfying the following condition:

> (*) Every non-small left R-module contains a non-zero injective
> submodule.

K. Oshiro named the ring a *left Harada ring* and studied it in [14] – [17]. The class of left Harada rings contains both QF-rings and serial rings. And both QF-rings and serial rings have self-dualities. So we naturally have the question: Whether left Harada rings have self-dualities or not. This problem was first considered in [10] and solved negatively by K. Koike in [11]. So we have another question: What kind of left Harada rings have self-dualities. In [10] J. Kado and K. Oshiro showed that every left Harada ring of homogeneous type has a weakly symmetric self-duality. But in general serial rings are not left Harada rings of homogeneous type. In this paper we define a certain kind of Harada rings which have weakly symmetric self-dualities and the class of which contains indecomposable serial rings.

By [8] and [15] a ring R is a left Harada ring if and only if R is an artinian ring with a complete set $\{e_{i,j}\}_{i=1,j=1}^{m,\ n(i)}$ of orthogonoal primitive idempotents of R such that $e_{i,1}R_R$ is injective and $e_{i,j}R_R \cong e_{i,j-1}R_R$ or $e_{i,j}R \cong e_{i,j-1}J_R$ for any $i = 1, \ldots, m$ and any $j = 2, \ldots, n(i)$ (see, for instance, [3]). We call the $\{e_{i,j}\}_{i=1,j=1}^{m,\ n(i)}$ a *left well indexed set* of R. (Symmetrically, we also define a *right well indexed set* for a right Harada ring. And a ring R is called a *Harada ring* if R is a left and right Harada ring.) In this paper, we call a ring R *a Harada ring of a component type* if R is a left Harada ring with a left well indexed set $\{e_{i,j}\}_{i=1,j=1}^{m,\ n(i)}$ which satisfies the following two conditions:

(A) For any $i = 1, \ldots, m-1$, there exists a left R-epimorphism $\theta_{i,n(i)}$: $Re_{i,n(i)} \to Je_{i+1,1}$ such that

$$\mathrm{Ker}\theta_{i,n(i)} = \begin{cases} S_{n(\kappa(i,n(i)))}(Re_{i,n(i)}) & \text{if } {}_RRe_{i,n(i)} \text{ is injective,} \\ 0 & \text{otherwise,} \end{cases}$$

where we let $\kappa(i,n(i)) \in \{1,\ldots,m\}$ with $Socle({}_RRe_{i,n(i)}) \cong {}_RRe_{\kappa(i,n(i)),1}/Je_{\kappa(i,n(i)),1}$, i.e., $(e_{\kappa(i,n(i)),1}R, Re_{i,n(i)})$ is an i-pair (we define the terminology i-pair later), and $S_{n(\kappa(i,n(i)))}(Re_{i,n(i)})$ means the $n(\kappa(i,n(i)))$-th socle of ${}_RRe_{i,n(i)}$.

(B) Either of the following two conditions holds:

(i) ${}_RRe_{m,n(m)}$ is injective.

(ii) There exists a left R-isomorphism $\theta_{m,n(m)}$: $Re_{m,n(m)} \to Je_{1,1}$.

Every Harada ring of a component type is a two sided Harada ring and every indecomposable serial ring is a Harada ring of a component type (see Theorems 3.1, 3.2). The purpose of this paper is to give the structure theorem of it and to show that it has a weakly symmetric self-duality.

2. Preliminaries

By M_R (resp. ${}_RM$) we stress M is a unitary right (resp. left) R-module. For an R-module M, we denote the injective hull, the Jacobson radical, the socle, the n-th socle, and the top $M/J(M)$ by $E(M)$, $J(M)$, $S(M)$, $S_n(M)$, and $T(M)$, respectively. And we put $M/S := M/S(M)$ and $M/S_n := M/S_n(M)$ for simplicity. Further, for a left R- right S- bimodule M, if $S({}_RM) = S(M_S)$ (resp. $S_n({}_RM) = S_n(M_S)$), we put $M/S := M/S({}_RM) = M/S(M_S)$ (resp. $M/S_n := M/S_n({}_RM) = M/S_n(M_S)$). In particular, for a local QF ring R, we put $\overline{R} := R/S({}_RR) = R/S(R_R)$.

Throughout this paper, we let R be a left Harada ring with a left well indexed set $\{e_{i,j}\}_{i=1,j=1}^{m,\ n(i)}$. Further we assume that R is a basic ring because the concepts of Harada rings and self-duality are Morita invariant.

We put $\mathbf{P} := \{(i,j)\}_{i=1,j=1}^{m,\ n(i)}$ and $E_{i,j} := E(T(_RRe_{i,j}))$ for any $(i,j) \in \mathbf{P}$.

Lemma 2.1. $_RRe_{\sigma(i),\rho(i)}/S_{j-1} \cong E_{i,j}$ *holds for any* $(i,j) \in \mathbf{P}$.

Let e, f be primitive idempotents of R. If both $S(eR_R) \cong T(fR_R)$ and $S(_RRf) \cong T(_RRe)$ hold, then we say that (eR, Rf) is an *injective pair* (abbreviated *i-pair*). By [1] we note that the following three conditions are equivalent:

(1) (eR, Rf) is an i-pair.
(2) eR_R is injective with $S(eR_R) \cong T(fR_R)$.
(3) $_RRf$ is injective with $S(_RRf) \cong T(_RRe)$.

And further we remark that, if (eR, Rf) is an i-pair, then $S(_{eRe}eRf) = S(eRf_{fRf})$ and it is simple both as a left eRe-module and as a right fRf-module.

Now we have maps $\sigma, \rho : \{1, \ldots, m\} \to \mathbf{N}$ such that $(e_{i,1}R, Re_{\sigma(i),\rho(i)})$ is an i-pair for any $i \in \{1, \ldots, m\}$ since $e_{i,1}R_R$ is injective.

In [12] K. Koike showed that, if R is a left Harada ring, then R' in the following lemma is also a left Harada ring. In this paper, we need describe this result in a detailed form to focus on left well indexed sets as follows:

Lemma 2.2. *Assume that there exists* $s \in \{1, \ldots, m\}$ *with* $n(s) \geq 2$. *Put* $R' := (1 - e_{s,n(s)})R(1 - e_{s,n(s)})$ *and*

$$n'(i) := \begin{cases} n(s) - 1 & \text{if } i = s, \\ n(i) & \text{otherwise.} \end{cases}$$

Then the following hold:

(1) *Suppose that* $_RRe_{s,n(s)}$ *is not injective. Then*

    (I) R' *is a left Harada ring with a left well indexed set* $\{e_{i,j}\}_{i=1,j=1}^{m,\ n'(i)}$, *and*

    (II) $(e_{i,1}R', R'e_{\sigma(i),\rho(i)})$ *is an* i-*pair for any* $i \in \{1, \ldots, m\}$.

(2) *Suppose that* $_RRe_{s,n(s)}$ *is injective but* $_RRe_{s,n(s)-1}$ *is not injective. Put*

$$(\sigma(i), \hat{\rho}(i)) := \begin{cases} (s, n(s) - 1) & \text{if } i = \kappa(s, n(s)), \quad \text{i.e., } (\sigma(i), \rho(i)) = (s, n(s)), \\ (\sigma(i), \rho(i)) & \text{otherwise.} \end{cases}$$

Then

(I) R' is a left Harada ring with a left well indexed set $\{e_{i,j}\}_{i=1,j=1}^{m,\ n'(i)}$, and

(II) $(e_{i,1}R', R'e_{\sigma(i),\hat{\rho}(i)})$ is an i-pair for any $i \in \{1,\ldots,m\}$.

(3) Suppose that both $_RRe_{s,n(s)}$ and $_RRe_{s,n(s)-1}$ are injective. We put $k := \kappa(s, n(s) - 1)$. And, for simplicity, we assume that $k \leq m - 1$ and $\kappa(s, n(s)) = k+1$, i.e., $(s, n(s)) = (\sigma(k+1), \rho(k+1))$. Then, for any $i \in \{1,\ldots, m-1\}$, further we put

$$\tilde{n}(i) := \begin{cases} n'(i) & \text{if } i \in \{1,\ldots,k-1\}, \\ n'(k) + n'(k+1) & \text{if } i = k, \\ n'(i+1) & \text{if } i \in \{k+1,\ldots,m-1\}, \end{cases}$$

and for any $j \in \{1,\ldots,\tilde{n}(i)\}$,

$$\tilde{e}_{i,j} := \begin{cases} e_{i,j} & \text{if } i \in \{1,\ldots,k-1\} \text{ and } j \in \{1,\ldots,\tilde{n}(i)\}, \\ e_{k,j} & \text{if } i = k \text{ and } j \in \{1,\ldots,n'(k)\}, \\ e_{k+1,j-n'(k)} & \text{if } i = k \text{ and } j \in \{n'(k)+1,\ldots,\tilde{n}(k)\}, \\ e_{i+1,j} & \text{if } i \in \{k+1,\ldots,m-1\} \text{ and } j \in \{1,\ldots,\tilde{n}(i)\}, \end{cases}$$

$$\sigma'(i) := \begin{cases} \sigma(i) & \text{if } i \in \{1,\ldots,k\}, \\ \sigma(i+1) & \text{if } i \in \{k+1,\ldots,m-1\}, \end{cases}$$

$$\tilde{\sigma}(i) := \begin{cases} \sigma'(i) & \text{if } \sigma'(i) \in \{1,\ldots,k\}, \\ \sigma'(i) - 1 & \text{if } \sigma'(i) \in \{k+1,\ldots,m\}, \end{cases}$$

$$\rho'(i) := \begin{cases} \rho(i) & \text{if } i \in \{1,\ldots,k\}, \\ \rho(i+1) & \text{if } i \in \{k+1,\ldots,m-1\}, \end{cases}$$

and

$$\tilde{\rho}(i) := \begin{cases} \rho'(i) & \text{if } \sigma'(i) \in \{1,\ldots,k\} \cup \{k+2,\ldots,m\}, \\ \rho'(i) + n'(k) & \text{if } \sigma'(i) = k+1. \end{cases}$$

Then

(I) R' is a left Harada ring with a left indexed set $\{\tilde{e}_{i,j}\}_{i=1,j=1}^{m-1,\ \tilde{n}(i)}$, and

(II) $(\tilde{e}_{i,1}R', R'\tilde{e}_{\tilde{\sigma}(i),\tilde{\rho}(i)})$ is an i-pair for any $i \in \{1,\ldots,m-1\}$.

3. Structure of Harada rings of a component type

First we give the following theorem:

Theorem 3.1. *Every indecomposable serial ring is a Harada ring of a component type.*

Proof. Let R be an indecomposable serial ring. Then R is a two sided Harada ring by, for instance, [3] . So we have a left well indexed set $\{e_{i,j}\}_{i=1,j=1}^{m\ \ n(i)}$ to satisfy $T(e_{i,n(i)}J_R) \cong T(e_{i+1,1}R_R)$ for any $i \in \{1,\ldots,m-1\}$. Then we note that, if there exists a primitive idempotent e of R with eR_R simple, then $e_{m,n(m)} = e$. Moreover, $e_{m,n(m)}R$, $e_{m,n(m)-1}R$, $\ldots$, $e_{m,1}R$, $e_{m-1,n(m-1)}R$, $\ldots$, $e_{1,1}R$ is a Kupisch series. And $Re_{1,1}$, $Re_{1,2}$, $\ldots$, $Re_{1,n(1)}$, $Re_{2,1}$, $\ldots$, $Re_{m,n(m)}$ is a Kupisch series by, for instance, [1] . So, for any $i \in \{1,\ldots,m-1\}$, $T(_RRe_{i,n(i)}) \cong T(_RJe_{i+1,1})$ holds. Suppose that $_RRe_{i,n(i)}$ is not injective. Then we have a left R-isomorphism $\theta_{i,n(i)} : Re_{i,n(i)} \to Je_{i+1,1}$ since R is a right Harada ring. Suppose that $_RRe_{i,n(i)}$ is injective. Then $_RRe_{i,n(i)}/S_{t-1} \cong E_{\kappa(i,n(i)),t}$ for any $t \in \{1,\ldots,n(\kappa(i,n(i)))\}$ by Lemma 2.1. And $_RRe_{i,n(i)}/S_{n(\kappa(i,n(i)))}$ is not injective since $\{Re_{\sigma(i),\rho(i)}/S_{j-1}\}_{i=1,j=1}^{m\ \ n(i)}$ is a basic set of indecomposable injective left R-modules by Lemma 2.1. So $E(_RRe_{i,n(i)}/S_{n(\kappa(i,n(i)))})$ is projective by [3] . Therefore $_RRe_{i,n(i)}/S_{n(\kappa(i,n(i)))} \cong {}_RJe_{i+1,1}$ since R is a right Harada ring and $T(_RRe_{i,n(i)}) \cong T(_RJe_{i+1,1})$, i.e., we have a left R-epimorphism $\theta_{i,n(i)} : Re_{i,n(i)} \to Je_{i+1,1}$ with $\mathrm{Ker}\theta_{i,n(i)} = S_{n(\kappa(i,n(i)))}(_RRe_{i,n(i)})$. The condition (A) is satisfied. Suppose that $e_{m,n(m)}R_R$ is not simple. Then $T(e_{m,n(m)}J_R) \cong T(e_{1,1}R_R)$ holds and so $T(_RRe_{m,n(m)}) \cong T(_RJe_{1,1})$ also holds by [1] . Therefore there exists a left R-isomorphism $\theta_{m,n(m)} : Re_{m,n(m)} \to Je_{1,1}$ as above. Suppose that $e_{m,n(m)}R_R$ is simple. Then $e_{m,n(m)}R_R \cong S(e_{m,1}R_R)$ by the definition of a left well indexed set. So $(e_{m,1}R, Re_{m,n(m)})$ is an i-pair since $e_{m,1}R_R$ is injective. Therefore $_RRe_{m,n(m)}$ is injective. The condition (B) also holds. $\square$

From now on, throughout this paper, we let R be a basic Harada ring of a component type with a left well indexed set $\{e_{i,j}\}_{i=1,j=1}^{m\ \ n(i)}$.

Lemma 3.1.

(1) $_RJe_{i,j}/J^2e_{i,j} \cong T(_RRe_{(i,j)-1})$ *for any* $(i,j) \in \mathbf{P} - \{(1,1)\}$.
   And, if $_RRe_{m,n(m)} \cong {}_RJe_{1,1}$, *then* $_RJe_{1,1}/J^2e_{1,1} \cong T(_RRe_{m,n(m)})$ *also holds.*

(2) $_RRe_{i,j}/J^{\nu(i,j)}e_{i,j}$ *is uniserial for any* $(i,j) \in \mathbf{P}$.

(3) *Suppose that* $_RJe_{1,1}/J^2e_{1,1}$ *is simple. Then* R *is a serial ring.*

So, in particular, if $_RRe_{m,n(m)} \cong {}_RJe_{1,1}$, *then* R *is a serial ring.*

Proof. (1), (2) It is clear by [17] and the condition (A).

(3) Suppose that $_RJe_{1,1}/J^2e_{1,1}$ is simple. Then R is a left serial ring by (1). On the other hand, since R is a left Harada ring, every indecomposable injective left R-module is a factor module of an indecomposable projective left R-module. Hence R is a serial ring. $\qquad\square$

We define a total order in $\mathbf{P}$ by

$$(i,j) \leq (s,t) \text{ if and only if either (i) } i < s, \text{ or (ii) } i = s \text{ and } j \leq t \text{ holds.}$$

And we consider the order preserving bijection

$$\nu : \mathbf{P} \to \{1, \ldots, \textstyle\sum_{i=1}^{m} n(i)\}.$$

For any $(i,j) \in \mathbf{P}$ we denote $(u,v) \in \mathbf{P}$ with

$$\nu(u,v) \equiv \nu(i,j) + 1 \quad (\text{mod } \textstyle\sum_{s=1}^{m} n(s))$$

by $(i,j)_{+1}$. And, for any $(1,1) \neq (s,t) \in \mathbf{P}$ we denote $(u,v) \in \mathbf{P}$ with

$$\nu(u,v) = \nu(s,t) - 1$$

by $(s,t)_{-1}$.

Now we have a bijection

$$\xi : \{1, \ldots, m\} \to \{(\sigma(i), \rho(i))\}_{i=1}^{m}$$

such that $\xi(1) > \xi(2) > \cdots > \xi(m)$. Then put

$$f_{i,1} := e_{\xi(i)}$$

and

$$n\langle i \rangle := \nu(\xi(i)) - \nu(\xi(i+1))$$

for any $i \in \{1, \ldots, m\}$, where we let

$$\nu(\xi(m+1)) = \begin{cases} 0 & \text{if } {}_RRe_{m,n(m)} \text{ is injective,} \\ \nu(\xi(1)) - \sum_{i=1}^{m} n(i) & \text{otherwise.} \end{cases}$$

And further we put

$$f_{i,j} := e_{u_{i,j}, v_{i,j}}$$

for any $i \in \{1, \ldots, m\}$ and any $j \in \{1, \ldots, n\langle i\rangle\}$, where we take $(u_{i,j}, v_{i,j}) \in \mathbf{P}$ to satisfy

$$\nu(u_{i,j}, v_{i,j}) \equiv \nu(\xi(i)) - j + 1 \quad (\text{mod } \textstyle\sum_{i=1}^{m} n(i)).$$

Then we see the following theorem which shows that every Harada ring of a component type is a two sided Harada ring.

Theorem 3.2. *R is a right Harada ring with a right well indexed set* $\{f_{i,j}\}_{i=1,j=1}^{m}{}^{n\langle i\rangle}$.

Proof. For any $(i, j) \neq (m, n(m))$ with $_RRe_{i,j}$ not injective,

$$_RRe_{i,j} \cong \begin{cases} _RJe_{i+1,1} & \text{if } j = n(i), \\ _RJe_{i,j+1} & \text{otherwise,} \end{cases}$$

by [17] and the condition (A). Further, if $_RRe_{m,n(m)}$ is not injective, $_RJe_{11} \cong {}_RRe_{m,n(m)}$ by the condition (B). Therefore, by the definition of $\{f_{i,j}\}_{i=1,j=1}^{m}{}^{n\langle i\rangle}$, we see that

(1) $_RRf_{i,1}$ is injective, and
(2) $_RJf_{i,j} \cong {}_RJ^{j-1}f_{i,1}$

for any $i \in \{1, \ldots, m\}$ and any $j \in \{1, \ldots, n\langle i\rangle\}$. The statement holds. $\square$

The following theorem is very important to show that every Harada ring of a component type has a weakly symmetric self-duality. To show it we need many lemmas. So we omit the proof in this paper. Please see [6] with respect to the detail.

Theorem 3.3. *Suppose that R is a Harada ring of a component type. Then R satisfies either the following (1) or (2).*

(1) $(\sigma(1), \rho(1)) < (\sigma(2), \rho(2)) < \cdots < (\sigma(m), \rho(m))$.

(2) R is a serial ring.

Therer is a possiblity that R has more than one left well indexed set. If the condition (1) of this theorem holds with respect to some left well indexed set $\{e_{i,j}\}_{i=1,j=1}^{m}{}^{n(i)}$ of R, we call it a Harada ring of a *linear* component type (with respect to $\{e_{i,j}\}_{i=1,j=1}^{m}{}^{n(i)}$).

Example 3.1. We give a simple example of a Harada ring of a component type. Let Q be a local QF ring. Then

$$\begin{bmatrix} Q & \overline{Q} & \overline{Q} & \overline{Q} \\ J(Q) & \overline{Q} & \overline{Q} & \overline{Q} \\ J(Q) & J(Q) & \overline{Q} & \overline{Q} \\ J(Q) & J(Q) & J(Q) & Q \end{bmatrix}$$

is a Harada ring of a componet type. But it is not a serial ring in general.

4. Self-duality of Harada ring of a component type

We say that an artinian ring R *has a (Morita) self-duality* if there exists a Morita duality D between the category of finitely generated left R-modules and the category of finitely generated right R-modules. It is well known that R has a self-duality if and only if there exist a finitely generated injective cogenerator $_RE$ and a ring isomorphism $\phi : R \to \text{End}(_RE)$ (which induces a right R-stucture on E via $x \cdot r = x\phi(r)$ for $x \in E$ and $r \in R$) such that the dualities D and $\text{Hom}_R(\ ?\ ,\ _RE_R)$ are naturally equivalent.

Further we say that R *has a weakly symmetric self-duality* if there exists a left R- right R- bimodule E which defines a Morita duality and satisfies the following condition:

$$\text{Hom}_R(T(_RRg),\ _RE)_R \cong T(gR_R) \text{ for any primitive idempotent } g \text{ of } R.$$

In this section, we show that a Harada ring of a component type has a weakly symmetric self-duality.

Lemma 4.1. *Suppose that R is a Harada ring of a linear component type and there exists $s \in \{1,\ldots,m\}$ with $n(s) \geq 2$. We put $R' := (1 - e_{s,n(s)})R(1 - e_{s,n(s)})$. Then R' is also a Harada ring of a linear component type.*

Proof. We put

$$m' := \begin{cases} m - 1 & \text{if } _RRe_{m,n(m)} \text{ is injective,} \\ m & \text{otherwise.} \end{cases}$$

By Lemma 2.2, R' is a left Harada ring. And we consider the following three cases as in Lemma 2.2 and use the same terminologies as in it.

  (1) $_RRe_{s,n(s)}$ is not injective.
  (2) $_RRe_{s,n(s)}$ is injective, but $_RRe_{s,n(s)-1}$ is not injective.

(3) Neither $_R Re_{s,n(s)-1}$ nor $_R Re_{s,n(s)}$ is injective.

Case 1: By Lemma 2.2 (1), R' is a left Harada ring with a left well indexed set $\{e_{i,j}\}_{i=1,j=1}^{m, \; n'(i)}$ and $(e_{i,1}R', R'e_{\sigma(i),\rho(i)})$ is an i-pair for any $i \in \{1,\ldots,m\}$. And

$$(\sigma(1),\rho(1)) < (\sigma(2),\rho(2)) < \cdots < (\sigma(m),\rho(m))$$

holds since R is a Harada ring of a linear component type. Further $_{R'}R'e_{m,n'(m)}$ is injective iff $_R Re_{m,n(m)}$ is so. And we put

$$\theta'_{i,n'(i)} := \begin{cases} \theta_{i,n(i)}|_{R'e_{i,n(i)}} & \text{if } i \neq s, \\ \theta_{s,n(s)-1}\theta_{s,n(s)}|_{R'e_{s,n(s)-1}} & \text{if } i = s, \end{cases}$$

for any $i \in \{1,\ldots,m'\}$. Then it is obvious that $\mathrm{Im}\theta'_{i,n'(i)} = J(R')e_{i+1,1}$ and

$$\mathrm{Ker}\theta'_{i,n'(i)} = \begin{cases} S_{n'(\kappa(i,n(i)))}(Re_{i,n'(i)}) & \text{if } _{R'}R'e_{i,n'(i)} \text{ is injective,} \\ 0 & \text{otherwise.} \end{cases}$$

The conditions (A), (B) hold.

Case 2: By Lemma 2.2 (2), R' is a left Harada ring with a left well indexed set $\{e_{i,j}\}_{i=1,j=1}^{m, \; n'(i)}$ and $(e_{i,1}R', R'e_{\sigma(i),\hat{\rho}(i)})$ is an i-pair for any $i \in \{1,\ldots,m\}$. Then

$$(\sigma(1),\hat{\rho}(1)) < (\sigma(2),\hat{\rho}(2)) < \cdots < (\sigma(m),\hat{\rho}(m))$$

holds and $_{R'}R'e_{s,n'(s)}$ is injective with $(e_{\kappa(s,n(s)),1}R', R'e_{s,n'(s)})$ an i-pair by the definition of $\hat{\rho}$ since R is a Harada ring of a linear component type. Now we consider $\theta'_{i,n'(i)}$ as in case 1. Then $\mathrm{Ker}\theta'_{s,n'(s)} = \mathrm{Ker}(\theta_{s,n(s)-1}\theta_{s,n(s)}|_{R'e_{s,n(s)-1}}) = (1 - e_{s,n(s)})\mathrm{Ker}(\theta_{s,n(s)-1}\theta_{s,n(s)}) = (1 - e_{s,n(s)})(S_{n(\kappa(s,n(s)))}(_R Re_{s,n(s)}))\theta_{s,n(s)-1}^{-1} = (1 - e_{s,n(s)})S_{n(\kappa(s,n(s)))}(_R Re_{s,n(s)-1}) = S_{n'(\kappa(s,n(s)))}(_{R'}R'e_{s,n'(s)})$ since $\theta_{s,n(s)-1}$ is monic and $\mathrm{Ker}\theta_{s,n(s)} = S_{n(\kappa(s,n(s)))}(_R Re_{s,n(s)})$. Hence the conditions (A), (B) hold because the remainder is obvious.

Case 3: We put $k := \kappa(s,n(s)-1)$. Then $k \leq m-1$ and $\kappa(s,n(s)) = k+1$ since R is a Harada ring of a linear component type. So, by Lemma 2.2 (3), R' is a left Harada ring with a left indexed set $\{\tilde{e}_{i,j}\}_{i=1,j=1}^{m-1, \; \tilde{n}(i)}$ and $(\tilde{e}_{i,1}R', R'\tilde{e}_{\tilde{\sigma}(i),\tilde{\rho}(i)})$ is an i-pair for any $i \in \{1,\ldots,m-1\}$. And

$$(\tilde{\sigma}(1),\tilde{\rho}(1)) < (\tilde{\sigma}(2),\tilde{\rho}(2)) < \cdots < (\tilde{\sigma}(m-1),\tilde{\rho}(m-1))$$

holds by the definitions of $\tilde{\sigma}$ and $\tilde{\rho}$ since R is a Harada ring of a linear component type. Further, we note that, if $s = m$, then $_{R'}R'\tilde{e}_{m-1,\tilde{n}(m-1)}$ is

10

injective. Furthermore, we put

$$\theta''_{i,\tilde{n}(i)} := \begin{cases} \theta'_{i,n'(i)} & \text{if } i \in \{1,\ldots,k-1\}, \\ \theta'_{i+1,n'(i+1)} & \text{if } i \in \{k,\ldots,m'-1\}. \end{cases}$$

In the case that $s = k$,

$$\theta''_{i,\tilde{n}(i)} := \begin{cases} \theta_{i,n(i)}|_{R'e_{i,n(i)}} & \text{if } i \in \{1,\ldots,k-1\}, \\ \theta_{i+1,n(i+1)}|_{R'e_{i+1,n(i+1)}} & \text{if } i \in \{k,\ldots,m'-1\}. \end{cases}$$

So clearly the conditions (A), (B) hold. In the case that $s \neq k$. Suppose that $s < k$. Then $_{R'}R'\tilde{e}_{s,\tilde{n}(s)}$ is injective with $(\tilde{e}_{k,1}R', R'\tilde{e}_{s,\tilde{n}(s)})$ an i-pair. And $\theta''_{s,\tilde{n}(s)} = \theta'_{s,n'(s)} = \theta_{s,n(s)-1}\theta_{s,n(s)}|_{R'e_{s,n(s)-1}}$. So $\mathrm{Ker}\theta''_{s,\tilde{n}(s)} = \mathrm{Ker}(\theta_{s,n(s)-1}\theta_{s,n(s)}|_{R'e_{s,n(s)-1}}) = (1 - e_{s,n(s)})\mathrm{Ker}(\theta_{s,n(s)-1}\theta_{s,n(s)}) = (1 - e_{s,n(s)})(S_{n(k+1)}(_RRe_{s,n(s)}))\theta^{-1}_{s,n(s)-1} =$

$(1 - e_{s,n(s)})((\oplus_{j=1}^{n(k)}S(e_{k,j}Re_{s,n(s)-1})) \oplus (\oplus_{j=1}^{n(k+1)}S(e_{k+1,j}Re_{s,n(s)-1}))) = S_{\tilde{n}(k)}(_{R'}R'\tilde{e}_{s,\tilde{n}(s)})$ by the proof of Lemma 2.2 (3) since $\mathrm{Ker}\theta_{s,n(s)-1} = S_{n(k)}(_RRe_{s,n(s)-1}) = \oplus_{j=1}^{n(k)}S(e_{k,j}Re_{s,n(s)-1})$ and $\mathrm{Ker}\theta_{s,n(s)} = S_{n(k+1)}(_RRe_{s,n(s)}) = \oplus_{j=1}^{n(k+1)}S(e_{k+1,j}Re_{s,n(s)})$. Suppose that $s > k$. Then $_{R'}R'\tilde{e}_{s-1,\tilde{n}(s-1)}$ is injective with $(\tilde{e}_{k,1}R', R'\tilde{e}_{s-1,\tilde{n}(s-1)})$ an i-pair. And $\theta''_{s-1,\tilde{n}(s-1)} = \theta'_{s,n'(s)} = \theta_{s,n(s)-1}\theta_{s,n(s)}|_{R'e_{s,n(s)-1}}$. So we see $\mathrm{Ker}\theta''_{s-1,\tilde{n}(s-1)} = S_{\tilde{n}(k)}(_{R'}R'\tilde{e}_{s,\tilde{n}(s)})$ by the same way as the case $s < k$. Hence the conditions (A), (B) hold because the remainder is obvious. $\square$

Using Lemma 4.1 and [12] , we can easily show the following main theorem:

Theorem 4.1. *Every Harada ring of a component type has a weakly symmetric self-duality.*

Proof. By [7] a serial ring has a weakly symmetric self-duality. So we assume that R is not a serial ring. Then R is a Harada ring of a linear component type by Theorem 3.3.

Now we construct a QF ring R_l from R by the same way as in [12] . Then R_l is a QF Harada ring of a linear component type by Lemma 4.1. That is, R is a QF ring with the identity Nakayama permutation (i.e., a QF ring satisfing that (gR_l, R_lg) is an i-pair for any primitive idempotent g of R_l. So it is clear that R_l has a weakly symmetric self-duality. Hence R also has a weakly symmetric self-duality by [12] . $\square$

Let R be a left Harada ring with a left well indexed set $\{e_{i,j}\}_{i=1,j=1}^{m,\ n(i)}$. We say that R is *homogeneous type* if σ is the identity permutation of

$\{1, \ldots, m\}$. In [10] J. Kado and K. Oshiro show that, if a left Harada ring is homogeneous type, then it has a self-duality. Last we remark that the ring given in Example 3.1 is not homogeneous type.

Acknowledgements

The author thanks to Prof. K. Koike for valuable discussion about [12] and to Prof. J. Clark for his terminological advice.

References

1. F. W. Anderson and K. R. Fuller, *Rings and categories of modules (second edition)*, Graduate Texts in Math. **13** (Springer-Verlag, 1991).
2. Y. Baba, Injectivity of quasi-projective modules, projectivity of quasi-injective modules, and projective cover of injective modules, *J. Algebra* **155, 2** (1993), 415-434
3. Y. Baba and K. Iwase, On quasi-Harada rings, *J. Algebra* **185** (1996) 544–570.
4. Y. Baba, Some classes of QF-3 rings, *Comm. in Alg.* **28 (6)** (2000) 2639–2669.
5. Y. Baba, On Harada rings and quasi-Harada rings with left global dimension at most 2, *Comm. in Alg.* **28 (6)** (2000) 2671–2684.
6. Y. Baba, Self-duality of Harada ring of a component type, *Preprint*
7. F. Dischinger and W. Müller, Einreihing zerlegbare artinsche Ringe sind selfstdual, *Arch. Math.* **43 (2)** (1984) 132–136.
8. M. Harada, *Non-small modules and non-cosmall modules*, Ring Theory , Proceedings of 1978 Antwerp Conference (F. Van Oystaeyen, Ed.), (Dekker, New York 1979) 669–690.
9. M. Harada, *Factor categories with applications to direct decomposition of modules*, Lecture Note in Pure and Appl. Math., **88** (Dekker, New York 1983)
10. J. Kado and K. Oshiro, Self-duality and Harada rings, *J. Algebra* **211** (1999) 354–408.
11. K. Koike, Examples of QF rings without Nakayama automorphism and H-rings without self-duality, *J. Algebra* **241 (2)** (2001) 731-744.
12. K. Koike, Almost self-duality and Harada rings, *J. Algebra* **254 (2)** (2002) 336-361.
13. K. Koike, Good self-duality of quasi-Harada rings and locally distributive rings, *Preprint*.
14. K. Oshiro, Lifting modules, extending modules and their applications to QF-rings, *Hokkaido Math. J.* **13** (1984) 310–338.
15. K. Oshiro, On Harada ring I, *Math. J. Okayama Univ.* **31** (1989) 161–178.
16. K. Oshiro, On Harada ring II, *Math. J. Okayama Univ.* **31** (1989) 179–188.
17. K. Oshiro, On Harada ring III, *Math. J. Okayama Univ.* **32** (1990) 111–118.

RING HULLS OF EXTENSION RINGS

GARY F. BIRKENMEIER

Department of Mathematics
University of Louisiana at Lafayette
Lafayette, LA 70504-1010, USA
E-mail: gfb1127@louisiana.edu

JAE KEOL PARK

Department of Mathematics
Busan National University
Busan 609-735, South Korea
E-mail: jkpark@pusan.ac.kr

S. TARIQ RIZVI

Department of Mathematics
Ohio State University
Lima, OH 45804-3576, USA
E-mail: rizvi.1@osu.edu

In this survey, we provide some results and examples on the behavior of the quasi-Baer and the right FI-extending right ring hulls. We focus on these ring hulls for various ring extensions including group ring extensions, full and triangular matrix ring extensions, and infinite matrix ring extensions. We also establish connections between the right FI-extending right ring hulls of semiprime homomorphic images of R and the subrings of $eQ(R)e$, where $e = e^2 \in Q(R)$.

Throughout this paper all rings are associative with unity and R denotes such a ring. Subrings and overrings preserve the unity of the base ring. Ideals without the adjective "right" or "left" mean two-sided ideals. All modules are unital and for an Abelian group M, we use M_R to denote a right R-module. If N_R is a submodule of M_R, then N_R is *essential* (resp., *dense* also called *rational*) in M_R if for any $0 \neq x \in M$, there exists $r \in R$ such that $0 \neq xr \in N$ (resp., for any $x, y \in M$ with $0 \neq x$, there exists $r \in R$ such that $xr \neq 0$, and $yr \in N$). We use $N_R \leq^{\mathrm{ess}} M_R$ to denote that N_R is an essential R-submodule of M_R.</p>

12

Recall that a *right ring of quotients T of R* is an overring of R such that R_R is dense in T_R. The maximal right ring of quotients of R is denoted by $Q(R)$. The right injective hull of R is denoted by $E(R_R)$.

We say that T is a *right essential overring* of a ring R if T is an overring of R such that R_R is essential in T_R [8]. Note that, for an overring T of a ring R, if R_R is dense in T_R, then R_R is essential in T_R. Thereby a right essential overring T of a ring R can be considered as a "generalized version" of a right ring of quotients of R (in fact, such rings were called right quotient rings in [18]).

A ring R is: right (*FI-*) *extending* if every (ideal) right ideal of R is essential in a right ideal generated by an idempotent, (*quasi-*) *Baer* if the right annihilator of every (ideal) nonempty subset of R is an idempotent generated right ideal, right (*p.q.-Baer*) *PP* if the right annihilator of a (principal ideal) singleton set of R is a right ideal generated by an idempotent. Note that the notion of a right PP ring is equivalent to every principal right ideal of R being projective (these rings are also called right *Rickart* rings). These classes have their roots in Operator Theory, especially in the study of von Neumann algebras, and in the study of right self-injective rings (see [3], [27], [26], [21], [5], [15], [37], [38], [17], [20], [40], [6], and [7]).

One of the important aspects of investigations in Ring Theory has been for a given ring R, the quest to find a more "well behaved" (i.e., with better properties than R) overring Q such that a rich information transfer between R and Q can take place. We develop methods which allow us to take any specific class $\mathfrak{K}$ of rings and determine existence and uniqueness results for right essential overrings of a given ring R, which are in some sense "minimal" with respect to belonging to the class $\mathfrak{K}$. Let S be a right essential overring of R and $\mathfrak{K}$ be a specific class of rings. We say S is a $\mathfrak{K}$ *right ring hull* of R if S is minimal among right essential overrings of R belonging to the class $\mathfrak{K}$ (i.e., $S \in \mathfrak{K}$ and if T is a right essential overring of R with T a subring of S and $T \in \mathfrak{K}$, then $T = S$).

Our methods provide a general setting to study various ring hulls. A number of results, examples, and applications of the concept of ring hulls are included in our papers [8], [9], and [10]. We apply these concepts to concrete classes of rings which satisfy properties such as the (FI-) extending, the (quasi-) Baer (or other generalizations of Baer or injective) properties. To the best of our knowledge, there is no information available about the (FI-) extending or quasi-Baer ring hulls for a given ring R and the existence of Baer hulls has only been known in the case when R is either commutative semiprime [29] or when R is a reduced right Utumi ring [23].

In this expository paper, we include our results from [8], [9], and [10] on the behavior of the quasi-Baer and the right FI-extending right ring hulls. Our focus here is on such hulls for various ring extensions including group ring extensions, full and triangular matrix ring extensions, and infinite matrix ring extensions. Some examples are included to clarify the concepts and delimit the results. We also include an interesting result which shows that for semiprime rings R and S, if R is Morita equivalent to S, then so are the quasi-Baer right ring hulls $\widehat{Q}_{q\mathfrak{B}}(R)$ and $\widehat{Q}_{q\mathfrak{B}}(S)$. We also survey connections (see [8]) between the right FI-extending right ring hulls of semiprime homomorphic images of R and the subrings of $eQ(R)e$, where $e = e^2 \in Q(R)$.

Let $\mathrm{Mat}_n(R)$ and $T_n(R)$ denote the n-by-n matrix ring and the n-by-n upper triangular matrix ring over R, respectively, where n is a positive integer. For a nonempty subset X of a ring R, we use $\langle X \rangle_R$ to denote the subring of R generated by X.

For the proofs of the results in this paper, detailed discussions of the various ring hull concepts and further examples, the reader is advised to see [8], [9], and [10].

We start with the following definition of ring hulls.

Definition 1. *Let $\mathfrak{K}$ denote a class of rings. For a ring R, let S, T, and U be right essential overrings of R. Consider the following conditions*:
 (i) $S \in \mathfrak{K}$.
 (ii) *If $T \in \mathfrak{K}$ and T is a subring of S, then $T = S$.*
 (iii) *If S and T are subrings of U and $T \in \mathfrak{K}$, then S is a subring of T.*
 (iv) *If $T \in \mathfrak{K}$, then S is a subring of T.*

If S satisfies (i) and (ii), then we say S is a $\mathfrak{K}$ *right ring hull* of R, denoted by $\widetilde{Q}_{\mathfrak{K}}(R)$. If S satisfies (i) and (iii), then we say S is the $\mathfrak{K}$ *absolute to U right ring hull* of R; for the $\mathfrak{K}$ absolute to $Q(R)$ right ring hull, we use the notation $\widehat{Q}_{\mathfrak{K}}(R)$. If S satisfies (i) and (iv), then we say S is the $\mathfrak{K}$ *absolute right ring hull* of R, denoted by $Q_{\mathfrak{K}}(R)$. Observe that if $Q(R) = E(R_R)$, then $\widehat{Q}_{\mathfrak{K}}(R) = Q_{\mathfrak{K}}(R)$.

We use

$$\mathfrak{B}, \quad q\mathfrak{B}, \quad pq\mathfrak{B}, \quad \mathfrak{E}, \quad \mathfrak{FI}, \quad q\mathfrak{C}$$

to denote the class of Baer rings, quasi-Baer rings, right p.q.-Baer rings, right extending rings, right FI-extending rings, right quasi-continuous rings, respectively.

Let $\mathbf{B}(Q(R))$ and $RB(Q(R))$ denote the set of central idempotents of $Q(R)$ and the idempotent closure of R (i.e., the subring of $Q(R)$ generated by R and $\mathbf{B}(Q(R))$, see [2]).

Theorem 2. ([9, Theorem 3.5]) *Let R be a semiprime ring. Then:*
(i) $RB(Q(R)) = \widehat{Q}_{q\mathfrak{B}}(R) = \widehat{Q}_{\mathfrak{F}\mathfrak{I}}(R)$.
(ii) *Let T be a right essential overring of R such that $RB(Q(R))$ is a subring of T. Then T is right extending and quasi-Baer.*

If R is a semiprime ring, then it can be seen that the Martindale symmetric ring of quotients, $Q^s(R)$, of R is a left ring of quotients of R [27, p.385]. Since $RB(Q(R))$ is a subring of $Q^s(R)$, $RB(Q(R))$ is also a left ring of quotients of R. From this, one can obtain a left sided version of Theorem 2 in which the idempotent closure of R in $Q^\ell(R)$ coincides with that in $Q(R)$ (i.e., $RB(\mathrm{End}(E(R_R)) = RB(Q(R)) = RB(Q^\ell(R)))$. For a ring R, $Q^m(R)$ denotes the Martindale right ring of quotients of R. Let C be the extended centroid of R (i.e., C is the center of $Q(R)$) and $N = \{x \in Q(R) \mid xR = Rx\}$. Then $R \cdot C\,(= \langle R \cup C \rangle_{Q(R)})$ and $R \cdot N\,(= \langle R \cup N \rangle_{Q(R)})$ are called the *central closure* and the *normal closure* of R, respectively. Since $C \subseteq N$, it follows that $R \cdot C \subseteq R \cdot N \subseteq Q^s(R)$ (see [27, p.395] for more detail on the central closure and the normal closure).

Corollary 3. ([9, Corollary 3.6]) (i) *If R is a semiprime ring, then $R \cdot C$, $R \cdot N$, $Q^m(R), Q^s(R)$, and $Q(R)$ are all quasi-Baer and right (and left) FI-extending.*

(ii) *If R is a semiprime ring, then there exists a right essential overring that is maximal with respect to being quasi-Baer (or right FI-extending).*

(iii) *If $Q(R)$ is semiprime, then $Q(R)$ is quasi-Baer and right (and left) FI-extending.*

In [19, p.1516], Ferrero has shown that the Martindale symmetric ring of quotients of a semiprime ring is quasi-Baer. There is a semiprime ring R for which both $Q^m(R)$ and $Q^s(R)$ are not Baer. There is also an example given by Zalesski and Neroslavskii [22] of a simple ring R which is not a domain and 0, 1 are its only idempotents. Then $Q^m(R) = R$ (and hence $Q^s(R) = R$). In this case, $Q^m(R)$ is not a Baer ring.

In [31] and [32, p.412] Osofsky poses the open question: *If $E(R_R)$ has a ring multiplication which extends its right R-module scalar multiplication, must $E(R_R)$ be a right self-injective ring?* To the authors' knowledge, this problem is still open [34]. Since the right FI-extending property generalizes the right self-injective property, our previous results allow us to provide the following partial answer to this question.

Corollary 4. ([9, Corollary 3.7]) *Let R be a ring such that $E(R_R)$ has a ring multiplication which extends its right R-module scalar multiplication. If R has a right FI-extending right essential overring which is a subring of $E(R_R)$, then $E(R_R)$ is right FI-extending. In particular, if either R or $Q(R)$ is semiprime, then $E(R_R)$ is a right FI-extending ring.*

In [34], Osofsky also posed the following question: *If $E(R_R)$ has a ring multiplication which extends its R-module scalar multiplication, must it be unique?* In [11], we answer her question in the negative, by exhibiting a ring R such that $E(R_R)$ has exactly four distinct ring multiplications which extend its R-module scalar multiplication and $E(R_R)$ is a right self-injective ring under each of these ring multiplications. Moreover, these right self-injective rings are all ring isomorphic.

Recently, there has been a series of papers [25], [1], and [35] on right extending group rings. In the next two results we consider applications of Theorem 2 to group rings. These results not only describe some quasi-Baer group rings, but they also enlarge the class of known right extending group rings.

Corollary 5. ([9, Corollary 3.8]) (i) *Assume that $R[G]$ is a semiprime right FI-extending (or quasi-Baer) group ring of a group G over a ring R. Then the order $|N|$ of N is invertible in R for any finite normal subgroup N of G.*

(ii) *Assume that $R[G]$ is a semiprime group ring of a finite group G over a ring R. If $R[G]$ is quasi-Baer (resp., Baer), then $|G|$ is invertible in R and R is quasi-Baer (resp., Baer). Also if $R[G]$ is right extending, then $|G|$ is invertible and R is right extending.*

(iii) *Assume that R is a subdomain of $\mathbb{Q}$ and G is a finite group, where $\mathbb{Q}$ is the field of rational numbers. Then the group ring $R[G]$ is quasi-Baer if and only if $|G|$ is invertible in R. In addition, if G is Abelian, then $R[G]$ is extending if and only if $|G|$ is invertible in R.*

In [35, Theorem 2.9], Parmenter and Zhou show the sufficiency of the result in the following corollary. However our result provides both necessity and sufficiency by Theorem 2.

Corollary 6. ([9, Corollary 3.9]) *Assume that R is a commutative domain* *with no 2-torsion and G is the group of order 2. Then the group ring $R[G]$ is extending if and only if 2 is invertible in R.*

Open Question. ([9]) If R is semiprime, is $RB(Q(R))$ a right FI-extending absolute right ring hull (i.e., does $RB(Q(R)) = Q_{\mathfrak{FJ}}(R)$)?

Our Corollary 5 is related to the question posed by Hirano in [24]: *Let R be a quasi-Baer ring and G a finite group. If $|G|$ is invertible in R, then is the group ring $R[G]$ quasi-Baer?* Corollary 5(ii) gives a partial answer to the question. However, in the following example, we see that the Baer ring version of Hirano's question does not hold.

Example 7. (see also [9]) Let $\mathbb{C}$ be the field of complex numbers and let $R = \mathbb{C}[x, y]$. Thus R is Baer and right (and left) extending. Take $G = S_3$, the symmetric group on $\{1, 2, 3\}$. Then we see that

$$\mathbb{C}[G] \cong \mathbb{C} \oplus \mathbb{C} \oplus \mathrm{Mat}_2(\mathbb{C}).$$

Thus

$$R[G] \cong \mathbb{C}[G][x, y] \cong \mathbb{C}[x, y] \oplus \mathbb{C}[x, y] \oplus \mathrm{Mat}_2(\mathbb{C}[x, y]).$$

Therefore $R[G]$ is quasi-Baer and right FI-extending. But note that $\mathrm{Mat}_2(\mathbb{C}[x, y])$ is neither Baer nor right extending since the domain $\mathbb{C}[x, y]$ is not Prüfer (see [26, p.17, Exercise 3] and [16, p.109, Corollary 12.10]). Therefore $R[G]$ is neither Baer nor right extending.

For a ring R, let $Z(R_R)$ denote the right singular ideal of R. The next example exhibits a nonsemiprime ring R with $Z(R_R) \neq 0$ which is not right FI-extending but does possess a right FI-extending right ring hull which, in general, is not right extending. Moreover $Q(R) = E(R_R)$ and, in general, these rings have no right essential overrings which are quasi-Baer.

Example 8. ([9, Example 3.12]) Let A be a QF ring which is not semisimple Artinian. Assume that A is right strongly FI-extending, and A has nontrivial central idempotents while the subring of A generated by 1_A (the unity of A) contains no nontrivial idempotents (e.g., $A = \mathbb{Q} \oplus \mathrm{Mat}_2(\mathbb{Z}_4)$, where $\mathbb{Z}_4$ is the ring of integers modulo 4). Let $1_{\prod_{i=1}^{\infty} A_i}$ denote the unity of $\prod_{i=1}^{\infty} A_i$, where $A_i = A$. Take R to be the subring of $\prod_{i=1}^{\infty} A_i$ generated by $1_{\prod_{i=1}^{\infty} A_i}$ and $\bigoplus_{i=1}^{\infty} A_i$. Observe that $Q(R) = \prod_{i=1}^{\infty} A_i = E(R_R)$ by [39, 2.1].

Now R has the following properties:
  (i) R is neither semiprime nor right FI-extending.
  (ii) $RB(Q(R)) = Q_{\mathfrak{FJ}}(R)$.

(iii) In general, $R\mathbf{B}(Q(R))$ is neither right extending nor right p.q.-Baer.

A ring R is called *reduced* if R has no nonzero nilpotent element. The following theorem is shown in [10].

Theorem 9. ([10, Theorem 6.6]) Every reduced ring has a Baer absolute right ring hull.

Theorem 9 generalizes results on the existence of Baer hulls by Mewborn for commutative semiprime rings [29], and by Hirano, Hongan and Ohori for reduced Utumi rings [23]. Part (ii) of the next example shows that the existence of Baer hulls is not guaranteed even for prime PI rings.

Example 10. ([10, Example 4.20]) (i) If R is a prime ring then $R = Q_{\mathfrak{F}\mathfrak{J}}(R) = Q_{\mathfrak{q}\mathfrak{B}}(R)$, and if $Z(R_R) \neq 0$ ([33], [28], and [12, Examples 4.3 and 4.4]) then R has no right essential overring which is Baer (hence R has no Baer absolute right ring hull).

(ii) For a field F and a positive integer $k > 1$, let $R = \mathrm{Mat}_k(F[x,y])$, where $F[x,y]$ is the polynomial ring over F with two indeterminates x and y. Then R has the following properties (observe that $Q(R) = E(R_R)$, hence $\widehat{Q}_{\mathfrak{K}}(R) = Q_{\mathfrak{K}}(R)$ for any class $\mathfrak{K}$ of rings):

    (1) R is quasi-Baer and right FI-extending.
    (2) $Q_{\mathfrak{B}}(R)$ does not exist.
    (3) $Q_{\mathfrak{E}}(R)$ does not exist.
    (4) $Q_{\mathfrak{q}\mathfrak{E}}(R)$ exists, and it is Baer.

For a ring R, let

$$\mathbf{B}_p(Q(R)) = \{e \in \mathbf{B}(Q(R)) \mid \text{there exists } x \in R \text{ with } RxR_R \leq^{\mathrm{ess}} eR_R\}.$$

In [10], we have proved that for a semiprime ring R, the p.q.-Baer absolute to $Q(R)$ right ring hull $\widehat{Q}_{\mathfrak{p}\mathfrak{q}\mathfrak{B}}(R)$ exists and is the subring $R\mathbf{B}_p(Q(R))$ of $Q(R)$ generated by R and $\mathbf{B}_p(Q(R))$.

Recall from [8] that a ring R is *principally right FI-extending* (resp., *finitely generated right FI-extending*) if every principal (resp., finitely generated ideal) of R is essential as a right R-module in a right ideal of R generated by an idempotent. We use $\mathfrak{p}\mathfrak{F}\mathfrak{J}$ (resp., $\mathfrak{f}\mathfrak{g}\mathfrak{F}\mathfrak{J}$) to denote the class of principally (resp., finitely generated) right FI-extending rings.

In [13] Burgess and Raphael study ring extensions of regular rings with bounded index. In particular, for a (von Neumann) regular ring R with bounded index, they obtain a closely related unique smallest overring, $R^{\#}$, which is "almost biregular" (see [13, p.76 and Theorem 1.7]). The next result shows that their ring $R^{\#}$ is precisely our principally right FI-extending right ring hull of a (von Neumann) regular ring R with bounded index.

Theorem 11. ([10, Theorem 4.8]) *Let R be a semiprime ring. Then:*

(i) R is p.q.-Baer if and only if $\mathbf{B}_p(Q(R)) \subseteq R$. Moreover, a right ring of quotient S of R is p.q.-Baer if and only if $\widehat{Q}_{\mathfrak{pqB}}(R) \subseteq S$.

(ii) $\widehat{Q}_{\mathfrak{pqB}}(R) = \widehat{Q}_{\mathfrak{pFJ}}(R) = \widehat{Q}_{\mathfrak{fgFJ}}(R)$.

In the following theorem we show connections between the right FI-extending right ring hulls of semiprime homomorphic images of a ring R and the subrings of $eQ(R)e$, where $e = e^2 \in Q(R)$.

Theorem 12. ([10, Theorem 5.5]) *Assume that $Q(R) = E(R_R)$ and I is a proper ideal of R such that I_R is closed in R_R. Then:*

(i) There exists $e \in \mathbf{I}(Q(R))$ such that $I_R \leq^{ess} (1-e)Q(R)_R$ and $I = R \cap (1-e)Q(R)$.

(ii) $eR = eRe$ and $R(1-e) = (1-e)R(1-e)$.

(iii) R/I is ring isomorphic to eRe.

(iv) $E(eRe_{eRe}) = eQ(R)e$ and $eQ(R)e = Q(eRe)$.

(v) If $\mathbf{P}(R) \subseteq I$, then R/I is semiprime and moreover $\widehat{Q}_{\mathfrak{FJ}}(R/I) \cong \langle eRe \cup \mathbf{B}(eQ(R)e) \rangle_{eQ(R)e}$.

A monoid G is called a *u.p.-monoid* (unique product monoid) if for any two nonempty finite subsets $A, B \subseteq G$ there exists an element $x \in G$ uniquely presented in the form ab, where $a \in A$ and $b \in B$. The class of u.p.-monoids is quite large and important (see [36] and [30]). For example, this class includes the right or left ordered monoids, submonoids of a free group, and torsion-free nilpotent groups. Every u.p.-monoid is cancellative, and every u.p.-group is torsion-free.

Theorem 13. ([10, Theorem 7.1]) *Let $R[G]$ be a semiprime monoid ring of a monoid G over a ring R. Then:*

(i) $\widehat{Q}_{\mathfrak{qB}}(R)[G] \subseteq \widehat{Q}_{\mathfrak{qB}}(R[G])$ and $\widehat{Q}_{\mathfrak{pqB}}(R)[G] \subseteq \widehat{Q}_{\mathfrak{pqB}}(R[G])$.

(ii) If G is a u.p.-monoid, then we have that $\widehat{Q}_{\mathfrak{qB}}(R)[G] = \widehat{Q}_{\mathfrak{qB}}(R[G])$ and $\widehat{Q}_{\mathfrak{pqB}}(R)[G] = \widehat{Q}_{\mathfrak{pqB}}(R[G])$.

From Theorem 13, we get the following immediately.

Corollary 14. ([10, Corollary 7.3]) *Let R be a semiprime ring and X a nonempty set of not necessarily commuting indeterminates. Then:*

(i) $\widehat{Q}_{\mathfrak{qB}}(R[x, x^{-1}]) = \widehat{Q}_{\mathfrak{qB}}(R)[x, x^{-1}]$, $\widehat{Q}_{\mathfrak{qB}}(R[X]) = \widehat{Q}_{\mathfrak{qB}}(R)[X]$, and $\widehat{Q}_{\mathfrak{qB}}(R[[X]]) = \widehat{Q}_{\mathfrak{qB}}(R)[[X]]$.

(ii) $\widehat{Q}_{\mathfrak{pqB}}(R[x, x^{-1}]) = \widehat{Q}_{\mathfrak{pqB}}(R)[x, x^{-1}]$, and also $\widehat{Q}_{\mathfrak{pqB}}(R[X]) = \widehat{Q}_{\mathfrak{pqB}}(R)[X]$.

Example 15. ([10, Example 7.4]) (i) Let $\mathbb{Z}[G]$ be the group ring of the group $G = \{1, g\}$ over the ring $\mathbb{Z}$ of integers. Then the group ring $\mathbb{Z}[G]$ is semiprime. But $\mathbb{Z}[G] = \widehat{Q}_{\mathfrak{qB}}(\mathbb{Z})[G] \subsetneq \widehat{Q}_{\mathfrak{qB}}(\mathbb{Z}[G])$. Thus the "u.p.-monoid" condition is not superfluous in Theorem 13(ii).

(ii) Let F be a field. Then $F[x]$ is a semiprime u.p.-monoid ring and $F[x] = Q(F)[x] \neq Q(F[x]) = F(x)$, where $F(x)$ is the field of fractions of $F[x]$. Thus "Q" cannot replace "$\widehat{Q}_{\mathfrak{qB}}$" in Theorem 13(ii).

(iii) From the fact that $\widehat{Q}_{\mathfrak{qB}}(R[[X]]) = \widehat{Q}_{\mathfrak{qB}}(R)[[X]]$ in Corollary 14(i), one might expect that $\widehat{Q}_{\mathfrak{pqB}}(R[[X]]) = \widehat{Q}_{\mathfrak{pqB}}(R)[[X]]$ holds. However, in [5, Example 2.3], there is a commutative von Neumann regular ring R (hence right p.q.-Baer), but the ring $R[[x]]$ is not right p.q.-Baer. Thus $\widehat{Q}_{\mathfrak{pqB}}(R) = R$, so $\widehat{Q}_{\mathfrak{pqB}}(R)[[x]] = R[[x]]$. Since $R[[x]]$ is not right p.q.-Baer, $\widehat{Q}_{\mathfrak{pqB}}(R[[x]]) \neq \widehat{Q}_{\mathfrak{pqB}}(R)[[x]]$.

From the following, one can see that the idempotent closure of $\mathrm{Mat}_n(R)$ is the matrix ring of n-by-n matrices over the idempotent closure of R and similarly for $T_n(R)$. Let 1_n denote the unity of $\mathrm{Mat}_n(R)$.

Proposition 16. ([10, Lemma 7.5]) *Let* $\delta \subseteq \mathbf{B}(Q(R))$ *and* $\Delta = \{1_n c \mid c \in \delta\}$. *Then*:
(i) $\mathrm{Mat}_n(\langle R \cup \delta \rangle_{Q(R)}) = \langle \mathrm{Mat}_n(R) \cup \Delta \rangle_{Q(\mathrm{Mat}_n(R))}$.
(ii) $Q(T_n(R)) = Q(\mathrm{Mat}_n(R)) = \mathrm{Mat}_n(Q(R))$.
(iii) $T_n(\langle R \cup \delta \rangle_{Q(R)}) = \langle T_n(R) \cup \Delta \rangle_{Q(\mathrm{Mat}_n(R))}$.

Using Proposition 16, the following can be shown (see [10]).

Theorem 17. ([10, Theorem 7.6]) *Assume that* R *is a semiprime ring and* n *is a positive integer. Then* $\widehat{Q}_{\mathfrak{K}}(\mathrm{Mat}_n(R)) = \mathrm{Mat}_n(\widehat{Q}_{\mathfrak{K}}(R))$, *where* $\mathfrak{K} = \mathfrak{qB}$, $\mathfrak{FJ}$, $\mathfrak{pqB}$, $\mathfrak{pFJ}$, *or* $\mathfrak{fgFJ}$.

We note that the statement of Theorem 17 does not hold true when $\mathfrak{K} = \mathfrak{qC}$. To show this, let K be a field and n be a positive integer such that $n > 1$. Then as shown in [9], the ring $\mathrm{Mat}_n(K[x]) = \mathrm{Mat}_n(K)[x]$ is not right quasi-continuous. So $Q_{\mathfrak{qC}}(\mathrm{Mat}_n(K)[x]) \neq \mathrm{Mat}_n(K)[x] (= Q_{\mathfrak{qC}}(\mathrm{Mat}_n(K))[x])$. Thus $Q_{\mathfrak{qC}}(\mathrm{Mat}_n(K)[x]) \neq Q_{\mathfrak{qC}}(\mathrm{Mat}_n(K))[x]$. Also we have that $Q_{\mathfrak{qC}}(\mathrm{Mat}_n(K[x])) \neq \mathrm{Mat}_n(K[x])(= \mathrm{Mat}_n(Q_{\mathfrak{qC}}(K[x])))$. Hence $Q_{\mathfrak{qC}}(\mathrm{Mat}_n(K[x])) \neq \mathrm{Mat}_n(Q_{\mathfrak{qC}}(K[x]))$.

Also we see that Theorem 17 does not hold for the case when $\mathfrak{K} = \mathfrak{B}$

or $\mathfrak{K} = \mathfrak{E}$. To show this, let $R = F[x,y]$, the polynomial ring over a field F with two indeterminates x and y. Then $\widehat{Q}_{\mathfrak{B}}(R) = R$, but $\widehat{Q}_{\mathfrak{B}}(\mathrm{Mat}_n(F[x,y]))$ does not exist for $n > 1$ as in Example 10(ii).

Theorem 17 suggests that we consider Morita equivalence.

Theorem 18. ([9, Theorem 5.7]) *Let R a semiprime ring. If R and a ring S are Morita equivalent, then $\widehat{Q}_{q\mathfrak{B}}(R)$ and $\widehat{Q}_{q\mathfrak{B}}(S)$ are Morita equivalent.*

Theorem 18 does not hold for the case of Baer absolute to $Q(R)$ right ring hulls. Let $R = F[x,y]$ be the polynomial ring over a field F and $S = \mathrm{Mat}_n(R)$ with $n > 1$. Then $\widehat{Q}_{\mathfrak{B}}(R) = R$, but $\widehat{Q}_{\mathfrak{B}}(S)$ does not exist as in Example 10(ii).

Theorem 19. ([10, Theorem 7.8]) *Let R be a semiprime ring. Then:*
   (i) $\widehat{Q}_{q\mathfrak{B}}(T_n(R)) = T_n(\widehat{Q}_{q\mathfrak{B}}(R)) = T_n(R\mathbf{B}(Q(R)))$.
   (ii) $T_n(R\mathbf{B}(Q(R))) = \widetilde{Q}_{\mathfrak{FJ}}(T_n(R))$.
   (iii) $\widehat{Q}_{\mathfrak{pq}\mathfrak{B}}(T_n(R)) = T_n(\widehat{Q}_{\mathfrak{pq}\mathfrak{B}}(R))$.

Theorem 19 provides us with examples of quasi-Baer right ring hulls and right FI-extending right ring hulls of nonsemiprime rings (recall Theorem 2).

For a ring R and a nonempty ordered set Γ, $\mathrm{CFM}_\Gamma(R), \mathrm{RFM}_\Gamma(R)$, and $\mathrm{CRFM}_\Gamma(R)$ denote the column finite, the row finite, and the column and row finite matrix rings over a ring R indexed by the set Γ, respectively. In [14, Theorem 1], it was shown that $\mathrm{CRFM}_\Gamma(R)$ is a Baer ring for all infinite index sets Γ if and only if R is semisimple Artinian. Our next result shows that the quasi-Baer property is always preserved by infinite matrix rings.

Theorem 20. ([9, Theorem 5.11]) (i) *If R is a quasi-Baer ring, then $\mathrm{CFM}_\Gamma(R), \mathrm{RFM}_\Gamma(R)$, and $\mathrm{CRFM}_\Gamma(R)$ are quasi-Baer rings.*

Conversely, if one of $\mathrm{CFM}_\Gamma(R), \mathrm{RFM}_\Gamma(R)$, and $\mathrm{CRFM}_\Gamma(R)$ is quasi-Baer, then so is R.

   (ii) *Assume that R is a semiprime ring. Then it follows that* $\widehat{Q}_{q\mathfrak{B}}(\mathrm{CFM}_\Gamma(R)) \subseteq \mathrm{CFM}_\Gamma(\widehat{Q}_{q\mathfrak{B}}(R))$, $\widehat{Q}_{q\mathfrak{B}}(\mathrm{RFM}_\Gamma(R)) \subseteq \mathrm{RFM}_\Gamma(\widehat{Q}_{q\mathfrak{B}}(R))$, *and* $\widehat{Q}_{q\mathfrak{B}}(\mathrm{CRFM}_\Gamma(R)) \subseteq \mathrm{CRFM}_\Gamma(\widehat{Q}_{q\mathfrak{B}}(R))$.

In [14, p.445] it is also shown that for any ring $R, \mathrm{CRFM}_\Gamma(R)$ is never right extending when Γ is countably infinite. For a semiprime ring R, Theorem 2, [7, Theorem 4.7], and Theorem 20(i) yield that $\mathrm{CRFM}_\Gamma(\widehat{Q}_{q\mathfrak{B}}(R))$ exists and is right FI-extending. Hence with each semiprime ring we can associate a right FI-extending ring which is not right extending. For a given

22

nonempty set Γ, we remark that R is quasi-Baer if and only if the column finite $\Gamma \times \Gamma$ upper triangular matrix ring is quasi-Baer as was shown in [16].

From Theorems 17 and 20, one might *expect* that either

$$\widehat{Q}_{q\mathfrak{B}}(\mathrm{CFM}_\Gamma(R)) = \mathrm{CFM}_\Gamma(\widehat{Q}_{q\mathfrak{B}}(R)) \text{ or } \widehat{Q}_{q\mathfrak{B}}(\mathrm{RFM}_\Gamma(R)) = \mathrm{RFM}_\Gamma(\widehat{Q}_{q\mathfrak{B}}(R)),$$

or

$$\widehat{Q}_{q\mathfrak{B}}(\mathrm{CRFM}_\Gamma(R)) = \mathrm{CRFM}_\Gamma(\widehat{Q}_{q\mathfrak{B}}(R)).$$

However, our next example shows that there is a commutative von Neumann regular ring R such that *none* of these equalities holds.

Example 21. ([9, Example 5.12]) Let $F_n = \mathbb{Z}_2$, the field of two elements, for $n = 1, 2, \ldots$, and

$$R = \left\{ (\gamma_n)_{n=1}^{\infty} \in \prod_{n=1}^{\infty} F_n \mid \gamma_n \text{ is eventually constant} \right\},$$

which is a subring of $\prod_{n=1}^{\infty} F_n$. Then the ring R is a commutative von Neumann regular ring. Let $\Gamma = \prod_{n=1}^{\infty} \mathbb{Z}_2$ as a set. We have shown in [8] that

$$\widehat{Q}_{q\mathfrak{B}}(\mathrm{CFM}_\Gamma(R)) \subsetneqq \mathrm{CFM}_\Gamma(\widehat{Q}_{q\mathfrak{B}}(R)), \quad \widehat{Q}_{q\mathfrak{B}}(\mathrm{RFM}_\Gamma(R)) \subsetneqq \mathrm{RFM}_\Gamma(\widehat{Q}_{q\mathfrak{B}}(R)),$$

and

$$\widehat{Q}_{q\mathfrak{B}}(\mathrm{CRFM}_\Gamma(R)) \subsetneqq \mathrm{CRFM}_\Gamma(\widehat{Q}_{q\mathfrak{B}}(R)).$$

Theorems 17 and 20 motivate the following questions: (1) *Is the right p.q.-Baer property preserved under the various infinite matrix ring extensions?* (2) *Does $\widehat{Q}_{pq\mathfrak{B}}(R)$ of a ring R have behavior similar to that of $\widehat{Q}_{q\mathfrak{B}}(R)$ for the various infinite matrix ring extensions?* Our next example provides negative answers to both of these questions.

Example 22. ([10, Example 7.11]) Let R be the ring defined as in Example 21 except that now we take $F_n = F$, where F is a fixed but arbitrary field. The ring R is a commutative von Neumann regular ring as in Example 21. Hence R is a right p.q.-Baer ring. Let $S = \mathrm{CFM}_\Gamma(R)$, where $\Gamma = \{1, 2, \ldots\}$. Take

$$a_1 = (0, 1, 0, 0, \ldots), \ a_2 = (0, 1, 0, 1, 0, 0, \ldots), \ a_3 = (0, 1, 0, 1, 0, 1, 0, 0, \ldots),$$

and so on, in R.

Let x be the element in S with a_n in the (n,n)-position for $n = 1, 2, \ldots$ and 0 elsewhere, and let

$$e = (q_n)_{n=1}^{\infty} \in Q(R) = \prod_{n=1}^{\infty} F_n$$

such that $q_{2n} = 1$ and $q_{2n-1} = 0$ for $n = 1, 2, \ldots$. Then $e = e^2 \in \mathbf{B}(Q(R))$, hence $eI \in \mathrm{CFM}_\Gamma(\widehat{Q}_{\mathfrak{q}\mathfrak{B}}(R)) \subseteq Q(S)$. Moreover,

$$eI \in \mathbf{B}_p(Q(S))$$

since $SxS_S \leq^{\mathrm{ess}} (eI)S_S$, where I is the unity matrix of S. But note that $eI \notin S$. Note that R is right p.q.-Baer, so $\widehat{Q}_{\mathfrak{p}\mathfrak{q}\mathfrak{B}}(R) = R$. Thus

$$\widehat{Q}_{\mathfrak{p}\mathfrak{q}\mathfrak{B}}(\mathrm{CFM}_\Gamma(R)) \nsubseteq \mathrm{CFM}_\Gamma(\widehat{Q}_{\mathfrak{p}\mathfrak{q}\mathfrak{B}}(R))$$

because $\mathrm{CFM}_\Gamma(\widehat{Q}_{\mathfrak{p}\mathfrak{q}\mathfrak{B}}(R)) = \mathrm{CFM}_\Gamma(R)$. Also

$$\widehat{Q}_{\mathfrak{p}\mathfrak{q}\mathfrak{B}}(\mathrm{CRFM}_\Gamma(R)) \nsubseteq \mathrm{CRFM}_\Gamma(\widehat{Q}_{\mathfrak{p}\mathfrak{q}\mathfrak{B}}(R))$$

and

$$\widehat{Q}_{\mathfrak{p}\mathfrak{q}\mathfrak{B}}(\mathrm{RFM}_\Gamma(R)) \nsubseteq \mathrm{RFM}_\Gamma(\widehat{Q}_{\mathfrak{p}\mathfrak{q}\mathfrak{B}}(R)).$$

Acknowledgments. The second author appreciates the partial support from Busan National University, 2003-2007, and the partial support from the Fund for the Promotion of International Scientific Research B-2, 2003, Aomori, Japan. The third author is thankful for the partial support received from OSU-Lima and MRI, OSU-Columbus.

References

1. A. Behn, Polycyclic group rings whose principal ideals are projective, *J. Algebra* **232** (2000), 697–707.
2. K. Beidar and R. Wisbauer, Strongly and properly semiprime modules and rings, Ring Theory, Proc. Ohio State-Denison Conf. (S. K. Jain and S. T. Rizvi (eds.)), World Scientific, Singapore, 1993, 58–94.
3. S. K. Berberian, Baer *-Rings, Springer-Verlag, Berlin-Heidelberg-New York, 1972.
4. G. F. Birkenmeier, H. E. Heatherly, J. Y. Kim and J. K. Park, Triangular matrix representations, *J. Algebra* **230** (2000), 558–595.
5. G. F. Birkenmeier, J. Y. Kim and J. K. Park, On quasi-Baer rings, Algebras and Its Applications (D. V. Huynh, S. K. Jain and S. R. López-Permouth (eds.)), *Contemp. Math.* 259, Amer. Math. Soc., Providence (2000), 67–92.

6. G. F. Birkenmeier, J. Y. Kim and J. K. Park, Principally quasi-Baer rings, *Comm. Algebra* 29 (2001), 639–660.

7. G. F. Birkenmeier, B. J. Müller and S. T. Rizvi, Modules in which every fully invariant submodule is essential in a direct summand, *Comm. Algebra* 30 (2002), 1395–1415.

8. G. F. Birkenmeier, J. K. Park and S. T. Rizvi, Ring hulls and application, Preprint.

9. G. F. Birkenmeier, J. K. Park and S. T. Rizvi, Ring hulls determined by central idempotents, Preprint.

10. G. F. Birkenmeier, J. K. Park and S. T. Rizvi, Ring hulls and their applications, Preprint.

11. G. F. Birkenmeier, J. K. Park and S. T. Rizvi, An injective hull with distinct ring structures, Preprint.

12. K. A. Brown, The singular ideals of group rings, *Quart. J. Math. Oxford* 28 (1977), 41–60.

13. W. D. Burgess and R. M. Raphael, On extensions of regular rings of finite index by central elements, Advances in Ring Theory (S. K. Jain and S. T. Rizvi (eds.)), Trends in Math., Birkhäuser, Boston (1997), 73–86.

14. V. P. Camillo, F. J. Costa-Cano and J. J. Simon, Relating properties of a ring and its ring of row and column finite matrices, *J. Algebra* 244 (2001), 435–449.

15. W. E. Clark, Twisted matrix units semigroup algebras, *Duke Math. J.* 34 (1967), 417–424.

16. J. Doh, H. L. Jin and J. K. Park, Quasi-Baer rings with essential prime radicals, Preprint

17. N. V. Dung, D. V. Huynh, P. F. Smith and R. Wisbauer, Extending Modules, Longman, Harlow, 1994.

18. C. Faith and Y. Utumi, Maximal quotient rings, *Proc. Amer. Math. Soc.* 16 (1965), 1084–1089.

19. M. Ferrero, Closed submodules of normalizing bimodules over semiprime rings, *Comm. Algebra* 29 (2001), 1513–1550.

20. K. R. Goodearl, Ring Theory: Nonsingular Rings and Modules, Marcel Dekker, New York, 1976.

21. K. R. Goodearl, Von Neumann regular rings: Connections with functional analysis, *Bull. Amer. Math. Soc. (N.S.)* 4 (1981), 125–134.

22. K. R. Goodearl, Simple Noetherian rings not isomorphic to matrix rings over domains, *Comm. Algebra* 12 (1984), 1412–1434.

23. Y. Hirano, M. Hongan and M. Ohori, On right P.P. rings, *Math. J. Okayama Univ.* 24 (1982), 99–109.

24. Y. Hirano, Open Problems, International Symposium on Ring Theory (G. F. Birkenmeier, J. K. Park and Y. S. Park (eds.)), Trends in Math., Birkhäuser, Boston (2001), 441–446.

25. S. K. Jain, P. Kanwar, S. Malik and J. B. Srivastava, KD_∞ is a CS-algebra, *Proc. Amer. Math. Soc.* 128 (2000), 397–400.

26. I. Kaplansky, Rings of Operators, Benjamin, New York, 1968.

27. T. Y. Lam, Lectures on Modules and Rings, Springer-Verlag, Berlin-

Heidelberg-New York, 1999.

28. J. Lawrence, A singular primitive ring, *Proc. Amer. Math. Soc.* 45 (1974), 59–62.
29. A. C. Mewborn, Regular rings and Baer rings, *Math. Z.* 121 (1971), 211–219.
30. J. Okniński, Semigroup Algebras, Marcel Dekker, New York, 1991.
31. B. L. Osofsky, Homological Properties of Rings and Modules, Doctoral Dissertation, Rutgers University, 1964
32. B. L. Osofsky, On ring properties of injective hulls, *Canad. Math. Bull.* 7 (1964), 405–413.
33. B. L. Osofsky, A non-trivial ring with non-rational injective hull, *Canad. Math. Bull.* 10 (1967), 275–282.
34. B. L. Osofsky, Personal Communication to S. T. Rizvi, Summer, 2003.
35. M. M. Parmenter and Y. Zhou, Finitely Σ-CS property of excellent extensions of rings, *Algebra Colloq.* 10 (2003), 17–21.
36. D. S. Passman, The Algebraic Structure of Group Rings, Wiley, New York, 1977.
37. A. Pollingher and A. Zaks, On Baer and quasi-Baer rings, *Duke Math. J.* 37 (1970), 127–138.
38. C. E. Rickart, Banach algebras with an adjoint operation, *Ann. Math.* 47 (1946), 528–550.
39. Y. Utumi, On quotient rings, *Osaka Math. J.* 8 (1956), 1–18.
40. Y. Utumi, On continuous rings and selfinjective rings, *Trans. Amer. Math. Soc.* 118 (1965), 158–173.

CONSTRUCTING MORPHIC RINGS

JIANLONG CHEN

Department of Mathematics
Southeast University
Nanjing, P.R.China 210096
E-mail: jlchen@seu.edu.cn

YUANLIN LI

Department of Mathematics
Brock University
St. Catharines, Canada L2S 3A1
E-mail: yli@spartan.ac.BrockU.CA

YIQIANG ZHOU

Department of Mathematics and Statistics
Memorial University of Newfoundland
St.John's A1C 5S7, Canada
E-mail: zhou@math.mun.ca

A ring R is called left morphic, if for every $a \in R$, $R/Ra \cong l(a)$ where $l(a)$ denotes the left annihilator of a in R. The ring R is called strongly left morphic if every matrix ring $\mathbb{M}_n(R)$ is left morphic. (Strongly) right morphic rings are defined analogously. For a subring C of a ring D, let $R[D,C] = \{(d_1, \cdots, d_n, c, c, \cdots) : d_i \in D, c \in C, n \geq 1\}$. A sufficient and necessary condition is obtained for $R[D,C]$ to be a left morphic ring. As consequences, a strongly left and right morphic, semiprimitive ring which is not regular is constructed. This example answered two questions both in the negative raised by Nicholson and Sánchez Campos in [2] and [4]. The example is also a counter-example to two questions on regular rings raised by Yue Chi Ming [5] and [6].

§1. All rings here are associative rings with identity. By the fundamental homomorphism theorem of modules, for any element a in a ring R, $R/l(a) \cong Ra$ where $l(a)$ denotes the left annihilator of a in R. An element a in a ring R is called left morphic if $R/Ra \cong l(a)$; equivalently, $a \in R$ is left morphic if and only if there exists $b \in R$ such that $Ra = l(b)$ and $Rb = l(a)$ (see [2, Lemma 1]). By Erlich [1], an element $a \in R$ is unit regular if and only if a is both (von Neumann) regular and left morphic. A ring R is called left morphic if every element of R is left morphic, and strongly left morphic if every matrix ring $\mathbb{M}_n(R)$ is left morphic. Right morphic rings and strongly right morphic rings are defined analogously. A left and right morphic ring is called a morphic ring. A strongly morphic ring means a strongly left and strongly right morphic ring. Left morphic rings were first introduced by Nicholson and Sánchez Campos [2] and were discussed in great detail in [2], [3] and [4]. The goal of this paper is to construct new examples of morphic rings. For a subring C of a ring D, let $R[D,C] = \{(d_1,\cdots,d_n,c,c,\cdots) : d_i \in D, c \in C, n \geq 1\}$. A sufficient and necessary condition is obtained for $R[D,C]$ to be a left morphic ring. As consequences, a strongly morphic, semiprimitive ring which is not regular is constructed. This example answered two questions both in the negative raised by Nicholson and Sánchez Campos in [2] and [4]. The example is also a counter-example to two questions on regular rings raised by Yue Chi Ming [5] and [6].

We use $J(R), Z_l(R)$ and $Z_r(R)$ to denote the Jacobson radical, left singular ideal and right singular ideal of the ring R respectively. The $n \times n$ matrix ring over R is denoted by $\mathbb{M}_n(R)$. By a subring of a ring R, we shall always mean a subring containing the identity of R. For a subring S of a ring R and $a \in R$, we let $\mathbf{l}_S(a) = \{s \in S : sa = 0\}$ and $\mathbf{l}_R(a) = \{r \in R : ra = 0\}$. Right annihilators are defined analogously. Sometimes, we simply write $\mathbf{l}(a)$ for $\mathbf{l}_R(a)$ and $\mathbf{r}(a)$ for $\mathbf{r}_R(a)$. A ring R is called unit regular if, for any $a \in R$, $a = aua$ for some unit of R. Regular rings here mean von Neumann regular rings.

§**2.** Let D be a ring and C be a subring of D. We set
$$R[D,C] = \{(d_1,\cdots,d_n,c,c,\cdots) : d_i \in D, c \in C, n \geq 1\}.$$
With addition and multiplication defined componentwise, $R[D,C]$ is a ring.

Theorem 0.1. *$R[D,C]$ is a left morphic ring if and only if the following hold:*

(1) D is a left morphic ring.

(2) For any $x \in C$ there exists $y \in C$ such that $\mathbf{l}_C(x) = Cy$, $\mathbf{l}_C(y) = Cx$, $\mathbf{l}_D(x) = Dy$, and $\mathbf{l}_D(y) = Dx$.

Proof. Write $R = R[D,C]$.

"$\Rightarrow$". Suppose R is a left morphic ring. Let $a_1 \in D$ and $a = (a_1,0,0,\cdots) \in R$. Then there exists $b = (b_1,b_2,\cdots) \in R$ such that $\mathbf{l}_R(a) =$

Rb, and $l_R(b) = Ra$. It follows that $l_D(a_1) \supseteq Db_1$ and $l_D(b_1) \supseteq Da_1$. For $s \in l_D(a_1)$, let $d = (s, 0, 0, \cdots) \in R$. Then $d \in l_R(a) = Rb$, showing that $s \in Db_1$. Therefore, $l_D(a_1) = Db_1$. Similarly, $l_D(b_1) = Da_1$. So D is a left morphic ring.

To show condition (2), let $x \in C$ and let $a = (x, x, \cdots) \in R$. Then there exists $b = (b_1, \cdots b_n, y, y, \cdots) \in R$ such that $l_R(a) = Rb$, and $l_R(b) = Ra$. It follows that $l_C(x) \supseteq Cy$, $l_C(y) \supseteq Cx$, $l_D(x) \supseteq Dy$, and $l_D(y) \supseteq Dx$.

If $s \in l_C(x)$, let $d = (s, s, \cdots) \in R$ and then $d \in l_R(a) = Rb$, showing that $s \in Cy$; thus $l_C(x) = Cy$.

If $t \in l_C(y)$, let $c = (c_i) \in R$ with $c_1 = \cdots = c_n = 0$ and $c_j = t$ for $j > n$. Then $c \in l_R(b) = Ra$, showing that $t \in Cx$; hence $l_C(y) = Cx$.

If $u \in l_D(x)$, let $d = (d_i) \in R$ with $d_1 = \cdots = d_{n+1} = u$ and $d_j = 0$ for $j > n + 1$. Then $d \in l_R(a) = Rb$, showing that $u \in Dy$; thus $l_D(x) = Dy$.

If $v \in l_D(y)$, let $c = (c_i) \in R$ with $c_1 = \cdots = c_n = 0$, $c_{n+1} = v$ and $c_j = 0$ for $j > n + 1$. Then $c \in l_R(b) = Ra$, showing that $v \in Dx$; hence $l_D(y) = Dx$.

Therefore, (2) holds.

"$\Leftarrow$". Suppose that conditions (1) and (2) hold. To show R is left morphic, let $a = (a_1, \cdots, a_n, x, x, \cdots) \in R$. Then, by (1) and (2), there exist $b_i \in D$ such that $l_D(a_i) = Db_i$ and $l_D(b_i) = Da_i$ for $i = 1, \cdots, n$ and there exists $y \in C$ satisfying

$$l_C(x) = Cy, \quad l_C(y) = Cx, \quad l_D(x) = Dy, \quad \text{and} \quad l_D(y) = Dx.$$

Let $b = (b_1, \cdots, b_n, y, y, \cdots) \in R$. We next show that $l_R(a) = Rb$ and $l_R(b) = Ra$. Clearly, we see that $l_R(a) \supseteq Rb$ and $l_R(b) \supseteq Ra$.

If $c = (c_i) \in l_R(a)$, then $0 = (c_1 a_1, \cdots, c_n a_n, c_{n+1} x, \cdots)$, showing that $c_i \in l_D(a_i) = Db_i$ for $i = 1, \cdots, n$ and $c_j \in l_D(x) = Dy$ for $j > n$. Thus, $c_i = d_i b_i$ with $d_i \in D$ for $i = 1, \cdots, n$. There exists $m > n$ such that $c_{m+1} = c_{m+2} = \cdots = u \in C$. Thus, $u \in l_C(x) = Cy$, so $u = zy$ for some $z \in C$. Moreover, $c_j = d_j y$ with $d_j \in D$ for $j = n + 1, \cdots, m$. Therefore, $c = (d_1, \cdots, d_m, z, z, \cdots)b \in Rb$. So $l_R(a) = Rb$.

If $c' = (c'_i) \in l_R(b)$, then $0 = (c'_1 b_1, \cdots, c'_n b_n, c'_{n+1} y, \cdots)$, showing that $c'_i \in l_D(b_i) = Da_i$ for $i = 1, \cdots, n$ and $c'_j \in l_D(y) = Dx$ for $j > n$. Thus, $c'_i = d'_i a_i$ with $d'_i \in D$ for $i = 1, \cdots, n$. There exists $m > n$ such that $c'_{m+1} = c'_{m+2} = \cdots = u' \in C$. Thus, $u' \in l_C(y) = Cx$, so $u' = z'x$ for some $z' \in C$. Moreover, $c'_j = d'_j x$ with $d'_j \in D$ for $j = n + 1, \cdots, m$. Therefore, $c' = (d'_1, \cdots, d'_m, z', z', \cdots)a \in Ra$. So $l_R(b) = Ra$. Hence, R is left morphic. $\qquad\square$

Corollary 0.1. $R[D, D]$ *is a left morphic ring if and only if D is a left morphic ring.*

Example 0.1. Let $R = R[D, C]$ where $D = \mathbb{M}_2(\mathbb{Z}_2)$ and $C = \{\left(\begin{smallmatrix} x & y \\ 0 & x \end{smallmatrix}\right) : x, y \in \mathbb{Z}_2\}$. Then the following hold:

(1) R is a morphic ring.
(2) R is semiprimitive.
(3) R is left and right nonsingular.
(4) R is not regular.

Proof. (1). $C = \{\left(\begin{smallmatrix} 0 & 0 \\ 0 & 0 \end{smallmatrix}\right), \left(\begin{smallmatrix} 1 & 0 \\ 0 & 1 \end{smallmatrix}\right), \left(\begin{smallmatrix} 1 & 1 \\ 0 & 1 \end{smallmatrix}\right), \left(\begin{smallmatrix} 0 & 1 \\ 0 & 0 \end{smallmatrix}\right)\}$. If $a = \left(\begin{smallmatrix} 0 & 0 \\ 0 & 0 \end{smallmatrix}\right)$, let $b = \left(\begin{smallmatrix} 1 & 0 \\ 0 & 1 \end{smallmatrix}\right)$; If $a = \left(\begin{smallmatrix} 1 & 0 \\ 0 & 1 \end{smallmatrix}\right)$ or $\left(\begin{smallmatrix} 1 & 1 \\ 0 & 1 \end{smallmatrix}\right)$, let $b = \left(\begin{smallmatrix} 0 & 0 \\ 0 & 0 \end{smallmatrix}\right)$; If $a = \left(\begin{smallmatrix} 0 & 1 \\ 0 & 0 \end{smallmatrix}\right)$, let $b = \left(\begin{smallmatrix} 0 & 1 \\ 0 & 0 \end{smallmatrix}\right)$.

In either case, we have
$$\mathbf{l}_C(a) = Cb, \ \mathbf{l}_C(b) = Ca, \ \mathbf{l}_D(a) = Db, \ \mathbf{l}_D(b) = Da$$
and
$$\mathbf{r}_C(a) = bC, \ \mathbf{r}_C(b) = aC, \ \mathbf{r}_D(a) = bD, \ \mathbf{r}_D(b) = aD.$$
Since $\mathbb{M}_2(\mathbb{Z}_2)$ is morphic, by Theorem 1, R is morphic.

(2). Let $a = (a_1, \cdots, a_n, x, x, \cdots) \in J(R)$. For any $r \in D$, let $b = (b_i) \in R$ with $b_1 = \cdots = b_n = r$ and $b_j = 0$ for $j > n$, and let $c = (c_i) \in R$ with $c_i = 0$ for $i = 1, \cdots, n$ and $c_{n+1} = r$ and $c_j = 0$ for $j > n + 1$. Then $1 - ba$ and $1 - ca$ are units in R. It follows that $1 - ra_i$ and $1 - rx$ are units in D for $i = 1, \cdots, n$. So, $a_i, x \in J(D)$ for $i = 1, \cdots, n$. But $J(D) = 0$, so $a = 0$.

(3). By (1), R is left and right morphic. Hence $Z_r(R) = Z_l(R) = J(R)$ by [2, Theorem 24], so R is left and right nonsingular by (2).

(4). R is not regular because its image C is not regular. $\quad\square$

Remark 0.1. By Erlich [1], every unit regular ring is morphic, and every regular left morphic ring is unit regular. Noting that $\mathbb{Z}_4$ is a morphic ring which is not unit regular, Nicholson and Sánchez Campos [2, Question, p.393] raised the question whether a morphic ring R with $J(R) = 0$ is necessarily regular. This is shown to be false in general by Example 3. In Yue Chi Ming [5, Question 5, p.41] and [6, Question 2, p.232], it was asked whether a semiprime ring R such that every principal one-sided ideal is the annihilator of an element of R is regular or whether a left nonsingular ring R such that every principal one-sided ideal is the annihilator of an element of R is regular. Example 3 clearly settled these questions both in the negative.

Furthermore, we prove that the ring R in Example 3 is strongly morphic.

Lemma 0.1. *Let C be a subring of a ring D. Then $\mathbb{M}_n(R[D,C]) \cong R[\mathbb{M}_n(D), \mathbb{M}_n(C)]$*

Proof. The map $\theta : \mathbb{M}_n(R[D,C]) \to R[\mathbb{M}_n(D), \mathbb{M}_n(C)]$ defined by

$$\begin{bmatrix} (x_{11}^{(i)}) & (x_{12}^{(i)}) & \cdots & (x_{1n}^{(i)}) \\ (x_{21}^{(i)}) & (x_{22}^{(i)}) & \cdots & (x_{2n}^{(i)}) \\ \vdots & \vdots & \vdots & \\ (x_{n1}^{(i)}) & (x_{n2}^{(i)}) & \cdots & (x_{nn}^{(i)}) \end{bmatrix} \mapsto \left(\begin{bmatrix} x_{11}^{(i)} & x_{12}^{(i)} & \cdots & x_{1n}^{(i)} \\ x_{21}^{(i)} & x_{22}^{(i)} & \cdots & x_{2n}^{(i)} \\ \vdots & \vdots & \vdots & \\ x_{n1}^{(i)} & x_{n2}^{(i)} & \cdots & x_{nn}^{(i)} \end{bmatrix} \right)$$

is the required ring isomorphism. $\qquad\square$

For convenience, we introduce the following definition.

Definition 0.1. Let C be a subring of a ring D. An element $x \in C$ is called left $[D,C]$-morphic if there exists $y \in C$ such that $l_C(x) = Cy$, $l_C(y) = Cx$, $l_D(x) = Dy$, and $l_D(y) = Dx$. In this case, we say that x is left $[D,C]$-morphic to y.

Lemma 0.2. *Let C be a subring of a ring D and let $x \in C$ be a left $[D,C]$-morphic element. Then for any unit u of C, ux and xu are left $[D,C]$-morphic.*

Proof. Suppose u is a unit of C and $x, y \in C$. If x is left $[D,C]$-morphic to y, then it can easily be verified that ux is left $[D,C]$-morphic to yu^{-1} and xu is left $[D,C]$-morphic to $u^{-1}y$. $\qquad\square$

Lemma 0.3. *Let C be a subring of a ring D. If $x_i \in C$ $(i = 1, \cdots, n)$ are left $[D,C]$-morphic. Then $\begin{bmatrix} x_1 & 0 & \cdots & 0 \\ 0 & x_2 & \cdots & 0 \\ \vdots & \vdots & \vdots & \\ 0 & 0 & \cdots & x_n \end{bmatrix} \in \mathbb{M}_n(C)$ is left $[\mathbb{M}_n(D), \mathbb{M}_n(C)]$-morphic.*

Proof. If $x_i, y_i \in C$ such that x_i is left $[D,C]$-morphic to y_i for $i = 1, \cdots, n$. Then $\begin{bmatrix} x_1 & 0 & \cdots & 0 \\ 0 & x_2 & \cdots & 0 \\ \vdots & \vdots & \vdots & \\ 0 & 0 & \cdots & x_n \end{bmatrix}$ is left $[\mathbb{M}_n(D), \mathbb{M}_n(C)]$-morphic to

$$\begin{bmatrix} y_1 & 0 & \cdots & 0 \\ 0 & y_2 & \cdots & 0 \\ \vdots & \vdots & & \vdots \\ 0 & 0 & \cdots & y_n \end{bmatrix}.$$

$\square$

Example 0.2. Let $D = \mathbb{M}_2(\mathbb{Z}_2)$ and $C = \{\left(\begin{smallmatrix} x & y \\ 0 & x \end{smallmatrix}\right) : x, y \in \mathbb{Z}_2\}$. Then $R[D, C]$ is a strongly morphic ring.

Proof. We prove that $R[D, C]$ is strongly left morphic; it is similar to show that $R[D, C]$ is strongly right morphic. By Lemma 5, we only need to show that $R[\mathbb{M}_n(D), \mathbb{M}_n(C)]$ is left morphic for all $n \geq 1$. Since $\mathbb{M}_n(D)$ is clearly left morphic, it suffices to show that every $0 \neq A = (a_{ij}) \in \mathbb{M}_n(C)$ is left $[\mathbb{M}_n(D), \mathbb{M}_n(C)]$-morphic by Theorem 1. Note that $C = \{\left(\begin{smallmatrix} 0 & 0 \\ 0 & 0 \end{smallmatrix}\right), \left(\begin{smallmatrix} 1 & 0 \\ 0 & 1 \end{smallmatrix}\right), \left(\begin{smallmatrix} 1 & 1 \\ 0 & 1 \end{smallmatrix}\right), \left(\begin{smallmatrix} 0 & 1 \\ 0 & 0 \end{smallmatrix}\right)\}$.

If a_{ij} is a unit of C for some i and j, interchanging the 1th and ith rows and interchanging the 1th and jth columns will bring a_{ij} to the $(1, 1)$-entry. Assume now that a_{11} is a unit of C. Let $k > 1$. Now subtract the first row times $a_{11}^{-1} a_{k1}$ from the kth row and subtract the first column times $a_{11}^{-1} a_{1k}$ from the kth column. These transformations change A to

$$B = \begin{bmatrix} b_{11} & 0 & \cdots & 0 \\ 0 & b_{22} & \cdots & b_{2n} \\ \vdots & \vdots & & \vdots \\ 0 & b_{n2} & \cdots & b_{nn} \end{bmatrix}.$$

If none of a_{ij} is a unit of C, then a_{ij} is equal to 0 or $\left(\begin{smallmatrix} 0 & 1 \\ 0 & 0 \end{smallmatrix}\right)$ and $a_{ij} = \left(\begin{smallmatrix} 0 & 1 \\ 0 & 0 \end{smallmatrix}\right)$ for some i and j. As above, we can bring this a_{ij} to $(1, 1)$-entry by elementary transformations. Assume that $a_{11} = \left(\begin{smallmatrix} 0 & 1 \\ 0 & 0 \end{smallmatrix}\right)$. Let $k > 1$. Now subtracting the first row from the kth row when $a_{k1} \neq 0$ and subtracting the first column from the kth column when $a_{1k} \neq 0$ will change A to a matrix of the same form as B above.

Thus, continuing in this way, we can change A to a diagonal by elementary transformations. Therefore, there exist units U and V of $\mathbb{M}_n(C)$ such that $UAV = \begin{bmatrix} a_1 & 0 & \cdots & 0 \\ 0 & a_2 & \cdots & 0 \\ \vdots & \vdots & & \vdots \\ 0 & 0 & \cdots & a_n \end{bmatrix}$, where $a_i \in C$ for $i = 1, \cdots, n$. (In fact, U and V are products of certain elementary matrices over C.) But, $R[D, C]$ is left morphic by Example 3. Thus, by Theorem 1, every element of C is left

$[D, C]$-morphic. So by Lemma 8, UAV is left $[\mathbb{M}_n(D), \mathbb{M}_n(C)]$-morphic. Therefore, by Lemma 7, A is left $[\mathbb{M}_n(D), \mathbb{M}_n(C)]$-morphic. $\qquad\square$

Remark 0.2. Since every unit regular ring is strongly left morphic, it is raised in [4, Question] whether a strongly left morphic, semiprimitive ring is necessarily unit regular. The answer is "No" by Example 9.

ACKNOWLEDGEMENTS

The research was carried out during a visit by the first author to Memorial University of Newfoundland and Brock University. He would like to gratefully acknowledge the financial support and kind hospitality from both institutes. The first author was supported by the National Natural Science Foundation of China (No. 10171011) and the Teaching and Research Award Program for Outstanding Young Teachers in Higher Education Institutes of MOE, P.R.C. The second author was supported by NSERC of Canada, and the third by NSERC (Grant OGP0194196) and a grant from the Office of Dean of Science, Memorial University.

References

1. G. Erlich, Units and one-sided units in regular rings, *Trans.A.M.S.* **216**(1976), 81-90.
2. W.K.Nicholson and E.Sánchez Campos, Rings with the dual of the isomorphism theorem. *J. Algebra* **271**(2004), 391-406.
3. W.K.Nicholson and E.Sánchez Campos, Principal rings with the dual of the isomorphism theorem, *Glasgow Math. J.* **46**(2004), 181-191.
4. W.K.Nicholson and E.Sánchez Campos, Morphic modules, Preprint, 2004.
5. R.Yue Chi Ming, On p-injectivity, YJ-injectivity and quasi-Frobeniusean rings, *Comment. Math. Univ. Carolinae* **43**(1)(2002), 33-42.
6. R.Yue Chi Ming, On injectivity and p-injectivity, IV. *Bull. Korean Math. Soc.*, **40**(2)(2003), 223-234.

SOME PROPERTIES OF ADDITIVE ENDOMORPHISMS AND MAPS ON GROUPS

YONG UK CHO

Department of Mathematics
Silla University
Pusan 617-736, Korea
E-mail: yucho@silla.ac.kr

In this expository paper, for any right R-module M, we introduce a concept of GM module and some characterizations of GM modules. Also, for any near-ring R, we can define the centralizer near-ring of S and a unitary R-group G which is a more general concept then old centralization, and we introduce an MR group and some properties of MR groups.

1. Introduction

Throughout this paper, all rings or all near-rings R are associative, all modules are right R-modules and for a near-ring R, we consider representations of R as R-groups.

For any group G and a nonempty subset S of $End(G)$, we know the centralizer of S and G as

$$C(S; G) = \{f \in M(G) \mid \alpha f = f\alpha \ \forall \alpha \in S\}.$$

Also, for a nonempty subset S of the distributive elements on G, we can define the centralizer near-ring of S and a unitary R-group G.

Next, for any right R module M, we define a new concept GM module and investigate some characterizations of GM modules. Also, for any near-ring R, we introduce an R-group with MR-property and some properties of MR groups as analogous properties of GM modules. Furthermore, we will survey that the commutativity of ring under faithful GM modules and faithful MR groups.

A near-ring R with $(R, +)$ abelian is called *abelian*. Consider the following notations: Given a near-ring R, $R_0 = \{a \in R \mid 0a = 0\}$ is called the *zero symmetric part* of R, and

$$R_d = \{a \in R \mid a \ \ is \ \ distributive\}$$

is called the *distributive part* of R. We note that R_0 is a subnear-ring of R, but R_d is a subsemigroup of R under multiplication.

Let $(G, +)$ be a group (not necessarily abelian). We will use right operations in the near-ring case to distinguish from left operations in the ring case in this paper. In the set

$$M(G) := \{f \mid f : G \longrightarrow G\}$$

of all self maps of G, if we define the sum $f + g$ of any two mappings f, g in $M(G)$ by the rule $x(f + g) = xf + xg$ for all $x \in G$ and the product $f \cdot g$ by the rule $x(f \cdot g) = (xf)g$ for all $x \in G$, then $(M(G), +, \cdot)$ becomes a near-ring. It is called the *self map near-ring* of the group G. Also, if we denote the set

$$M_0(G) := \{f \in M(G) \mid of = o\}$$

for the additive group G with identity o, then $(M_0(G), +, \cdot)$ is a zero symmetric near-ring.

Let R be any near-ring and G an additive group. Then G is called an *R-group* if there exists a near-ring homomorphism

$$\theta : (R, +, \cdot) \longrightarrow (M(G), +, \cdot).$$

Such a homomorphism θ is called a *representation* of R on G, we write xr for $x(\theta_r)$ for all $x \in G$ and $r \in R$. If R is unitary and $x1 = x$ for all $x \in G$, then R-group G is called *unitary*. Note that R itself is an R-group called the *regular group*.

Naturally, every group G has an $M(G)$-group structure by applying the $f \in M(G)$ to the $x \in G$ as a scalar multiplication xf.

An R-group G with the property that for each x, $y \in G$ and $a \in R$, $(x + y)a = xa + ya$ is called a *distributive R-group*, and also an R-group G with $(G, +)$ is abelian is called an *abelian R-group*. For example, if $(G, +)$ is abelian, then $M(G)$ is an abelian near-ring and moreover, G is an abelian $M(G)$-group. On the other hand, every distributive near-ring R is a distributive R-group.

A near-ring R is called *distributively generated* (briefly, *D.G.*) by S if

$$(R, +) = gp < S >= gp < R_d >$$

where S is a semigroup of distributive elements in R, in particular, $S = R_d$. This D.G. near-ring R which is generated by S is denoted by (R, S).

On the other hand, the set of all distributive elements of $M(G)$ are obviously the set $End(G)$ of all endomorphisms of the group G, that is,

$$(M(G))_d = End(G)$$

which is a semigroup under composition, but not yet a near-ring. Here we denote that $E(G)$ is the D.G. near-ring generated by $End(G)$, that is,

$$E(G) = gp < End(G) > .$$

Obviously, $E(G)$ is a subnear-ring of $(M_0(G), +, \cdot)$. Thus we say that $E(G)$ is the *endomorphism near-ring* of the group G.

For the remainder basic concepts and results on ring and near-ring case, we refer to [1], [8] and [9].

2. Results

Hereafter, we can introduce similar notions of AE rings [10] for right R-modules and R-groups. First, we introduce the concepts of GM-property of a right R-module and MR-property of an R-group, and then investigate their properties.

For any ring R, right R-modules M and N, the set of all R-module homomorphisms from M to N is denoted by $Hom_R(M, N)$ and the set of all group homomorphisms from M to N is $Hom(M, N) := Hom_{\mathbb{Z}}(M, N)$, in particular we denote that $End_R(M) := Hom_R(M, M)$ and $End(M) := End_{\mathbb{Z}}(M)$, In this case, M is called a GM *module* over R if every group homomorphism of M is an R-module homomorphism, that is,

$$End(M) = End_R(M).$$

In particular, R is called a GM *ring* if R is a GM module as a right R-module, that is, for all $f \in End_{\mathbb{Z}}(R)$, x, $r \in R$, we have $f(xr) = f(x)r$.

Examples 2.1. (1) $\mathbb{Z}$ *and* $\mathbb{Q}$ *are* GM *modules because* $End(\mathbb{Z}) = \mathbb{Z} = End_{\mathbb{Z}}\mathbb{Z}$ *and* $End(\mathbb{Q}) = \mathbb{Q} = End_{\mathbb{Q}}\mathbb{Q}$.

(2) *Every subgroup of* $(\mathbb{Q}, +)$ *is a* GM *module, for example,* $\mathbb{Z}_{(p)} = \{m/p^n \mid m, n \in \mathbb{Z}, n \geq 1\}$ *is a* GM *module, where* p *is a prime.*

(3) *For a multiplicatively closed set* S *of* $\mathbb{Z}$, *localization* $\mathbb{Z}_S$ *is a* GM *module.*

Proposition 2.2 [2]. *Let* $\{M_i \mid i \in \Lambda\}$ *be any family of right R-modules. Then each M_i is a GM module for all $i \in \Lambda$ if and only if $M := \oplus M_i$ is a GM module.*

Proposition 2.3. *Let R be a GM ring. Then for any $x \in R$, xR is a GM ring. Furthermore, this xR is also a GM module as an R-module.*

Proof. Let $f \in End(xR)$, and $g : R \longrightarrow R$ be defined by $g(a) = f(xa)$ for all $a \in R$. Then $g \in End_{\mathbb{Z}}(R)$. This implies that $g(axb) = g(a)xb$, because $End_{\mathbb{Z}}(R) = End_R(R)$. So we have

$$f(xaxb) = g(axb) = g(a)xb = f(xa)xb.$$

Hence, for any $x \in R$, xR is a GM ring. Obviously, we can check that xR is a GM module as an R-module. $\blacksquare$

Applying Propositions 2.2 and 2.3, we obtain the following:

Corollary 2.4. *Let R be a GM unitary ring. Then all finitely generated right ideals and all direct sums of principle right ideals are GM rings.*

From the faithful GM-property, we get a commutativity of rings.

Proposition 2.5 [2]. *Let M be a right R-module. If M is a faithful GM module, then R is a commutative ring.*

Next, we shall treat a D.G. near-ring R generated by S, and a faithful R-group G, furthermore, there is a module like concept as follows: Let (R, S) be a D.G. near-ring. Then an additive group G is called a *D.G. (R, S)-group* if there exists a D.G. near-ring homomorphism

$$\theta \ : \ (R, S) \longrightarrow (E(G), End(G))$$

such that $S\theta \subset End(G)$. If we write that xr instead of $x(\theta_r)$ for all $x \in G$ and $r \in R$, then an D.G. (R, S)-group is an additive group G satisfying the following conditions:

$$x(rs) = (xr)s, \quad x(r + s) = xr + xs, \quad (x + y)s = xs + ys,$$

for all x, $y \in G$ and all r, $s \in S$.

Such a homomorphism θ is called a *D.G. representation* of (R, S) on G. This D.G. representation is said to be *faithful* if $Ker\theta = \{0\}$. In this case, we say that G is called a *faithful D.G. (R, S)-group* [3], [7], [9].

Let G and T be two R-groups. Then the mapping $f : G \longrightarrow T$ is called a *R-group homomorphism* if for all x, $y \in G$ and $a \in R$, (i) $(x+y)f = xf+yf$ and (ii) $(xa)f = (xf)a$. In this paper, we call that the mapping $f : G \longrightarrow T$ with the condition $(xa)f = (xf)a$ is an *R-homogeneous map* (or simply, *R-map*) [6]. We define the set

$$M_R(G, \ T) := \{f \in M(G, \ T) \mid (xr)f = (xf)r, \ \forall \ x \in G, \ r \in R\}$$

of all R-homogeneous maps from G to T.

For any near-ring R and R-group G, we write the set

$$M_R(G) := \{f \in M(G) \mid (xr)f = (xf)r, \ \forall \, x \in G, \ \ r \in R\}$$

of all R-homogeneous maps on G as defined previously.

On the other hand, an element $a \in R$ is said to *distributive on G* if $(x + y)a = xa + ya$ for all $x, \ y \in G$.

Putting $D_R(G)$ the set of all distributive elements on G, $D_R(G)$ becomes a ring whenever G is abelian. In particular, every unitary abelian near-ring contains a unitary ring.

The following two statements are motivation of MR-property of R-groups.

Lemma 2.6. *Assume that G is an abelian D.G. (R, S)-group. Then the set $M_R(G) := \{f \in M(G) \mid (xr)f = (xf)r, \ \forall \, x \in G, \ \ r \in R\}$ is a subnear-ring of $M(G)$.*

On the other hand, for a group G and a nonempty subset S of $End(G)$, we define the *centralizer* of S in G as following:

$$C(S; G) = \{f \in M(G) \mid \alpha f = f\alpha \ \forall \, \alpha \in S\},$$

which is a subnear-ring of $M(G)$, we say that $C(S; G)$ is the *centralizer near-ring* of S and G. This is an extended concept of centralizer a near-ring which is introduced in [5, 6], at there, S is a subsemigroup of $End(G)$.

Also, for any endomorphism α of G, the centralizer of α in G is $C(\{\alpha\}; G)$ we denote it simply by $C(\alpha; G)$.

Note that obviously, $C(\alpha; G)$ is a subnear-ring of $M(G)$ and

$$C(S; G) \ = \ \cap_{\alpha \in S} C(\alpha; G).$$

Also, we see that $C(1_G; G) \ = \ M(G)$ and $C(0; G) \ = \ M_0(G)$.

In ring and module theory, we obtain the following important structure for near-ring and R-group theory:

Considering each element $a \in R$ is an endomorphism of V and

$$M_R(V) := \{f \in M(V) \mid af = fa, \ \forall \, a \in R\}$$

we see that

$$M_R(V) \ = \ C(R; \ V)$$

is the centralizer near-ring of R and V. Also

$$M_R(V) \ = \ \cap_{a \in R} M_a(V).$$

Proposition 2.7 [2]. *Let R be a semisimple ring with unity 1 and let M be a right R-module. Then $M_R(M)$ is a semisimple near-ring.*

Now we get a more general concept then centralization which is known till now.

Proposition 2.8. *Let R be a near-ring with unity 1 and G a unitary R-group. Then for any nonempty subset S of $D_R(G)$,*

$$M_S(G) := C(S;G) = \{f \in M(G) \mid af = fa, \ \forall\, a \in S\}$$

is a centralizer subnear-ring of $M(G)$ and

$$M_S(G) \ = \ \bigcap_{a \in S} M_{\{a\}}(G).$$

Moreover, we see that $M_{\{1\}}(G) \ = \ M(G)$ and $M_{\{0\}}(G) \ = \ M_0(G)$.

In Proposition 2.8, $M_S(G)$ is called the *centralizer near-ring* of S and G which is a generalization of centralizer near-rings in [4, 5, 6]. We denote $M_{\{a\}}(G)$ by $M_a(G)$ for convenance. Then

$$M_S(G) \ = \ \bigcap_{a \in S} M_a(G).$$

Corollary 2.9. *([6]) Let R be a ring with unity 1 and V a unitary right R-module. Then $M_R(V) := \{f \in M(V) \mid (xa)f = (xf)a, \ for \ all \ x \in V, \ a \in R\}$ is a subnear-ring of $M(V)$.*

Lemma 2.10. *([9]) Let G be a faithful R-group. Then we have the following conditions:*
(1) If $(G, +)$ is abelian, then $(R, +)$ is abelian.
(2) If G is distributive, then R is distributive.

Applying Lemma 2.10, we get the following:

Proposition 2.11. *If G is a distributive abelian faithful R-group, then R is a ring.*

The following statement which is obtained from Lemma 2.10 and property of faithful D.G. (R, S)-group is a generalization of the Proposition 2.11.

Proposition 2.12. *Let (R, S) be a D.G. near-ring. If G is an abelian faithful D.G. (R, S)-group, then R is a ring.*

Finally, we also introduce the *MR*-property of *R*-group, which is motivated by the Lemma 2.6. An *R*-group G is called an *MR group* over near-ring *R*, provided that every mapping on G is an *R*-homogeneous map of G, that is,

$$M(G) = C(R; G)$$

From now on, we introduce two characterizations of *MR* groups in the following propositions 2.13 and 2.15.

Proposition 2.13 *Let G be an R-group. Then G is an MR group if and only if G has the condition that $ar = a$ for all $a \in G$ and $r \in R$.*

Proof. Suppose G has the condition that $ar = a$ for all $a \in G$ and $r \in R$. Let $f : G \longrightarrow G$ be any given mapping on G. Then by hypothesis,

$$(ar)f = (a)f = (a)fr$$

for all $a \in G$ and $r \in R$. Thus the 'if part' is proved.

Now we will prove 'only if part'. Assume that G is an *MR* group and assume to the contrary that there is a in G and r in R such that $ar \neq a$. Define a mapping $f : G \longrightarrow G$ given by $(ar)f = a$ and $(x)f = x$ for all $x \in G$ which is not equal to ar. Then clearly, $f \in M(G)$, however,

$$(ar)f = a \neq ar = (a)fr.$$

This implies that f is not an *R*-homogeneous map, a contradiction. ∎

Examples 2.14. (1) *An additive group G with multiplication on G: $ab = a$ for all a, $b \in G$ (Example 1.4, (b) in [9]). We call these near-rings left thread near-rings. Every regular R-group which is left thread is an MR-group.*

(2) *If $M(V)$ is a centralizer near-ring determined by R and V then R-module V is an MR group.*

From the Proposition 2.13, we can directly obtain a characterization of *MR* groups for direct sum whose proof is different from the proof of the Proposition 2.2 for *GM*-property of *R*-module as following.

Proposition 2.15. *Let $\{G_i \mid i \in \Lambda\}$ be any family of R-groups. Then each G_i is an MR group if and only if $G := \oplus \, G_i$ is an MR group.*

A similar property of Proposition 2.5 for *MR* group is obtained, using the variables on the right side of maps on *R*-group as defined previously, together with Proposition 2.11. Thus we have the following:

Proposition 2.16. *Let G be an R-group.*

(1) If G is a faithful MR group, then R is a commutative near-ring.

(2) If G is a faithful distributive abelian MR group, then R is a commutative ring.

Proof. Let a, $b \in R$. Define a mapping $f : G \longrightarrow G$ given by $xf = xa$, for all $x \in G$. Then clearly, $f \in M(G)$. Since G is an MR group, $f \in C(R;G)$. Thus we have the equalities: $(xb)f = (xb)a = x(ba)$ and since $f \in M(G) = C(R;G)$,

$$(xb)f = (xf)b = (xa)b = x(ab).$$

Since G is a faithful R-group, these two equalities implies that $ab = ba$. Hence R is a commutative near-ring. ∎

From the Propositions 2.10 and 2.16, we get the following statement.

Corollary 2.17. *If G is an abelian faithful MR group over near-ring R, then R becomes a commutative ring.*

References

1. F. W. Anderson and K. R. Fuller, Rings and Categories of Modules, Springer-Verlag, New York, Heidelberg, Berlin, 1974.
2. Y. U. Cho, R-homomorphisms and R-homogeneous maps, *J. Korean Math. Soc.* to appear, (2005).
3. C. G. Lyons and J. D. P. Meldrum, Characterizing series for faithful D.G. near-rings, *Proc. Amer. Math. Soc.* 72 (1978), 221–227.
4. C. J. Maxson and K. C. Smith, The centralizer of a group endomorphism, *J. Algebra* 57 (1979), 441–448.
5. C. J. Maxson and K. C. Smith, Simple near-ring centralizers of finite rings, *Proc. Amer. Math. Soc.* 75 (1979), 8–12.
6. C. J. Maxson and A. B. Van der Merwe, Forcing linearity numbers for modules over rings with nontrivial idempotents, *J. Algebra* 256 (2002), 66–84.
7. J. D. P. Meldrum, Upper faithful D.G. near-rings, *Proc. Edinburgh Math. Soc.* 26 (1983), 361–370.
8. J. D. P. Meldrum, Near-rings and Their Links with Groups, Pitman, Boston, London, Melbourne, 1985.
9. G. Pilz, Near-rings, North Holland, Amsterdam, New York, 1983.
10. R. P. Sullivan, Research problem No. 23, *Period. Math. Hungar.* 8 (1977), 313–314.

LOCALLY SEMI-T-NILPOTENT FAMILIES OF MODULES

JOHN CLARK

Department of Mathematics and Statistics,
University of Otago, PO Box 56,
Dunedin, New Zealand
E-mail: jclark@maths.otago.ac.nz

We present a brief survey of the local semi-T-nilpotency condition on families of modules, with some indication of its uses and connections with other module properties.

1. The definition and introduction

The local semi-T-nilpotency condition evolved in the 1970's in a series of papers, including [23], [24], [25], [28], [30], by M. Harada and his coauthors, particularly H. Kanbara and Y. Sai. They used it initially in their study of projective modules, in particular perfect and semiperfect modules, and were clearly motivated by the T-nilpotency property introduced by Bass in his seminal study of perfect rings [8]. The definition of the condition is as follows.

Definition 1.1. A family of modules $\{M_i : i \in I\}$ over a ring R is said to be *locally semi-T-nilpotent* if, for any countably infinite set of non-isomorphisms $\{f_n : M_{i_n} \to M_{i_{n+1}} \mid n \in \mathbb{N}\}$ where all the i_n are distinct indices from I, given any $x \in M_{i_1}$ there is a $k \in \mathbb{N}$ (depending on x) for which $f_k \cdots f_1(x) = 0$. The condition is frequently abbreviated as "lsTn", where "T" denotes "transfinite".

In this survey we shall look at how this condition interacts with various properties associated with the decomposition of modules into direct summands. Indeed we will see that it is frequently equivalent to such conditions. Moreover, as phrased in S. H. Mohamed and B. J. Müller's text [42], although it may appear at first sight to be somewhat technical, *"it is usually the one condition that can be explicitly verified"*.

2. Decompositions which complement summands

Much of the success of the lsTn condition has been achieved when the modules M_i in the family are all indecomposable, and more particularly are *LE-modules*, i.e. have local endomorphism rings. We now attempt to trace some of the early development in this area, beginning with a reminder of some of the key ingredients. The first of these is due F. W. Anderson and K. R. Fuller [2] and more information can be found in their text [3].

Definition 2.1. Let $M = \oplus_{i \in I} M_i$ be a decomposition of the module M into nonzero summands M_i.

(i) This decomposition is said to *complement direct summands* if, whenever A is a direct summand of M, there is a subset J of I for which $M = (\oplus_{j \in J} M_j) \oplus A$.

(ii) The decomposition is said to *complement maximal direct summands* if, whenever A_1, A_2 are submodules of M for which $M = A_1 \oplus A_2$ and A_1 is indecomposable, then $M = M_i \oplus A_2$ for some $i \in I$.

(iii) Given a second decomposition $M = \oplus_{j \in J} N_j$ of M, the two decompositions are said to be *equivalent* or *isomorphic* if there is a bijection $\sigma : I \to J$ such that $M_i \simeq N_{\sigma(i)}$ for each $i \in I$.

To place our subsequent discussion in context, we record an early milestone in the study of indecomposable decompositions, namely the following theorem of Azumaya [6] which generalized the classical Krull-Schmidt Theorem on the decomposition of modules of finite length (see [3], §12).

Theorem 2.1. (Azumaya) *Let* $M = \oplus_{i \in I} M_i$ *be an LE-decomposition, i.e. each M_i is an LE-module. Then*

(i) *every nonzero direct summand of M has an indecomposable direct summand,*

(ii) *the decomposition $M = \oplus_{i \in I} M_i$ complements maximal direct summands, and consequently*

(iii) *the decomposition is equivalent to every indecomposable decomposition of M.*

Moreover, if I is finite then the decomposition complements direct summands.

Another useful tool in decomposition theory is given by the next definition.

Definition 2.2. An internal direct sum $\oplus_{i \in I} A_i$ of submodules of a module M is called a *local (direct) summand of M* if, given any finite subset F of the index set I, the direct sum $\oplus_{i \in F} A_i$ is a direct summand of M.

If, moreover, the direct sum $\oplus_{i \in I} A_i$ is itself a summand of M, then we say that the local direct summand $\oplus_{i \in I} A_i$ is also a (direct) summand of M. This (somewhat verbose) phrasing emphasises that local direct summands are not always summands.

It was observed by Oshiro in [46] that if every local summand of M is also a summand of M then M has an indecomposable decomposition. Moreover, every local summand of M is also a summand of M if and only if the union of any chain of summands in M is also a summand in M (see, for example, Lemma 2.16, Theorem 2.17 of [42]).

We now come to the interconnection between these ideas and local semi-T-nilpotency. The following major result is due to Harada [26], Theorems 7.3.15 and 8.2.1. However, earlier partial results appeared in papers by Yamagata [48], [49], Ishii [33], and Kanbara [36], as well as previous papers by Harada himself.

Theorem 2.2. (Harada) *Let $M = \oplus_{i \in I} M_i$ be an LE-decomposition, $S = End_R(M)$ and $J(S)$ denote the Jacobson radical of S. Then the following statements are equivalent.*

(a) *The radical factor ring $S/J(S)$ is (von Neumann) regular and idempotents lift modulo $J(S)$, i.e. S is a* semiregular *ring.*
(b) *Every local summand of M is a summand.*
(c) *The decomposition complements direct summands.*
(d) *$\{M_i : i \in I\}$ is locally semi-T-nilpotent.*

The proof given by Harada in [26] relies heavily on his theory of factor categories and is not for the faint-hearted. However, one can circumvent his arguments as we now explain. We first give the definition of the total of a ring, a concept pioneered by F. Kasch.

Definition 2.3. Given any ring R, an element $r \in R$ is called *partially invertible* if there is an $s \in R$ for which sr is a nonzero idempotent in R. Then the *total* of R is defined to be the set given by

$$\mathrm{Tot}(R) = \{r \in R : r \text{ is not partially invertible }\}.$$

Details of the total can be found in several publications by Kasch and his coauthors, including [38] and the recent text by him and A. Mader [37].

44

In these last two citations, one finds the following two theorems (proved by relatively elementary methods), which together recover part of Harada's Theorem.

Theorem 2.3. *Let $M = \oplus_{i \in I} M_i$ be an LE-decomposition with $S = End_R(M)$. Then $Tot(S)$ is an ideal of S and the factor ring $S/Tot(S)$ is isomorphic to a direct product of endomorphism rings of vector spaces over division rings (and so a regular ring).*

Theorem 2.4. *Let $M = \oplus_{i \in I} M_i$ be an LE-decomposition, with S and $J(S)$ as before. Then the following statements are equivalent.*

(a) $J(S) = Tot(S)$.
(b) The decomposition complements direct summands.
(c) $\{M_i : i \in I\}$ is locally semi-T-nilpotent.

We note that D. Khurana and R. N. Gupta [40] give an alternative approach to identifying $J(S)$ and $S/J(S)$ on the assumption that the LE-decomposition is lsTn.

Next we state a theorem due to N. V. Dung [15]. This result is important for two reasons. Firstly, it generalises part of Harada's Theorem by replacing the LE-decomposition hypothesis by the weaker complementing maximal summands condition. Secondly, his method of proof is quite different from that of Harada's, employing module theory techniques similar to those used in a paper by Zimmermann-Huisgen and Zimmermann [51] which we will feature later.

Theorem 2.5. (Dung) *Let $M = \oplus_{i \in I} M_i$ be an indecomposable decomposition of the module M which complements maximal direct summands. Then the following statements are equivalent.*

(a) The decomposition complements direct summands.
(b) Every nonzero summand of M contains an indecomposable direct summand and $\{M_i : i \in I\}$ is locally semi-T-nilpotent.
(c) Every local summand of M is a summand.

We note that in a forerunner to [26], Harada had an alternative to condition (a) of his theorem, namely that $J(S) = J' \cap \mathrm{End}(M)$ where J' is a set of non-isomorphisms closely allied to the total of S. In [52], A. Zöllner has given a different proof that the lsTn condition yields this description of $J(S)$. On the other hand, in [33] T. Ishii gives an alternative proof that this description of $J(S)$ forces the decomposition to complement summands.

Furthermore, as recorded in detail in Mohamed and Müller [42], Kasch and Zöllner have shown (unpublished) that *any* decomposition $M = \oplus_{i \in I} M_i$ which complements summands is locally semi-T-nilpotent, without the LE-decomposition assumption. (Dung uses this in the proof of his theorem.)

While Dung's theorem relaxes the LE-decomposition requirement, it is interesting to note that J. L. Gómez Pardo and P. A. Guil Asensio have established in [21] the following (where $M^{(I)}$ denotes the I-clone of M, namely the direct sum of I copies of M).

Theorem 2.6. *Let M be an indecomposable module and suppose that, for each index set I, every local summand of $M^{(I)}$ is a summand. Then M is an LE-module.*

3. The exchange property

The following definition is due to Crawley and Jónnson [12] who introduced it in the wider context of general algebra.

Definition 3.1. Let c be any cardinal number. A module M is said to have the *c-exchange property* if, for any module A and any decompositions

$$A = M' \oplus N = \oplus_{i \in I} A_i$$

for modules M', N, A_i where $M' \simeq M$ and $\mathrm{card}(I) \le c$, there always exist submodules $B_i \le A_i$ for each $i \in I$ such that

$$A = M' \oplus \left(\oplus_{i \in I} B_i \right).$$

If M has the n-exchange property for every positive integer n then M is said to have the *finite exchange property*.

If M has the c-exchange property for every cardinal number c then M is said to have the (*full* or *unrestricted*) *exchange property*.

Every module with the 2-exchange property has the finite exchange property but it remains a mystery as to whether finite exchange implies the full exchange property in general.

While the exchange property connections with local semi-T-nilpotency were recognised at an early stage by Japanese ring theorists, the following theorem, due to B. Zimmermann-Huisgen and W. Zimmermann [51], marks another milestone since it unified and generalized previous results and was proved using module-theoretic techniques in contrast to earlier categorical ones.

Theorem 3.1. (Zimmermann-Huisgen–Zimmermann) *Let* $M = \oplus_{i \in I} M_i$ *be a decomposition of* M *into indecomposable modules. Then the following statements are equivalent.*

(a) M has the exchange property.

(b) M has the finite exchange property.

(c) Each M_i is an LE-module and $\{M_i : i \in I\}$ is lsTn.

Prior to this result, the equivalence of (b) and (c) had been established by Harada, Sai, and Yamagata in [30] and [49], while the implication (c) $\Rightarrow$ (a) was shown by Harada, Ishii, and Yamagata ([27] and [47], [48]) under the assumptions that all the M_i are injective or they are all mutually isomorphic. A key step in the proof of Theorem 3.1 was showing that the exchange property for the module M can be checked by just taking the direct sum $\oplus_{i \in I} A_i$ in the definition above to be the I-clone $M^{(I)}$.

We now record two additional results of Zimmermann-Huisgen and Zimmermann which they deduced from their theorem. The first of these was established earlier in [27] and [47].

Theorem 3.2. *If R is a ring in which the identity is a finite sum of primitive orthogonal idempotents, the following statements are equivalent.*

(a) Every projective right R-module has the exchange property.

(b) The free right R-module $R^{(\mathbb{N})}$ has the finite exchange property.

(c) R is right perfect.

Theorem 3.3. *All strongly invariant submodules of an algebraically compact module have the exchange property.*

Here a submodule N of a module M is *strongly invariant* if $f(N) \subset N$ for any homomorphism $f : N \to M$. Examples are numerous and include all quasi-injective modules, all algebraically compact modules, and all linearly compact modules (and so all artinian modules) over a commutative ring.

The following striking result appears as the starting place of a recent paper by L. Angeleri-Hügel and M. Saorín [5]. The equivalence of conditions (a), (b), (e), and (f) is proved by Gómez Pardo and Guil Asensio in [21] as a consequence of their Theorem 2.6 above and the Zimmermann-Huisgen–Zimmermann Theorem. The equivalence of (b) and (c) is due to Huisgen-Zimmermann and Saorín and appears as part of Proposition E of [32], while that of (a), (b), and (c) is shown by Angeleri-Hügel in [4].

Here Add M is the class of modules consisting of the direct summands of clones $M^{(I)}$ of M. Also the right R-module M is said to be *coperfect* over its endomorphism ring S if the module $_SM$ satisfies the descending chain on its cyclic submodules.

Theorem 3.4. *The following statements are equivalent for a module M.*

(a) Every local summand of a module in $\operatorname{Add} M$ is a direct summand.

(b) Every module X in $\operatorname{Add} M$ has an LE-decomposition $X = \oplus_{i \in I} X_i$ where the family $\{X_i : i \in I\}$ is locally semi-T-nilpotent.

(c) M has an LE-decomposition and M is coperfect over its endomorphism ring.

(d) M has an LE-decomposition and $\operatorname{End}_R(A)$ is a semiregular ring for all A in $\operatorname{Add} M$.

(e) M has an indecomposable decomposition and every module in $\operatorname{Add} M$ has the exchange property.

(f) Every module in $\operatorname{Add} M$ has a decomposition that complements direct summands.

When a module M satisfies any of the conditions of Theorem 3.4, the authors of [5] say that M has a *perfect decomposition* and illustrate this concept with many examples. Note that taking M to be the R-module R_R recovers Theorem 3.2 and other characterizations of right perfect rings.

4. Extending modules

Much of the early use of the lsTn condition was in the investigation of the decomposition of injective modules, projective modules, and their generalizations. In this section we look at generalizations of injectivity.

As an entrée, we first mention an early result of Yamagata [48].

Theorem 4.1. *Let $M = \oplus_{i \in I} M_i$ be a decomposition of M into indecomposable injective modules, with $S = \operatorname{End}(M)$ and $J(S)$ as before. Then the following statements are equivalent.*

(a) M has the exchange property.

(b) M has the finite exchange property.

(c) $\{M_i : i \in I\}$ is locally semi-T-nilpotent.

(d) $J(S) = \{f \in S : \operatorname{Ker}(f)$ is essential in $M\}$.

Recall that, as a generalization of injectivity, a module M is an *extending* or a *CS-module* if each of its (essentially) closed submodules is a summand.

We also need to refer to the following chain condition, called (A_2) in [42], which arose in the study of the quasi-injectivity of direct sums.

48

Definition 4.1. A family of R-modules $\{M_i : i \in I\}$ is said to satisfy (A_2) if, given any countably infinite family of elements $x_n \in M_{i_n}$, where all the i_n are distinct indices from I, for which there is a $y \in M_j$ for some $j \in I$ such that $\text{ann}(y) \subseteq \cap_{n=1}^{\infty}\text{ann}(x_n)$, then the ascending chain $\cap_{k=n}^{\infty}\text{ann}(x_k)$, $(n \in \mathbb{N})$, becomes stationary.

Using his Theorem 2.5 above, Dung generalized earlier results of his in [13] and [14] by showing the following in [15].

Theorem 4.2. *Let $M = \oplus_{i \in I} M_i$ be a direct sum of uniform submodules M_i which complements maximal direct summands. Then the following statements are equivalent.*

(a) M is an extending module.
(b) $\oplus_{i \in H} M_i$ is an extending module for every countable subset H of I.
(c) $M_i \oplus M_j$ is an extending module for every distinct pair of indices $i, j \in I$ and $\{M_i : i \in I\}$ is lsTn and satisfies (A_2).

In this case, any local summand of M is a summand.

As Dung acknowledges in [15], some of the inspiration for Theorem 4.2 also comes from Harada and Y. Oshiro's [29] and M. A. Kamal and Müller's [35]. We also note that, with the same hypothesis as Theorem 12, further characterizations of when M is extending are given by J. Kado, Y. Kuratomi, and Oshiro in [34] using a variant of (A_2) and the concepts of generalized injectivity and the internal exchange property.

In some cases, the lsTn condition is equivalent to a weaker version of (A_2), as the following result due to Müller and S. T. Rizvi [44] shows.

Theorem 4.3. *Let $M = \oplus_{i \in I} M_i$ be a direct sum of uniform modules for which M_i is M_j-injective for all distinct $i, j \in I$. Then $\{M_i : i \in I\}$ is locally semi-T-nilpotent if and only if, given any countably infinite family of elements $x_n \in M_{i_n}$, where all the i_n are distinct indices from I, if the sequence $\text{ann}(x_k)$ is an ascending chain then it becomes stationary. In particular, this will be so if $\oplus_{i \in I} M_i$ is quasi-continuous.*

The following alternative characterization of the extending property for direct sums (in the special case of LE-modules) appears as Theorem 8.13 in the text [18] on extending modules by Dung, D. V. Huynh, P. F. Smith, and R. Wisbauer. Here a module M is *uniform extending* if every uniform submodule of M is essential in a direct summand of M. (Of course, the equivalence of conditions (b), (c) and (d) follows from Harada's Theorem.)

Theorem 4.4. *Let* $M = \oplus_{i \in I} M_i$ *be a direct sum of uniform LE-modules* M_i. *Then the following statements are equivalent.*

(a) M is an extending module.

(b) M is a uniform extending module and local summands of M are summands.

(c) M is uniform extending and $M = \oplus_{i \in I} M_i$ complements direct summands.

(d) M is uniform extending and $\{M_i : i \in I\}$ is locally semi-T-nilpotent.

(e) M is uniform extending and there is no infinite sequence of non-isomorphic monomorphisms $\{f_k : M_{i_k} \to M_{i_{k+1}}\}$ in which all $i_k \in I$ are distinct.

We note that the implication (a) $\Rightarrow$ (d) of Theorem 4.4 also appears in Y. Baba and Harada's [7].

Definition 4.2. If $M^{(I)}$ is extending for each index set I then the module M is said to be $\sum$-*extending*. If $M^{(\mathbb{N})}$ is extending then M is *countably* $\sum$-*extending*.

In [9] Dung and the author proved that if $M = \oplus_{i \in I} M_i$ is an indecomposable decomposition of a nonsingular extending module M then the family $\{M_i : i \in I\}$ is lsTn. They then used this to show that any nonsingular self-generator $\sum$-extending module is a direct sum of uniserial noetherian quasi-injective submodules. These results were extended in [11] to polyform modules. Moreover several conditions for a uniform module to be either $\sum$-extending or countably $\sum$-extending were considered by A. O. Al-attas and N. Vanaja in [1], including the non-isomorphic monomorphisms condition (e) of Theorem 4.4. They also show that an indecomposable module is $\sum$-extending if and only if it is $\sum$-quasi-injective. Gómez Pardo and Guil Asensio [22] use this to prove part (ii) of their following important result. (See also their earlier paper [20].)

Theorem 4.5. *Let M be a $\sum$-extending module. Then*

(i) M is a direct sum of uniform submodules,

(ii) these submodules are LE-modules, and consequently

(iii) all indecomposable decompositions of M are equivalent and complement summands.

5. Lifting, quasi-discrete and discrete modules

In this section we consider concepts which are dual to that of extending, quasi-continuous and continuous modules. We begin by recording the following definition taken from Mohamed and Müller [42].

Definition 5.1. A module M is said to satisfy property

(D_1) if, for every submodule N of M, there is a decomposition $M = M_1 \oplus M_2$ with $M_1 \leq N$ and $N \cap M_2 \ll M$ (where here $A \ll B$ means that A is a small submodule of B),

(D_2) if, whenever N is a submodule of M for which M/N is isomorphic to a summand of M, then N is a summand of M,

(D_3) if, whenever M_1 and M_2 are summands of M with $M_1 + M_2 = M$, then $M_1 \cap M_2$ is also a summand of M.

If M satisfies (D_1) it is called a *lifting* module. A lifting module which also satisfies (D_2) is called *discrete* while a lifting module satisfying (D_3) is called *quasi-discrete*.

Note that an indecomposable module M is lifting if and only if it is hollow, i.e. every submodule of M is small.

The first important result of the section establishes a nice decomposition for quasi-discrete modules due to Oshiro [45], (who called them *quasi-semiperfect* modules). (See also Theorem 4.15 of [42].)

Theorem 5.1. (Oshiro) *Any quasi-discrete module M decomposes as a sum $\oplus_{i \in I} H_i$ of hollow modules. Moreover this decomposition complements summands and so is unique up to isomorphism and is lsTn.*

The next result, due to Mohamed and Müller [42] (see also [41] and [43]) describes precisely when a direct sum of hollow modules is quasi-discrete.

Theorem 5.2. *Let $M = \oplus_{i \in I} H_i$ be a direct sum of hollow modules H_i. Then M is quasi-discrete if and only if the following conditions all hold*

(i) H_i is $\oplus_{j \neq i} H_j$-projective for each $i \in I$,
(ii) every local summand of M is a summand, and
(iii) $M = \oplus_{i \in I} H_i$ complements direct summands.

Using his Theorem 5 above, Dung then showed in [15] that condition (ii) of Theorem 5.2 is superfluous.

We mention one further result on quasi-discrete modules from Mohamed and Müller [42].

Theorem 5.3. *If $M = \oplus_{i \in I} M_i$ is a direct sum of pairwise mutually projective local modules M_i, the following statements are equivalent.*

    (a) M is quasi-discrete.
    (b) M has small radical.
    (c) Every proper submodule of M is contained in a maximal submodule.
    (d) $M = \oplus_{i \in I} M_i$ complements direct summands.
    (e) $\{M_i : i \in I\}$ is locally semi-T-nilpotent.

We next note that J. M. Zelmanowitz [50] gives a quick proof that if M is a discrete module with endomorphism ring S then $S/J(S)$ is isomorphic to a direct product of endomorphism rings of vector spaces over division rings (cf. Harada's Theorem and Theorems 2.3 and 2.4). For this he noted that, if $M = \oplus_{i \in I} M_i$ where M is a discrete module and the M_i pairwise have no isomorphic summands, then $\{M_i : i \in I\}$ is lsTn.

Now we turn our attention to lifting modules. We first mention that Harada and A. Tozaki [31] show that if $M = \oplus_{i \in I} H_i$ is a direct sum of hollow LE-modules where $\{H_i : i \in I\}$ is lsTn, then M is lifting if and only if every non-small submodule of M contains a nonzero indecomposable summand of M and this in turn can be characterised by a form of relative projectivity. Moreover Baba and Harada [7] show that if $\{M_i : i \in I\}$ is a family of LE-modules such that $\oplus_{i \in I} M_i$ is lifting then $\{M_i : i \in I\}$ is lsTn. (Their proof uses Harada's theory of factor categories but, in a private communication, N. Vanaja has indicated that she has a module-theoretic proof. The latter is scheduled to appear in a text by the author, C. Lomp, Vanaja, and Wisbauer [10].) We also note that D. Keskin and Lomp [39] have used Baba and Harada's results to determine when $M \oplus S$ is lifting when S is semisimple and M is a lifting module with an LE-decomposition.

6. Final remarks

We close with two further remarks on local semi-T-nilpotency.

The first is in connection with the Krull-Schmidt decomposition results which have been extensively investigated by, in particular, A. Facchini. While it has been shown that Krull-Schmidt fails for finitely presented modules over serial rings (and such modules are finite direct sums of uniserial modules) (see Chap. 9 of [19]), Dung and Facchini have shown the following in [17]. Here, for two modules A and B, we write $[A]_m = [B]_m$ if there are monomorphisms from A into B and B into A while we write $[A]_e = [B]_e$ if there are epimorphisms from A onto B and B onto A.

Theorem 6.1. *If $\{U_i : i \in I\}$ and $\{V_j : j \in J\}$ are two locally semi-T-nilpotent families of nonzero uniserial modules over an arbitrary ring R then $\oplus_{i \in I} U_i \simeq \oplus_{j \in J} V_j$ if and only if there are two bijections $\sigma, \tau : I \to J$ for which $[U_i]_m = [V_{\sigma(i)}]_m$ and $[U_i]_e = [V_{\tau(i)}]_e$.*

Lastly we note the recent appearance of a dual to lsTn, due to Huisgen-Zimmermann and Saorín [32]. Renaming the usual lsTn as *right semi-T-nilpotency*, they make the following definition, using it to show relationships between the structure of $M = \oplus_{i \in I} M_i$ over its endomorphism ring and the finiteness of the isomorphism classes of the M_i. (See also Dung [16].)

Definition 6.1. A family $\{M_i : i \in I\}$ of indecomposable R-modules is said to be *left semi-T-nilpotent* if, for any sequence $(i_n)_{n \in \mathbb{N}}$ of distinct indices in I, any family of non-isomorphisms $f_n \in \mathrm{Hom}_R(M_{i_{n+1}}, M_{i_n})$, and any finitely cogenerated factor module M_{i_1}/X of the R-module M_{i_1}, there exists an $n_0 \in \mathbb{N}$ such that $\mathrm{Im}(f_1 f_2 \cdots f_{n_0}) \subseteq X$.

Acknowledgments

Many thanks to Professors Chen Jianlong and Ding Nanqing and their band of helpers for a most enjoyable conference.

References

1. Al-attas, A. O. and Vanaja, N., *On Σ-extending modules*, Comm. Algebra **25**, 2365–2393 (1997).
2. Anderson, F. W. and Fuller, K. R., *Modules with decompositions that complement direct summands*, J. Algebra **22**, 241–253 (1972).
3. Anderson, F. W. and Fuller, K. R., *Rings and Categories of Modules*, Springer, Berlin (1974).
4. Angeleri-Hügel, L., *Covers and envelopes via endoproperties of modules*, Proc. London Math. Soc. **86**, 649–665 (2003).
5. Angeleri-Hügel, L. and Saorín, M., *Modules with perfect decompositions*, preprint, 2004.
6. Azumaya, G., *Corrections and supplementaries to my paper concerning Krull-Remak-Schmidt's theorem*, Nagoya Math. J. **1**, 117–124 (1950).
7. Baba, Y. and Harada, M., *On almost M-projectives and almost M-injectives*, Tsukuba J. Math. **14**, 53–69 (1990).
8. Bass, H. *Finitistic dimension and a homological generalization of semi-primary rings*, Trans. Amer. Math. Soc. **95**, 466–488 (1960).
9. Clark, J. and Dung, N. V., *On the decomposition of nonsingular CS-modules*, Canad. Math. Bull. **39**, 257–265 (1996).
10. Clark, J., Lomp, C., Vanaja, N., and Wisbauer, R., *Lifting Modules*, manuscript in preparation.

11. Clark, J. and Wisbauer, R., *Polyform and projective $\sum$-extending modules*, Algebra Colloq. **5**, 391–408 (1998).

12. Crawley, P. and Jónnson, B., *Refinements for infinite direct decompositions of algebraic systems*, Pacific J. Math. **91**, 249–261 (1980).

13. Dung, N. V., *On indecomposable decompositions of CS-modules*, J. Austral. Math. Soc. Ser. A **61**, 30–41 (1996).

14. Dung, N. V., *On indecomposable decompositions of CS-modules. II*, J. Pure Appl. Algebra **119**, 139–153 (1997).

15. Dung, N. V., *Modules with indecomposable decompositions that complement maximal direct summands*, J. Algebra **197**, 449–467, (1997).

16. Dung, N. V., *On the finite type of families of indecomposable modules*, J. Algebra Appl. **3**, 111–119 (2004).

17. Dung, N. V. and Facchini, A., *Weak Krull-Schmidt for infinite direct sums of uniserial modules*, J. Algebra **193**, 102–121, (1997).

18. Dung, N. V., Huynh, D. V., Smith, P. F., and Wisbauer, R., *Extending modules*, Longman Scientific & Technical, Harlow (1994).

19. Facchini, A., *Module Theory. Endomorphism rings and direct sum decompositions in some classes of modules*, Birkhäuser, Basel (1998).

20. Gómez Pardo, J. L. and Guil Asensio, P. A., *Indecomposable decompositions of $\aleph$-$\sum$-CS-modules*, Algebra and its applications, 467–473, Contemp. Math., **259**, Amer. Math. Soc., Providence, (2000).

21. Gómez Pardo, J. L. and Guil Asensio, P. A., *Big direct sums of copies of a module have well behaved indecomposable decompositions*, J. Algebra **232**, 86–93 (2000).

22. Gómez Pardo, J. L. and Guil Asensio, P. A., *Indecomposable decompositions of modules whose direct sums are CS*, J. Algebra **262**, 194–200 (2003).

23. Harada, M., *On categories of indecomposable modules. II*, Osaka J. Math. **8**, 309–321 (1971).

24. Harada, M., *Supplementary remarks on categories of indecomposable modules*, Osaka J. Math. **9**, 49–55 (1972).

25. Harada, M., *Small submodules in a projective module and semi-T-nilpotent sets*, Osaka J. Math. **14**, 355–364 (1977).

26. Harada, M., *Factor categories with applications to direct decomposition of modules*, Marcel Dekker, Inc., New York, 1983.

27. Harada, M. and Ishii, T., *On perfect rings and the exchange property*, Osaka J. Math. **12**, 483–491 (1975).

28. Harada, M. and Kanbara, H., *On categories of projective modules*, Osaka J. Math. **8**, 471–483 (1971).

29. Harada, M. and Oshiro, K., *On extending property on direct sums of uniform modules*, Osaka J. Math. **18**, 767–785 (1981).

30. Harada, M. and Sai, Y., *On categories of indecomposable modules. I*, Osaka J. Math. **7**, 323–344 (1970).

31. Harada, M. and Tozaki, A., *Almost M-projectives and Nakayama rings*, J. Algebra **122**, 447–474 (1989).

32. Huisgen-Zimmermann, B. and Saorín, M., *Direct sums of representations*

as modules over their endomorphism rings, J. Algebra **250**, 67–89 (2002).

33. Ishii, T., *On locally direct summands of modules*, Osaka J. Math. **12**, 473–482 (1975).

34. Kado, J., Kuratomi, Y., and Oshiro, K., *CS-property of direct sums of uniform modules*, International Symposium on Ring Theory (Kyongju, 1999), 149–159, Trends Math., Birkhäuser Boston, Boston, (2001).

35. Kamal, M. A. and Müller, B. J., *The structure of extending modules over Noetherian rings*, Osaka J. Math. **25**, 539–551 (1988).

36. Kanbara, H., *Note on Krull-Remak-Schmidt-Azumaya's theorem*, Osaka J. Math. **8**, 409–413 (1971).

37. Kasch, F. and Mader, A., *Rings, modules, and the total*, Birkhäuser Verlag, Basel, (2004).

38. Kasch, F. and Schneider, W., *The total of modules and rings*, Algebra Berichte, **69**, Verlag Reinhard Fischer, Munich (1992).

39. Keskin, D. and Lomp, Ch., *On lifting LE-modules*, Vietnam J. Math. **30**, 167–176 (2002).

40. Khurana, D. and Gupta, R. N., *Endomorphism rings of Harada modules*, Vietnam J. Math. **28**, 173–175 (2000).

41. Mohamed, S. H. and Müller, B. J., *Dual continuous modules over commutative Noetherian rings*, Comm. Algebra **16**, 1191–1207 (1988).

42. Mohamed, S. H. and Müller, B.J., *Continuous and Discrete Modules*, London Math. Soc. Lect. Notes Ser., **147**, Cambridge (1990).

43. Mohamed, S. H., Müller, B. J., and Singh, S., *Quasi-dual-continuous modules*, J. Aust. Math. Soc., Ser. A **39**, 287–299 (1985).

44. Müller, B. J. and Rizvi, S. T., *Direct sums of indecomposable modules*, Osaka J. Math. **21**, 365–374 (1984).

45. Oshiro, K., *Semiperfect modules and quasi-semiperfect modules*, Osaka J. Math. **20**, 337–372 (1983).

46. Oshiro, K., *Lifting modules, extending modules and their applications to QF-rings*, Hokkaido Math. J. **13**, 310–338 (1984).

47. Yamagata, K., *On projective modules with the exchange property*, Pacific J. Math. **55**, 301–317 (1974).

48. Yamagata, K., *The exchange property and direct sums of indecomposable injective modules*, Sci. Rep. Tokyo Kyoiku Daigaku Sect. A **12**, 39–48, (1974).

49. Yamagata, K., *On rings of finite representation type and modules with the finite exchange property*, Sci. Rep. Tokyo Kyoiku Daigaku Sect. A **13**, 347–365, 1–6. (1975).

50. Zelmanowitz, J. M., *On the endomorphism ring of a discrete module: a theorem of F. Kasch*, Advances in ring theory, 317–322, Birkhäuser, Boston, (1997).

51. Zimmermann-Huisgen, B. and Zimmermann, W., *Classes of modules with the exchange property*, J. Algebra **88**, 416–434 (1984).

52. Zöllner, A., *On modules that complement direct summands*, Osaka J. Math. **23**, 457–459 (1986).

SMOOTH ALGEBRAS AND THEIR APPLICATIONS

C.R.HAJARNAVIS

Mathematics Institute,
University of Warwick,
Coventry CV4 7AL,
England
E-mail: crh@maths.warwick.ac.uk

We give here an account of the author's joint work with A.Braun in the area of rings of finite global dimension. We consider smooth rings with trivial K_0 as a natural generalisation of commutative regular local rings. We show that this treatment yields rich results which apply to important classes such as characteristic p enveloping algebras. Detailed proofs will appear in [4].

1. Introduction

The theory of commutative regular local rings plays an extensive role in commutative algebra and geometry. In homological terms these rings can be characterised as Noetherian local rings of finite global dimension. We show that extending this idea appropriately to non-commutative rings pays rich dividends yielding a theory applicable to naturally occurring classes of rings such as universal enveloping algebras of finite dimensional Lie algebras over fields of characteristic p .

A famous theorem of Auslander and Buchbaum states that a regular local ring must be a unique factorisation domain. In the non-commutative case, as an analogue admitting practical applications, we may ask if the height one prime ideals in a smooth Noetherian PI ring with trivial K_0 are principal (as right ideals and left ideals). While we have been unable to settle this issue, we can show that a sufficiently high symbolic power of a height one prime ideal is principal and is even centrally generated (Theorem 4.2). This is enough to determine that the ideal class group of the centre of the universal enveloping algebra of a finite dimensional Lie algebra over a field of characteristic p must be a torsion group (Corollary 4.4).

2. Preliminaries and Background

All rings will be assumed to have an identity. Terms such as Noetherian will mean two-sided Noetherian.

Let R be a ring, I an ideal of R and M a right R-module. We denote $C(I) = \{c \in R \mid c + I$ regular in the ring $R/I\}$.

dim M = the uniform (or Goldie) dimension of M .

This is the maximal number of non-zero terms possible in a direct sum of submodules in M .

pd M = projective dimension of M .

gl. dim. R = sup M_R the right global dimension of R .
$\quad\quad${M}

But note that this also equals the left global dimension of R when R is a Noetherian ring.

A finitely generated module P is said to be *stably free* if there exists a finitely generated free module F such that P $\oplus$ F is free.

We say that *R has trivial K_0* if every finitely generated projective module P with dim P = n dim R (n $\geq$ 1) is stably free.

By [19, Theorem 12.3.4] the universal enveloping algebra of a finite dimensional Lie algebra satisfies the above condition.

We denote $F(R)$ = {M_R | M is a finitely generated torsion-free module with dim M = n dim R_R for some integer n} .

Recall that R is called a *polynomial identity* (PI) ring if there exists a monic polynomial f in the free algebra $\mathbb{Z} <x_1 , \ldots , x_n>$ such that $f(r_1, r_2, \ldots , r_n) = 0$ for all choices of $r_i \in R$.

Let R be a Noetherian ring with gl. dim. R = n < ∞. By a result of Bhatwadekar-Goodearl [7, Theorem 12.2], there exists a simple module S such that pd S = n . In general, of course, R will have other simple modules of projective dimension less than n . We define R to be a *smooth* ring if R is a Noetherian PI ring, gl. dim. R = n < ∞ and pd R/M is equal for all maximal ideals belonging to the same clique in R . Clearly, by above there exists a maximal ideal M with pd R/M = n .

For details of clique theory we refer the reader to [10, 15]. By [5, Corollary 1.10], the universal enveloping algebra of a finite dimensional Lie algebra over a field of characteristic p is a smooth ring. It is well-known that this ring is a finite module over its affine normal centre.

Let I be a non-zero ideal of a prime Noetherian ring R . Let Q be the quotient ring of R . We denote $I^* = \{q \in Q \mid qI \subseteq R\}$ and $I^\# = \{q \in Q \mid Iq \subseteq R\}$. Clearly, $I^*I \subseteq R$ and $II^\# \subseteq R$. We say that I is *left invertible* if $I^*I = R$ and *right invertible* if $II^\# = R$. Properties of one-sided invertible ideals are discussed in [11,12] and the symmetry result proved there is crucial to this theory.

Recall that R is called a *Krull-symmetric* ring if for each bimodule M such that $_RM$ and M_R are finitely generated, the Krull dimensions on the two sides are equal. It is now well-known [14, Theorem 2.3], that Noetherian PI rings are Krull symmetric. Goldie [9] defined symbolic powers for prime ideals in a noncommutative Noetherian ring. It was shown in [11] that these take the classical form in Krull-symmetric rings. Thus for a prime ideal P in such a ring, we have the n-th symbolic power $P^{(n)} = \{x \in R \mid xc \in P^n$ for some $c \in C(P)\} = \{x \in R \mid cx \in P^n$ for some $c \in C(P)\}$.

Let R, S be orders in a simple Artinian ring Q. The rings R and S are said to be *equivalent* if there exist units u, v, w, $t \in Q$ such that $uRv \subseteq S$ and $wSt \subseteq R$. The ring R is called a *maximal order* if there is no order in Q which is equivalent to R and strictly contains it. When R is commutative and Noetherian, this is equivalent to R being an integrally closed domain. A fractional ideal X of such a ring is called *reflexive* if $X^{**} = X$. Clearly, a principal ideal is reflexive. Let G be the set of all reflexive fractional ideals of R. We define a product for two reflexive ideals X and Y by $X{\cdot}Y = (XY)^{**}$. Then G is a group since R is integrally closed. Let K be the subgroup of G generated by principal ideals. The factor group G/K is called *the ideal class group* of R and is denoted by $cl(A)$. It is easily seen that R is a unique factorisation ring (UFD) if and only if $cl(A) = \{1\}$.

PI deg R will denote the PI degree of a PI ring R. (See [19, 13.3.6]).

$Z(R)$ will denote the centre of a ring R.

3 Stably Free Ideals

3.1 Lemma [3, Proposition 1.8]:

Let R be a prime Noetherian PI ring with $aR \supseteq Ra$ for some $a \in R$. Then $aR = Ra$.

An interesting consequence of the above is the following symmetry result.

3.2 Corollary [3, Proposition 1.14]

Let I be an ideal in a semi prime Noetherian PI ring. Then
$$I_R \text{ is stably free} \Leftrightarrow {_R}I \text{ is stably free}$$

3.3 Proposition:

Let R be a prime PI ring which is a maximal order. Let I be an ideal of R

satisfying

$$I \oplus I \oplus \ldots \oplus I \text{ (n times)} \cong R \oplus R \oplus \ldots \oplus R \text{ (n times)}$$

as right R-modules. Then $I^{dn} = cR$ where $c \in Z(R)$ and $d = PI \deg R$.

Proof:

This requires [17, Proposition 4] and Lemma 3.1 plays a key role. $\qquad\square$

3.4 Theorem:

Let R be a Noetherian prime PI ring which is a maximal order. Let I be an ideal of R such that I_R is stably free. Then $I^d = cR$ where $c \in Z(R)$ and $d = PI \deg R$.

Proof:

We may assume that I is non-zero. By [18, Theorem 1], there exists an integer t such that $I \oplus I \oplus \ldots \oplus I$ (n times) is free as a right R-module for all $n \geq t$. We have $I \oplus I \oplus \ldots \oplus I$ (n times) $\cong R \oplus R \oplus \ldots \oplus R$ (n times) since the two sides must have the same uniform dimension. Applying Proposition 3.3 successively to $n = t$ and $n = t + 1$ we have $I^{dt} = c_1 R$ and $I^{d(t+1)} = c_2 R$ where $c_1, c_2 \in Z(R)$ and c_1, c_2 are regular in R. So we obtain $I^d = cR$ where $c = c_2 c_1^{-1} \in Z(Q)$. But $c \in I^d$ and so $c \in Z(R)$. $\qquad\square$

We say that a module has FFR if it has a finite free resolution.

3.5 Lemma:

Let $P \in F(R)$ where R be a prime Noetherian ring. Suppose that P is projective and has FFR. Then P is stably free.

Proof:

We prove this by induction on the length of the finite free resolution. If P is free, the result is trivial. Now suppose that P has a FFR of length $n > 0$. Then we have $0 \rightarrow K \rightarrow F_0 \rightarrow P \rightarrow 0$ where F_0 is free and K is projective of length less than n. We have $F_0 \cong P \oplus K$ and so, in particular, $K \in F(R)$. By the induction hypothesis there exist finitely generated free modules G, H such that $K \oplus H \cong G$. Hence, $F_0 \oplus H \cong P \oplus K \oplus H \cong P \oplus G$ and so P is stably free. $\qquad\square$

Conversely, we have the following.

3.6 Lemma:

Let R be a prime Noetherian ring with trivial K_0 and let $M \in F(R)$ with pd $M < \infty$. Then M has FFR.

Proof:

This is similar to above using induction on the length of the projective resolution for M.

$\square$

The following Corollary is easy to deduce.

3.7 Corollary:

Let R be a prime Noetherian ring with trivial K_0 and let S be an Ore set in R. Let $M \in F(R)$ with pd $M < \infty$. If M_S is projective as an R_S-module then M_S is stably free.

We can now prove our first main result.

3.8 Theorem:

Let R be a smooth prime Noetherian PI ring with trivial K_0. Then R is a maximal order.

Proof:

By [20, Theorem 5.4], we have $R = \bigcap_x R_x$ where x runs over all the cliques of height one prime ideals of R and the intersection is taken in the quotient ring of R. Thus it suffices to show that each R_x is a maximal order. We note that by [20, Theorem 5.4], R is integral over its centre. So by [6, Theorem 3.5], R_x is a hereditary ring and thus P_x is a projective R_x-module. By Corollary 3.7, it follows that P_x is a stably free R_x-module. Using [18, Theorem 1], the proof of [3, Proposition 1.13] shows that P_x is invertible. It follows that P_x is localisable in R_x and hence P is a localisable prime ideal of R. Hence $x = \{P\}$ and R_P is a prime Noetherian local hereditary ring. Thus by [13, Proposition 1.3], R_P is a principal right and a principal left ideal ring. In particular, R_P is a maximal order. $\square$

4 Symbolic Powers

We require Kaplansky's trick of adjoining an indeterminate.

4.1 Lemma:

Let R be a smooth prime Noetherian PI ring. Then so is the polynomial ring R[t].

Proof:

This is routine. We do need the fact that R is integral over its centre. □

We can now state the main result of [4].

4.2 Theorem [4, Theorem13]:

Let R be a prime Noetherian smooth PI ring with trivial K_0 . Let P be a height one prime ideal of R . Then $P^{(d)} = cR$ where $c \in Z(R)$ and $d = PI \deg R$.

Proof:

We sketch the main steps of the argument given in [4]. First we move over to the polynomial ring R[t] . It is enough to show that $(P[t])^{(d)} = pR[t]$ for some $p \in Z(R[t]) = Z(R)[t]$. Adjoining the indeterminate allows us to use the fact that, as in the commutative case, a + bt is a prime element in Z(R)[t] for the regular sequence {a , b} . The argument proceeds by localising at S which consists of products of prime elements in Z(R)[t] . The localised ring is smooth and has global dimension which does not exceed two. The result is then deduced by analysing this situation. □

We do not yet know if P itself is principal. However it is worth noting that when R is commutative, we have d = 1 , and we recover the Auslander-Buchsbaum unique factorisation theorem in this case.

Theorem 4.2 allows us to obtain information on the class group of the centre.

4.3 Theorem [4, Theorem16]:

Let R be a smooth prime Noetherian PI ring with trivial K_0 . Then Cl(Z(R)) is a d-torsion group where d = PI \deg R .

Proof:

By [8, Proposition 6.8], it is enough to show that $p^{(k)} = cZ(R)$ for some $c \in Z(R)$ where p is a height one prime ideal of Z(R) and k is a divisor of d . This requirement can be deduced from Theorem 4.2. □

[4] gives examples where $d = 2$ and $Cl(Z(R)) \cong Z/2Z$.

Theorem 4.3, in particular, applies to the following.

4.4 Corollary [4, Theorem17]:

Let $\mathbf{g}$ be a finite dimensional Lie algebra over a field of finite characteristic and let $U_{\mathbf{g}}$ be its enveloping algebra. Then $cl(Z(U_{\mathbf{g}}))$ is a d-torsion group where $d = PI \deg U_{\mathbf{g}}$.

We do not know if $cl(Z(U_{\mathbf{g}}))$ is actually a finite group. [4] contains further results which apply to cross products and quantum enveloping algebras.

References

1. A.Braun and C.R.Hajarnavis, Finitely generated P.I. rings of global dimension two, *J. Algebra* **169** (1994), 587--604.
2. A. Braun and C.R.Hajarnavis, A structure theorem for Noetherian P.I. rings with global dimension two, *J. Algebra* **215** (1999), 248--289.
3. A. Braun and C.R.Hajarnavis, Generator ideals in Noetherian PI rings, *J. Algebra* **247** (2002), 134--152.
4. A.Braun and C.R.Hajarnavis, Smooth polynomial identity rings with almost factorial centres, *To appear.*
5. K.A.Brown and K.R.Goodearl, Homological aspects of Noetherian PI Hopf algebras and irreducible modules of maximal dimension, *J. Algebra* **198** (1997), 240--265.
6. K.A.Brown and C.R.Hajarnavis, Homologically homogeneous rings, *Trans. Amer. Math. Soc.* **281** (1984), 197--208.
7. A.W.Chatters and C.R.Hajarnavis, Rings with chain conditions, *Research notes in mathematics 44,* Pitman advanced publishing program, London (1980).
8. R.M.Fossum, The divisor class group of a Krull domain, Springer-Verlag 1973
9. A.W.Goldie, Localisation in non-commutative Noetherian rings, *J. Algebra* **5** (1967), 89--105.
10. K.R.Goodearl and R.B.Warfield, An introduction to non-commutative Noetherian rings, *London Math. Soc. Student Texts* **16** Cambridge University Press, Cambridge 1989.
11. C.R.Hajarnavis, One-sided invertibility and localisation, *Glasgow Math. J.* **34** (1992), 333--339.
12. C.R.Hajarnavis, One-sided invertibility and localisation II, *Glasgow Math. J.* **37** (1995), 15--19.
13. C.R.Hajarnavis and T.H.Lenagan, Localisation in Asano orders, *J. Algebra* **21** (1972), 441--449.

14. A.V.Jategaonkar, Jacobson's conjecture and modules over fully bounded Noetherian rings, *J. Algebra* **30** (1974), 103--121.

15. A.V.Jategaonkar, Localisation in Noetherian rings, *London Math. Soc. Lecture Notes Series* **98** Cambridge University Press, Cambridge 1986.

16. I. Kaplansky, Commutative algebra, Allyn and Bacon, Boston 1970.

17. M-A. Knus and M. Ojanguren, A note on the automorphisms of maximal orders, *J. Algebra* **22** (1972), 573--577.

18. T.Y.Lam, Series summation of stably free modules, *Quart. J. Math .* Oxford ser. (2) **27** (1976), 37--46.

19. J.C.McConnell and J.C.Robson, Non-commutative Noetherian rings, *Pure and Applied Mathematics,* Wiley-Interscience, New York 1987.

20. J.T.Stafford and J.J.Zhang, Homological properties of (graded) Noetherian PI rings, *J. Algebra* **168** (1994), 988--1026.

RINGS WHOSE SIMPLE MODULES HAVE SOME PROPERTIES

YASUYUKI HIRANO

Department of Mathematics, Okayama University,
Okayama 700-8530, Japan
E-mail: yhirano@math.okayama-u.ac.jp

In this paper we give a survey of results on rings whose simple modules have some propertites. We also mension some questions and conjectures.

Some important rings are charactrized by certain properties of their simple modules. For example, I. Kaplansky [40] proved that a commutative ring R is von Neumann regular if and only if every simple R-module is injective. Later, G. O. Michler and O. E. Villamayor [32] studied the rings whose simple modules are injective. In this survey, we state some results on rings whose simple modules have certain propertites. We consider some homological conditions on simple modules. We state projectivety, flatness and injectivety of simple modules. We also state some results on semiartinian rings and max rings. Finally we state some generalizations of V-rings.

Throughout this paper, all rings have identity and all modules are unital. For a ring R, $J(R)$ denotes the Jacobson radical of R. Let M be a left R-module, let N be a subset of M and let S be a subset of R. Then we set $Ann_R(N) = \{a \in R \mid aN = 0\}$ and $Ann_M(S) = \{m \in M \mid Sm = 0\}$.

1. Projectivity of simple modules

It is well-known that a ring R is semisimple Artinian if and only if every module is projective. Then what can we say about a ring R whose simple module are projective? The following result is also well-known, but for the convenience of readers, we give its proof.

Theorem 1.1. *For a ring R, the following statements are equivalent:*

(i) *R is a semisimple Artinian ring;*
(ii) *Every simle left R-module is projective.*

Proof. (i) $\Rightarrow$ (ii): This is trivial. (ii) $\Rightarrow$ (i): It suffices to show that every left ideal of R is a direct summand of R. So let L be a nonzero left ideal of R. Using Zorn's lemma, we can find a left ideal K of R which is maximal with respect to the property $L \cap K = 0$. If $L + K \neq R$, there exists a maximal left ideal M which contains $L + K$. By hypothesis, the short exact sequence $0 \to M \to R \to R/M \to 0$ must split. Then there exists a minimal left ideal S such that $R = M \oplus S$. Then $L \cap (K + S) = 0$ and $K + S \neq K$. This is a contradiction. ∎

A ring R is called *semiperfect* in case $R/J(R)$ is semisimple and idempotents lift modulo $J(R)$. The following theorem [1, Theorem 27.6] shows that a semiperfect ring is characterized using its simple modules.

Theorem 1.2. *For a ring R, the following statements are equivalent:*

(i) *R is semiperfect;*
(ii) *Every simple left R-module has a projective cover.*

2. Flatness of simple modules

A ring R is called a *left SF-ring* if every simple left R-module is flat. Ramamulthi [36] has conjectured that such rings are necessarily von Neumann regular. No counterexample is presently known. Some results about left SF-rings are presented in [7, 19, 23, 36, 38, 41, 44, 45, 47].

Conjecture 2.1. [36] *A left SF-ring is von Neumann regular.*

Proposition 2.1. *The following are equivalent:*

(i) *R is a left SF-ring;*
(ii) *For each maximal left ideal K of R, there holds that $u \in Ku$ for all $u \in K$.*

Theorem 2.1. [19] *Suppose that $R/Ann(M)$ is Artinian for any singular simple right R-module M. Then the following are equivalent:*

(i) *R is left SF-ring;*
(ii) *R is von Neumann regular.*

Corollary 2.1. *Let R be a ring with primitive factor rings Artinian. Then the following are equivalent:*

(i) *R is left SF-ring;*

(ii) *R is von Neumann regular.*

Z. Y. Huang and F. C. Vheng [24] generalized this corollary as follows.

Proposition 2.2. [24] *Let R be a ring, and I an ideal of R such that R/I is semisimple artinian. Then the left flat dimension of R/I is equal to the right injective dimension of R/I.*

A ring R is called a *right pp-ring* if xR is projective for all $x \in R$.

Theorem 2.2. [41] *A left SF right pp-ring is von Neumann regular.*

Remark 2.1. Let R be a left Noetherian left SF-ring. Since a finitely presented flat module is projective, every simple left R-module is projective. Hence R is semisimple artinian by Theorem 1.1.

Theorem 2.3. [23] *A ring R is von Neumann regular if and only if R is a right nonsingular right SF-ring and every principal right ideal is either a maximal right annihilator or a pprojective right annihilator of an element.*

3. Injectivity of simple modules

A ring R is called a *left V-ring* if every simple left R-module is injective. Some results about left V-rings are presented in [3, 4, 11, 12, 13, 25, 32, 33, 43]. G. O. Michler and O. E. Villamayor [32] studied V-rings and obtained many results on V-rings.

Theorem 3.1. [32, 33] *The following properties of a ring R are equivalent:*

  (i) *R is a left V-ring;*
  (ii) *Every left ideal of R is an intersection of maximal left ideals of R;*
  (iii) *Every left R-module has the property that zero is an intersection of maximal submodules;*
  (iv) *The category of left R-modules has a cogenerator which is a direct sum of simple R-modules.*

J. Cozzens [10] constructed an example of a non-regular Noetherian V-domain which has only one isomorphism class of simple modules (see also L. A. Koĭfman [29]). B. L. Osofsky [35] constructed an example of a Noetherian V-domain who has infinitely many nonisomorphic simple modules. R. D. Resco [39] constructed an example of a right Noetherian right V-domain T with a T-bimodule W which is the unique simple right T-module. This example was used by C. Faith and P. Menal [16] to construct a counterexample John's theorem [26]. A ring R is called a *right annihilator ring* if every right ideal of R is a right annihilator. A right Noetherian right annihilator ring is called a *right Johns ring*. The question has been raised whether every right Johns ring is right Artinian. A counter-example was given by C. Faith and P. Menal [16] using V-domain. If W is a left R-module, we say that W satisfies the *double annihilator condition with respect to right ideals* if $I = Ann_R Ann_W I$.

Theorem 3.2. [17] *A ring R is a right V-ring if and only if some semisimple modules satisfies the double annihilator condition with respect to right ideals.*

Corollary 3.1. *IF R is a right Johns ring, then R/J is a right V-ring.*

A ring R is called *left coherent* if any direct product of copies of R is flat as a right R-module.

Theorem 3.3. [8] *Let R be a right Johns and left coherent ring. Then R is right Artinian.*

A ring R is a *strongly right Johns* ring if $M_n(R)$ is right Johns for all positive integers n.

Question 3.1. Is a strongly Johns ring right Artinian?

I. Kaplansky proved the following theorem.

Theorem 3.4. [40] *A commutative ring R is a V-ring if and only if R is von Neumann regular.*

Many authors generalized this result to some noncommutative rings. Here we state some results of G. Baccella [3, 4].

A ring R is said to be *right weakly regular* if $I^2 = I$ for every right ideal I of R.

Theorem 3.5. [3] *Let R be a ring all of whose right primitive factor rings are artinian. Then the following are equivalent:*

   (i) *R is a right V-ring;*
   (ii) *R is von Neumann regular;*
   (iii) *R is right weakly regular.*

A ring R is called *left semi-artinian* if every nonzero left R-module has a non-zero socle.

Proposition 3.1. [4]

    (1) *The following properties of a ring R are equivalent:*

        (i) *R is a right semiartinian right V-ring;*

        (ii) *Every nonzero right R-module contains a nonzero injective submodule.*

    (2) *A right semiartinian right V-ring is von Neumann regular.*

The notion of a left V-ring was generalized in many ways. A ring R is called a *left GV-ring* if each simple left R-module is either projective or injective (or equivalently if each singular simple left R-module is injective).

Theorem 3.6. [2, 37] *The following properties of a ring R are equivalent:*

    (i) *R is a left GV-ring;*

    (ii) *Every essential left ideal of R is an intersection of maximal left ideals, and $Z(R) \cap J(R) = 0$;*

    (iii) *For each left R-module M, $Z(M) \cap J(M) = 0$ and every essential submodule of M is an intersection of maximal submodule;*

    (iv) *$Soc(_R R)$ is projective and $R/Soc(_R R)$ is a left V-ring;*

    (v) *For each left R-module M, $Z(M) \cap M(Soc_R R) = 0$ and every essential submodule of M is an intersection of maximal submodule.*

A right R-module M is called *P-injective* if, for any $0 \neq a \in R$, any right R-homomorphism of aR into M extends to one of R into M.

Proposition 3.2. [46] *If every simple left R-module is P-injective, then R is left weakly regular.*

Corollary 3.2. *Let R be a ring all of whose right primitive factor rings are artinian. Then R is a right V-ring if and only if every simple right R-module is P-injective.*

A right R-module M is called *GP-injective* if, for any $0 \neq a \in R$, there exists a positive integer n such that $a^n \neq 0$ and any right R-homomorphism of $a^n R$ into M extends to one of R into M. GP-injectivity of simple modules were investigated by [27, 28, 34, etc.]. A ring R is called *right quasi-duo* if every maximal right ideal of R is a two-sised ideal.

Theorem 3.7. [34] *Let R be a right quasi-duo ring. Then the following are equivalent:*

(i) *R is von Neumann regular;*

(ii) *R is a right (or left) V-ring;*

(iii) *Every simple right (or left) R-module is P-injective;*

(iv) *Every simple right (or left) R-module is GP-injective.*

4. Semiartinian rings and max rings

Let $\mathfrak{S}$ denote an irredundant set of representatives of the simple left R-modules and let S denote the direct sum of all modules in $\mathfrak{S}$.

Recall that a ring R is left semi-artinian if every nonzero left R-module has a non-zero simple submodule. So R is left semi-artinian if and only if $\mathrm{Hom}_R(S, M) \neq 0$ for each non-zero left R-module M. A module is called *semi-artinian* if every non-zero quotient has a non-zero socle. The Jacobson radical $J(R)$ is *left T-nilpotent* if, for every sequence $a_1, a_2 \ldots$, in $J(R)$ one has $a_n a_{n-1} \cdots a_1 = 0$. The following characterizations of a left semi-artinian ring are well-known.

Theorem 4.1. *The following are equivalent:*

(i) *R is left semi-artinian;*

(ii) *Every left R-module is semi-artinian;*

(iii) *Every non-zero left R-module has non-zero socle;*

(iv) *Every left R-module is an essential extension of its socle;*

(v) *$J(R)$ is left T-nilpotent and $R/J(R)$ is left semi-artinian.*

If R is a commutative semi-artinian ring, then $R/J(R)$ is von Neumann regular (cf. [4, p.591]). Hence we have the following.

Theorem 4.2. *Let R be a commutative ring. Then the following are equivalent:*

(i) *R is a semi-artinian ring;*

(ii) *$J(R)$ is T-nilpotent and $R/J(R)$ is semi-artinian and von Neumann regular.*

The dual notion of "semi-artinian" is "max". A ring R is called a *left max ring* if every nonzero left R-module has a maximal submodule. A ring R is left max if and only if $\mathrm{Hom}_R(M, S) \neq 0$ for each non-zero left R-module M.

This notion is extended to modules. A left R-module M is called *max* if every submodule of M has a maximal submodule. Hence a left R-module M is max if and only if every submodule has a simple homomorphic image.

Also a ring R is a left max ring if and only if every left R-module is max. Some results about left max rings are presented in [6, 14, 15, 18, 20, 30, 43].

A ring R is said to be *right perfect* if every right R-module has a projective cover. A right Artinian ring is a right max ring. More generally we have the following.

Theorem 4.3. [5] *A semilocal ring R is a right max ring if and only if it is a right perfect ring.*

Let R be a ring and let M be a left R-module. Then $E(M)$ denotes the injective hull of M. A submodule K of a left R-module M is *small in M*, in case for every submodule L of M $K + L = M$ implies $L = M$. The following characterizations of a left max ring are well-known.

Theorem 4.4. *Let R be a ring with Jacobson radical $J(R)$. Then the following conditions are equivalent:*

  (i) *R is a left max ring;*
 (ii) *For every non-zero left R-module M, $J(M)$ is small in M;*
(iii) *For every simple left R-module S, every submodule of $E(S)$ has a maximal submodule;*
 (iv) *There is a cogenerator C in R-Mod which is max;*
  (v) *$J(R)$ is left T-nilpotent and $R/J(R)$ is left max.*

A ring R is called a *π-regular ring* if for every element $a \in R$, there is an element $b \in R$ such that $a^n = a^n b a^n$ for some positiove integer n. If a right max ring R satisfies a polynomial identity, then $R/J(R)$ is π-regular.

Theorem 4.5. [30] *Let R be a PI-ring. Then the following are equivalent:*

  (i) *R is a right max ring;*
 (ii) *R is a left max ring;*
(iii) *$J(R)$ is right or left T-nilpotent and $R/J(R)$ is π-regular.*

Corresponding to Theorem 4.2, we have the following.

Theorem 4.6. [18] *Let R be a commutative ring. Then the following are equivalent:*

  (i) *R is a max ring;*
 (ii) *$J(R)$ is T-nilpotent and $R/J(R)$ is von Neumann regular.*

Corollary 4.1. *A commutative semi-artinian ring is a max ring.*

More generally we have the following.

Proposition 4.1. [6]

    (1) *Let R be a left semi-artinian ring with the maximum condition on (right and left) primitive ideals. Then R is a right max ring.*

    (2) *There exists a right and left semi-artinian ring which is not a right max ring.*

5. π-V rings

E. Matlis [31] proved that the injective hull of a simple module over a commutative Noetherian ring is Artinian. R. L. Snider [42] proved that if G is nilpotent-by-finite then the injective hull of each simple $\mathbf{Z}[G]$-module is Artinian.

The following examples show that the result of Matlis can not extend to right Noetherian rings.

Example 5.1. Let Z and Q denote the ring of integers and the field of rational numbers, respectively. Consider the ring $A = \begin{pmatrix} Z & Q \\ 0 & Q \end{pmatrix}$. Then A is a right Noetherian PI-ring. Clearly $K = \begin{pmatrix} 0 & Q \\ 0 & 0 \end{pmatrix}$ is a minimal right ideal of A. The right ideal $H = \begin{pmatrix} Z & Q \\ 0 & 0 \end{pmatrix}$ is an essential extension of K and H_A is not artinian.

Example 5.2. Let F be a field of characteristic zero and let $A_1(F)$ denote the first Weyl algebra over F, that is $A_1(F) = F[x][y; d/dx]$. It is well-known that $A_1(F)$ is a left and right Noetherian domain. We can easily see that $F[x]$ is a simple left $A_1(F)$-module. Let $\{a_1, a_2, \cdots\}$ be an infinite subset of F and set $T_j = \{x - a_i \mid i = j, j + 1, \cdots\}$. Let S_j denote the multiplicative subsemigroup of $F[x]$ generated by T_j and consider the localization $F[x]_{S_j}$ of $F[x]$ by S_j. Then $F[x]_{S_1}$ is an essetial extension of $F[x]$. Since $F[x]_{S_1} \supset F[x]_{S_2} \supset \cdots$ is a strictly descending chain of left $A_1(F)$-submodules of $F[x]_{S_1}$, the injective hull $E(F[x])$ is not artinian.

Problem. Characterize a ring R in which the injective hull of every simple right R-module is artinian.

Let R be a ring and let M be a left R-module. Then again $E(M)$ denotes the injective hull of M.

A ring R is called a *left π-V ring* if $E(S)$ is of finite length for every simple left R-module S. Let n be a positive integer. A ring R is called a *left n-V ring* if the length of $E(S)$ is equal to or less than n for every simple left R-module S. A 1-V ring is just a V ring.

Theorem 5.1. [40] *A left and right Artinian PI-ring is a left and right π-V ring.*

Example 5.3. By Cohn [9], given any integer $n > 1$, there exists a skew field extension S/T of left degree n and right degree ∞. Hence by Rosenberg and Zelinsky [40], there exists a right and left Artinian ring R which is not a left π-V ring.

Let M be a module. A submodule N of M is said to be *of finite co-length*, if the length of M/N is finite.

Theorem 5.2. [21] *Let R be a ring. Then the following conditions are equivalent:*

    (i) *R is a π-V ring;*
    (ii) *Every left R-module M of finite length has an injective hull of finite length;*
    (iii) *For every left R-module M, the intersection of all submodules of finite co-length is 0.*

We also have the following.

Theorem 5.3. [21] *Let n be a positive integer. Then the following conditions are equivalent for a ring R:*

    (i) *R is a left n-V-ring;*
    (ii) *For every left R-module M, the intersection of all submodules N with $Le_R M/N \leq n$ is zero.*

By R. M. Hamsher [18], we have the following.

Proposition 5.1. *A commutative ring R is a π-V-ring if and only if R_M is artinian for any maximal ideal M of R.*

Also from the proof of Hamsher, we have the following.

Proposition 5.2. *A commutative ring R is a n-V-ring if and only if R_M has length $\leq n$ as an R_M-module for any maximal ideal M of R.*

Proposition 5.3. *Let R be a π-V ring. Then R is a left max ring and hence the Jacobson radical $J(R)$ of R is left T-nilpotent.*

Question 5.1. Is a perfect PI-ring a left (and right) π-V ring?

Proposition 5.4. *Let n be a positive integer and let R be a left n-V ring. Then, for any left ideal I of R, $I^n = I^{n+1}$.*

Question 5.2. Let R be a left π-V ring and let I be an arbitrary left ideal of R. Is there a positive integer n such that $I^n = I^{n+1}$?

6. Rings whose modules of finite length are semisimple

If R is a left V-ring, then every left R-module of finite length is semisimple. Hence a ring whose left modules of finite length are semisimple, can be considered as a generalization of a left V-ring. The following is easily proved.

Proposition 6.1. *Let R be a ring. Then the following conditions are equivalent:*

    (i) *Every left R-module of finite length is semisimple;*
    (ii) *$Ext_R^1(S,T) = 0$ for all simple left R-modules S, T;*
    (iii) *For each simple left R-module S, $Soc(E(S)/S) = 0$.*

Corollary 6.1. *Let R be a left semi-artinian ring. Then the following are equivalent:*

    (i) *R is a left V-ring;*
    (ii) *Every left R-module of finite length is semisimple.*

In case every primitive factor ring of a ring R is Artinian, we have the following characterization.

Theorem 6.1. *Let R be a ring whose primitive factor rings are Artinian. Then the following conditions are equivalent:*

    (i) *Every left R-module of finite length is semisimple;*
    (ii) *For any two primitive ideals P, Q, there holds that $PQ = P \cap Q$.*

There is an example of a commutative ring R such that R is not a V-ring and all R-modules of finite length are semisimple.

Proposition 6.2. *Let R be a commutative ring. Then the following are equivalent:*

> (i) *R is a V-ring (,or equivalently, R is a von Neumann regular ring);*
> (ii) *R is a max ring and every left R-module of finite length is semisimple.*

Conjecture 6.1. *A ring R is a left V-ring if and only if R is left max and every left R-module of finite length is semisimple.*

We conclude this paper with a characterization of a simple principal ideal domain whose left modules of finite length are semisimple. To state it, we need the following characterizations of rings all of whose modules of finite length are cyclic.

Proposition 6.3. [22] *Let R be a ring. Then the following statements are equivalent:*

> (i) *Any left R-module of finite length is cyclic;*
> (ii) *There is a positive integer n such that any left R-module of finite length is generated by n elements;*
> (iii) *Every finitely cogenerated left R-module has an essential cyclic submodule;*
> (iv) *For any simple left R-module M and any positive integer n,the direct sum of n copies of $M^{(n)}$ of M is cyclic;*
> (v) *R has no left Artinian factor rings;*
> (vi) *R has no simple left Artinian factor rings;*
> (i')-(vi') *The left-right symmetric versions of (i)-(vi).*

We call a ring R a *FLC-ring* if R satisfies these equivalent conditions. Obviously a non-artinian simple principal ideal domain is a FLC-ring. Also it is known that for every nonzero left ideal L of a simple principla ideal domain R, R/L is of finite length. Hence we obtain the following.

Proposition 6.4. *Let R be a simple principal ideal domain. Then the following are equivalent:*

> (i) *Every left R-module of finite length is semisimple;*
> (ii) *For every nonzero left ideal L of R, R/L is semisimple.*

References

1. F. W. Anderson and K. R. Fuller, Rings and Categories of Modules, Second Edition, Springer-Verlag, New York-Heidelberg-Berlin, 1992.

2. G. Baccella, Generalized V-rings and von Neumann regular rings, *Rend. Sem. Mat. Univ. Padova*, 72 (1984), pp.117–133.

3. G. Baccella, Von Neumann regularity of V-rings with artinian primitive factor rings, *Proc. Amer. Math. Soc.*, 103 (1988), pp.747–749.

4. G. Baccella, Semiartinian V-rings and semiartinian von Neumann regular rings, *J. Algebra*, 173 (1995), pp.587–612.

5. H. Bass, Finitistic dimension and a homological generalizations of semi-primary rings, *Trans. Amer. Math. Soc.*, 95 (1960), pp.466–488.

6. V. P. Camilo and K.R. Fuller, A note on Loewy rings and chain conditions on primitive ideals, Lecture Notes in Math. Vol.700, Springer, 1979.

7. J. Chen, On von Neumann regular rings and SF-rings, *Math. Japon.*, 36 (1991), pp.1123–1127.

8. J. Chen, N. Ding and M.F. Yousif, On Noetherian rings with essential socle, *J. Austra. Math. Soc.*, 76 (2004), pp.39–49.

9. P. M. Cohn, Quadratic extensions of skew fields, *Proc. London Math. Soc.*, 11 (1961), pp.531–556.

10. J. Cozzens, Homological properties of the ring of differential polynomials, *Bull. Amer. Math. Soc.*, 76 (1970), pp.75–79.

11. N. V. Dung and P. F. Smith, On semi-artinian V-modules, *J. Pure and Appl. Algebra*, 82 (1992), pp.27–37.

12. C. Faith, Algebra: rings, modules and categories. Vol. I., Springer, 1973.

13. C. Faith, Modules finite over endomorphism ring. Lectures on rings and modules, Lecture Notes in Math. Vol.246, (1973), pp.145–189.

14. C. Faith, Locally perfect commutative rings are those whose modules have maximal submodules, *Comm. Algebra*, 23 (13) (1995), pp.4885–4886.

15. C. Faith, Rings whosee modules have maximal submodules, *Publ. Mat.*, 39 (1995), pp.201–214.

16. C. Faith and P. Menal, A counter-example to a conjecture of Johns, *Proc. Amer. Math. Soc.*, 116 (1992), pp.21–26.

17. C. Faith and P. Menal, A new duality theorem for semisimple modules and characterization of Villamayor rings, *Proc. Amer. Math. Soc.*, 123 (1995), pp.1635–1637.

18. R. M. Hamsher, Commutative rings over which every modules has a maximal submodule, *Proc. Amer. Math. Soc.*, 18 (1967), pp.1133–1137.

19. Y. Hirano, On rings all of whose simple modules are flat, *Canad. Math. Bull.*, 37 (1994), pp.361–364.

20. Y. Hirano, On rings over which each module has a maximal submodule, *Comm. Algebra*, 26 (1998), pp.3435–3445.

21. Y. Hirano, On injective hulls of simple modules, *J. Algebra*, 225 (2000), pp.299–308.

22. Y. Hirano, On rings all of whose modules of finite length are cyclic, *Bull. Austral. Math. Soc.*, 69 (2004), pp.137–140.

23. C. Y. Hong, J.Y. Kim and N.Y. Kim, On von Neumann regular rings, *Comm. Algebra*, 28 (2000), pp.791–801.

24. Z. Y. Huang and F. C. Vheng, On homological dimensions of simple modules over non-commutative rings, *Comm. in Algebra*, 24 (10) (1996), pp.3259–

3264.

25. D. V. Huynh, S. K. Jain and S. R. López-Permouth, On a class of non-Noetherian V-rings, *Comm. Algebra*, 24 (1996), pp.2839–2850.

26. B. Johns, Annihilator conditions in Noetherian rings, *J. Algebra*, 49 (1977), pp.222–224.

27. J. Y. Kim, H. S. Yang, N. K. Kim and S. B. Nam, Some comments on rings whose simple singular modules are GP-injective or flat, *Kyungpook Math. J.*, 41 (2001), pp.23–27.

28. N. K. Kim, S. B. Nam and J. Y. Kim, On simple singular GP-injective modules, *Comm. Algebra*, 27 (1999), pp.2087–2096.

29. L. A. Koĭfmann, Rings over which every module has a maximal submodule, *Mat. Zametki*, 7 (1970), pp.350–367 = *Math. Notes* 7 (1970), pp.215–219.

30. V. T. Markov, On B-rings with a polynomial identity, *Trudy Sem. Petrovsk.*, 7 (1981), pp.232–238.

31. E. Matlis, Injective modules over Noetherian rings, *Pacific J. Math.*, 8 (1959), pp.511–528.

32. G. O. Michler and O. E. Villamayor, On rings whose simple modules are injective, *J. Algebra*, 25 (1973), pp.185–201.

33. C. Nastasescu, Quelques remarques sur la dimension homologique des anneaux, *J. Algebra*, 18 (1971), pp.470–485.

34. S. B. Nam, N. K. Kim and J. Y. Kim, On simple GP-injective modules, *Comm. Algebra*, 23 (1995), pp.5437–5444.

35. B. L. Osofsky, On twisted polynomial rings, *J. Algebra*, 18 (1971), pp.597–607.

36. V. S. Ramamurthi, On the injectivity and flatness of certain cyclic modules, *Proc. Amer. Math. Soc.*, 48 (1975), pp.21–25.

37. V. S. Ramamurthi and K. M. Rangaswamy, Generalized V-rings, *Math. Scand.*, 31 (1972), pp.69–77.

38. M. B. Rege, On von Neumann regular rings ans SF-rings, *Math. Japon.*, 31 (1986), pp.927–936.

39. R. D. Resco, Division rings and V-domains, *Proc. Amer. Math. Soc.*, 99 (1987), pp.427–431.

40. A. Rosenberg and D. Zelinsky, Finiteness of the injective hull, *Math. Zeitschr.*, 70 (1959), pp.372–380.

41. A. Shamsuddin, Homological properties of SF rings, *Bull. Austral. Math. Soc.*, 55 (1997), pp.327–333.

42. R. L. Snider, Injective hulls of simple modules over group rings, Ring theory (Proc. Conf., Ohio Univ., Athens, Ohio, 1976), pp. 223–226. Lecture Notes in Pure and Appl. Math., Vol. 25, Dekker, New York, 1977.

43. A. Tuganbaev, Max rings and V-rings, Handbook of Algebra Vol 3, pp.567–584, Elsevir Science, 2003.

44. Y. Xiao, One sided SF rings with certain chain conditions, *Canad. Math. Bull.*, 37 (1994), pp.272–277.

45. Y. Xiao, SF rings and excellent extensions, *Comm. Algebra* 22 (1994), pp.2463–2471.

46. R. Yue Chi Ming, On simple p-injective modules *Math. Japonicae* 19 (1974),

pp.173–176.

47. Z. Zhang and X. Du, Von Neumann regularity of SF-rings, *Comm. Algebra*, 21 (1993), pp.2445-2451.

ON A FINITELY GENERATED P-INJECTIVE LEFT IDEAL

YASUYUKI HIRANO

Department of Mathematics, Okayama University
Okayama 700-8530, Japan
E-mail: yhirano@math.okayama-u.ac.jp

JIN YONG KIM

Department of Mathematics and Institute of Natural Sciences
Kyung Hee University, Suwon 449-701, South Korea
E-mail: jykim@khu.ac.kr

We study in this paper for rings containing a finitely generated P-injective left ideal. We prove that if R contains a finitely generated P-injective left ideal I such that R/I is completely reducible, and if every left semicentral idempotent of R is central, then R is a left P-injective ring. As a byproduct of this result we give a new characterization of a von Neumann regular ring with nonzero socle. Also we are able to find a necessary and sufficient condition for semiprime left Noetherian rings to be Artinian.

Throughout this paper, R denotes an associative ring with identity and all modules are unitary. Recall that an idempotent $e \in R$ is left (resp. right) semicentral if $xe = exe$ (resp. $ex = exe$), for all $x \in R$. The set of left (resp. right) semicentral idempotents of R is denoted by $S_\ell(R)$ (resp. $S_r(R)$). For the set of all central idempotents of R will be denoted by $B(R)$. Observe $S_r(R) \cap S_\ell(R) = B(R)$ and if R is semiprime then $S_r(R) = S_\ell(R) = B(R)$. We deal with rings containing a finitely generated P-injective left ideal I such that R/I is completely reducible. We show that if R contains a finitely generated P-injective left ideal I such that R/I is completely reducible, and satisfying $S_\ell(R) = B(R)$, then R is left P-injective. As a byproduct of this result we are able to give a new characterization of von Neumann regular rings with nonzero socle. Actually we prove that a ring R is a von Neumann regular ring with nonzero socle if and only if R is a left pp-ring containing a finitely generated P-injective proper left ideal I such that R/I is completely reducible, and satisfying $S_\ell(R) = B(R)$. And we are able to find a necessary and sufficient condition for semiprime left Noetherian rings to be Artinian.

Also a connection between GP-injective rings and $C2$-rings is investigated. Recall that a ring R is called a *left pp-ring* if every principal left ideal of R is projective. A left R-module M is called to be *left P-injective* [9] if every left R-homomorphism from a principal left ideal Ra to M extends to one from $_RR$ to M. A well-known theorem of Ikeda-Nakayama [1] asserts that R is a *left P-injective* ring if and only if every principal right ideal of R is a right annihilator. A left R-module M is called *generalized left prinicipally injective* (briefly *left GP-injective*) [4] if, for any $0 \neq a \in R$, there exists a positive integer n such that $a^n \neq 0$ and any left R-homomorphism of Ra^n into M extends to one of $_RR$ into M. Note that GP-injective modules defined here are also called YJ-injective modules in [11].

Lemma 1. *For an idempotent $e \in R$, the following conditions are equivalent*:

(i) $e \in S_r(R)$;
(ii) $eR(1 - e) = 0$;
(iii) Re *is an ideal of R*.

Proof. The proof is routine. ∎

The following lemma was proved by Ming [10, Lemma 1.2]. But we shall give an elementary proof here.

Lemma 2. *If I is a finitely generated P-injective left ideal of R, then I is a direct summand of R.*

Proof. Let $I = Ra_1 + Ra_2 + \cdots + Ra_n$ where $a_1, a_2, \cdots, a_n \in I$. Since I is left P-injective, the inclusion map $\varphi_1 : Ra_1 \hookrightarrow I$ can be extended by $\hat{\varphi}_1 : R \to I$. Then $a_1 = \varphi_1(a_1) = \hat{\varphi}_1(a_1) = a_1 e_1$ where $\hat{\varphi}_1(1) = e_1$. Consider the element $a_2 - a_2 e_1 \in I$ and the inclusion map $\varphi_2 : R(a_2 - a_2 e_1) \hookrightarrow I$. Similarly there exists an element $e_2 \in I$ such that $(a_2 - a_2 e_1)e_2 = a_2 - a_2 e_1$. Let $e' = e_1 + e_2 - e_1 e_2$. Then $a_1 e' = a_1$ and $a_2 e' = a_2$. Now we will show that there exists an element $f \in I$ such that $a_i f = a_i$ for $i = 1, 2, \cdots, n$. We go by induction on n. The cases $n = 1$ and $n = 2$ are already done. Also we have an element $f_n \in I$ such that $a_n f_n = a_n$. Consider the $n - 1$ elements $a_1 - a_1 f_n, a_2 - a_2 f_n, \cdots, a_{n-1} - a_{n-1} f_n$. By induction, there exists an element $f' \in I$ such that $(a_i - a_i f_n)f' = a_i - a_i f_n$ for $i = 1, 2, \cdots, n-1$. Let $f = f_n + f' - f_n f'$. Then $a_n f = a_n f_n + (a_n - a_n f_n)f' = a_n$. Therefore we have an element $f \in I$ such that $a_1 f = a_1, a_2 f = a_2, \cdots, a_n f = a_n$. Since $I = Ra_1 + Ra_2 + \cdots + Ra_n$, $xf = f$ for any $x \in I$. Hence $I = Rf$ and $f^2 = f$. ∎

Recall that a direct sum of modules is *P-injective* if and only if each direct summand is *P-injective*.

Theorem 3. *If R contains a finitely generated P-injective left ideal I such that R/I is completely reducible, and satisfying $S_\ell(R) = B(R)$, then R is a left P-injective ring.*

Proof. By Lemma 2, we have $R = I \oplus L$ where $L = Re, I = R(1 - e)$ and $e = e^2 \in L$. Since $R/I \cong L$, $L = L_1 \oplus \cdots \oplus L_n$ where L_i minimal left ideals. Say $L = Re = Re_1 \oplus \cdots \oplus Re_n$, then by [2, p.50, Proposition 2] $e = e_1 + e_2 + \cdots + e_n$ and $\{e_1, e_2, \cdots, e_n\}$ is a set of orthogonal idempotents. If $IRe_i = 0$ for all $i \in \{1, 2, \cdots, n\}$, then $IL = R(1 - e)Re = 0$. So $e \in S_\ell(R) = B(R)$, we have $R = I \oplus L$ as a direct sum of two rings. Since L is a semisimple Artinian ring, R is obviously a left P-injective ring. So we may assume that $IRe_i \neq 0$ for some $i \in \{1, 2, \cdots, n\}$. Without loss of generality we can write $IRe_k \neq 0$ for $1 \leq k \leq m$ and $IRe_{m+1} = \cdots = IRe_n = 0$. Since $IRe_k = Re_k$ is projective, Re_k is isomorphic to a direct summand of I for $1 \leq k \leq m$. Thus, Re_k is P-injective for $1 \leq k \leq m$. Hence $I \oplus Re_1 \oplus \cdots \oplus Re_k$ is a finitely generated P-injective left ideal of R. Again by Lemma 2, there exists an idempotent $f \in R$ such that $R(1 - f) = I \oplus Re_1 \oplus \cdots \oplus Re_k$ and $Rf = Re_{k+1} \oplus \cdots \oplus Re_n$. Now we will claim that $R(1 - f)Rf = 0$. If not, there exists positive integers i, j such that $Re_iRe_j = Re_j \neq 0$ where $1 \leq i \leq k$ and $k + 1 \leq j \leq n$. Thus there is a nonzero element $x \in Re_j$, so we have a nonzero map $f : Re_i \to Re_j$ defined by $f(a) = ax$ for all $a \in Re_i$. Hence Re_i is isomorphic to Re_j. It is a contradiction, because $0 \neq IRe_i \cong IRe_j = 0$. Therefore $R(1 - f)Rf = 0$, so $f \in S_\ell(R) = B(R)$. Hence $R = R(1 - f) \oplus Rf$ as a direct sum of two rings. Since Rf is a semisimple Artinian ring, R is a left P-injective ring. ∎

Corollary 4. *Let R be a semiprime ring or an abelian ring. If R contains a finitely generated P-injective left ideal I such that R/I is completely reducible, then R is a left P-injective ring.*

Proof. Note that any semiprime ring or an abelian ring satisfies the condition $S_\ell(R) = B(R)$. ∎

Corollary 5. *If R contains a finitely generated P-injective maximal left ideal, and satisfying $S_\ell(R) = B(R)$, then R is a left P-injective ring.*

Theorem 6. *For a ring R, the following statements are equivalent:*
  (i) *R is a von Neumann regular ring with nonzero socle;*
  (ii) *R is a left pp-ring containing a finitely generated P-injective proper left ideal I such that R/I is completely reducible, and satisfying $S_\ell(R) = B(R)$.*

Proof. (i) $\Rightarrow$ (ii): Suppose that R is a von Neumann regular ring with nonzero socle. Obviously, R is a semiprime left pp-ring, hence satisfies the condition $S_\ell(R) = B(R)$. If every maximal left ideal of R is essential, then the socle of R is contained in $J(R) = 0$. Since R has a nonzero socle, there exists a maximal left ideal M of R which is not essential. Therefore M is a direct summand of R. Note that R is von Neumann regular if and only if every cyclic left R-module is P-injective [9, Lemma 2]. Hence M is finitely generated P-injective and R/M is simple left R-module.

(ii) $\Rightarrow$ (i): Let I be a finitely generated P-injective proper left ideal of R such that R/I is completely reducible. Then by Lemma 2, $R = I \oplus L$, where L is completely reducible. Hence the left socle of R is nonzero. Also by Theorem 3, R is left P-injective. It is known that R is a von Neumann regular ring if and only if R is a left P-injective and left pp-ring [8, Theorem 3]. Therefore R is a von Neumann regular ring with nonzero socle. ∎

Corollary 7. *For a ring R, the following statements are equivalent:*
  (i) *R is a von Neumann regular ring with nonzero socle;*
  (ii) *R is an left pp-ring containing a finitely generated P-injective maximal left ideal, and satisfying $S_\ell(R) = B(R)$.*

Corollary 8. *For a ring R, the following statements are equivalent:*
  (i) *R is a strongly regular ring with nonzero socle;*
  (ii) *R is an abelian left pp-ring containing a finitely generated P-injective maximal left ideal.*
  (iii) *R is a reduced ring containing a finitely generated P-injective maximal left ideal.*

Theorem 9. *Let R be a ring containing a finitely generated P-injective left ideal I such that R/I is completely reducible, and satisfying $S_\ell(R) = B(R)$. If R has ACC on left annihilators, then R is right Artinian.*

Proof. Combine Theorem 3 with Rutter's Theorem [7, Theorem], R is right Artinian. ∎

Corollary 10. *Let R be a ring containing a finitely generated P-injective maximal left ideal, and satisfying $S_\ell(R) = B(R)$. If R has ACC on left annihilators, then R is right Artinian.*

The next Corollary 11 contains a necessary and sufficient condition for semiprime left Noetherian rings to be Artinian.

Corollary 11. *For a ring R, the following statements are equivalent:*
  (i) *R is a semisimple Artinian ring;*
  (ii) *R is a semiprime left Noetherian ring containing a P-injective maximal left ideal;*
  (iii) *R is a semiprime ring containing a finitely generated P-injective maximal left ideal, and satisfying ACC on left annihilators.*

A ring R is called a *left $C2$-ring* [5] if every left ideal isomorphic to a direct summand of R is itself a direct summand. It is known that every left P-injective ring is a left $C2$-ring, but not conversely [5, Example 4]. Hence every von Neumann regular ring is a left and right $C2$-ring. But we do not know whether or not a left GP-injective ring is a left $C2$-ring. We shall give a partial answer as follows.

Theorem 12. *The following statements are equivalent for a ring R containing a finitely generated P-injective maximal left ideal:*
  (i) *R is left P-injective;*
  (ii) *R is left $C2$-ring;*
  (iii) *R is left GP-injective.*

Proof. (i) $\Rightarrow$ (ii) : See [5, Example 4]. (i) $\Rightarrow$ (iii) : Obvious. Now let M be a finitely generated P-injective maximal left ideal of R. Then by Lemma 2, $R = M \oplus U$ where $M = Re$, $U = R(1 - e)$ and $e = e^2 \in R$. We will show that the minimal left ideal $_RU$ is P-injective. If $e \notin S_r(R)$, then $MU = ReR(1 - e) \neq 0$. So there exists nonzero element $u \in U$ such that $Mu \neq 0$. Let $f : M \to Mu = U$ such that $f(m) = mu$. Since $_RU$ is projective, $M \approx \mathrm{Ker} f \oplus N$ where $_RN \approx_R U$. Hence $_RU$ is P-injective, so R is left P-injective. Thus it remains only to consider the case $e \in S_r(R)$.

(ii) $\Rightarrow$ (i): Assume that $e \in S_r(R)$. Then $M = Re$ is a two-sided ideal of R and $MU = 0$. Now $eR \subseteq ReR = M = Re$. First we will show that $M_R \cap (1 - e)R = 0$. If not, then there exists $a \in R$ such that

$0 \neq (1-e)a \in M$. Consider $g : R(1-e)a \to R/M$ such that $g(r(1-e)a) = r + M$. Since $\mathrm{Ker} g = M(1-e)a = 0$, we have $R(1-e)a \approx_R (R/M) \approx_R U$. Now R is a left $C2$-ring, $R(1-e)a$ is a direct summand of ${}_R R$. Thus, $0 \neq (R(1-e)a)^2 \subseteq M(1-e)a = 0$. It is a contradiction. Hence we have $M_R \cap (1-e)R = 0$ and $R_R = M_R \oplus (1-e)R$. Now $(R/M)_R$ is projective (and hence flat) which implies ${}_R(R/M)$ is P-injective [6, Proposition 1.4]. Then ${}_R U$ is P-injective, so R is left P-injective.

(iii) $\Rightarrow$ (i): Let $e \in S_r(R)$. Then $M = Re$ is a two-sided ideal of R. Now $r(M) = (1-e)R$, we claim that $(1-e)R$ is a minimal right ideal of R. If not, there exists $0 \neq a \in (1-e)R$ such that $aR \subsetneq (1-e)R$. Since R is left GP-injective, there exists $n \in \mathbb{Z}^+$ such that $a^n R$ is a nonzero right annihilator [11, Lemma 3]. Hence $M = \ell(a^n R)$, so $a^n R = r(M) = (1-e)R$. It is absurd because $aR \subsetneq (1-e)R$. Therefore $(1-e)R$ is a minimal right ideal of R. Now $eR \subseteq ReR = Re = M$, so $R_R = M_R + (1-e)R$. Assume that there is $0 \neq b \in M \cap (1-e)R$. Then $(1-e)R = bR \subseteq M = Re$, it is a contradiction. Thus $R_R = M_R \oplus (1-e)R$, hence $(R/M)_R$ is projective. Therefore ${}_R(R/M)$ is P-injective which implies that $U = R(1-e)$ is P-injective. Thus R is left P-injective. $\blacksquare$

We conclude with the following question.

Question: Could we show that every left GP-injective ring is a left C2-ring?

ACKNOWLEDGMENTS

This paper was written while the second named author visited Okayama University in Japan under the Memorandum of Understanding between KOSEF and JSPS. He is grateful to the staffs of Department of Mathematics of Okayama University for their hospitality. The second named author was partially supported by the Grant No.R05-2002-000-00715-0 from the Basic Research Program of the Korea Science and Engineering Foundation.

References

1. M. Ikeda and T. Nakayama, On some characteristic properties of quasi-Frobenius and regular rings, *Proc. Amer. Math. Soc.*, 5 (1954), pp.15–19.
2. N. Jacobson, Structure of rings, Amer. Math. Soc., Reprinted 1968.
3. J. Y. Kim and N. K. Kim, On rings containing a p-injective maximal left ideal, Comm. Korean. Math. Soc. 18 (4) (2003), 629–633.

4. S.B. Nam, N.K. Kim and J.Y. Kim, On simple GP-injective modules, Comm. Algebra 23 (14), (1995), 5437–5444.

5. W. K. Nicholson and M. F. Yousif, C2-rings and the FGF-conjecture, *Contemporary Math.* 273 (2001), pp.245–251.

6. V.S. Ramamurthi, On the injective and flatness of certain cyclic modules, *Proc. Amer. Math. Soc.* 48 (1975), pp.21–26.

7. E. A. Rutter, Jr, Rings with the principal extension property, *Comm. in Algebra*, 3(3) (1975), pp.203–212.

8. W. M. Xue, On *pp*-rings, Kobe J. Math. 7(2) (1990), 77–80.

9. R. Yue Chi Ming, On (von Neumann) regular rings, Proc. Edinburgh Math. Soc. 19 (1974), 89–91.

10. R. Yue Chi Ming, On von Neumann regular rings, III, *Mh. Math.* 86 (1978), pp.251–257.

11. R. Yue Chi Ming, On regular rings and Artinian rings(II), *Riv. Mat. Univ. Parma.* 11 (1985), pp.101–109.

12. R. Yue Chi Ming, A note on YJ-injectivity, *Demonstratio Math.* 30 (1997), pp.551–556.

CROSSED PRODUCTS AND FULLY PRIME RINGS

LIUJIA HUANG

Department of Mathematics and Computer Science,
Guangxi University for Nationality,
Nanning, Guangxi, 530006
E-mail: huangliujia@126.com

ZHONG YI[*]

Department of Mathematics, Guangxi normal University,
Guilin, Guangxi, 541004
E-mail: zyi@mailbox.gxnu.edu.cn

Some equivalent characterizations for a crossed product to be a fully prime ring
(almost fully prime ring) are given.

In this paper, all rings are associative and have identity, and all modules
are unitary. Let R be a ring and let G be a multiplicative group. A
crossed product $R * G$ of G over R is an associative ring which contains R
and is a free $R-$module with an $R-$basis the set $\bar{G}$, a copy of G. Thus
$R * G = \bigoplus_{g \in G} \bar{g} R$. Addition of $R * G$ is as expected and multiplication is
determined by the two rules below:

$$\bar{g}\bar{h} = \overline{gh}\alpha(g, h)$$

for all $g, h \in G$, where $\alpha : G \times G \to U(R)$, the group of units of R, and

$$r\,\bar{g} = \bar{g}\,r^{t(g)}$$

for all $r \in R$ and $g \in G$, where $t : G \to Aut(R)$. If $\alpha(g, h) = 1$ for all
$g, h \in G$, then $R * G$ is called a *skew group ring*. For basic properties and
some well-known results of crossed products, see [1] for details.

---

[*]Supported by NSF of China (10271021), NSF of Guangxi (0135005), EYTP of MOE of
China (2002-40)

We recall that a ring R is called *fully idempotent* if every ideal of R is idempotent, and a ring R is called a *fully prime ring* if every ideal of R is prime(see [2]), in this case we briefly call R an FPR. The following is a basic result of an FPR:

Lemma 1[2] A ring R is an FPR if and only if it is fully idempotent and the set of ideals of R is linearly ordered under inclusion.

At first, we discuss the FPR properties between a ring R and a crossed product $R * G$.

Lemma 2 Let R be a ring such that the set of ideals of R is linearly ordered under inclusion, and let G be a finite group acting on R as automorphisms. If $I \lhd R$, then I is $G-$stable.

Proof. $\forall g \in G$, we have $I^g \lhd R$. Since the ideals of R are linearly ordered, we have $I^g \subseteq I$ or $I \subseteq I^g$. If $I^g \subseteq I$, then $I = I^{g^n} \subseteq I^{g^{n-1}} \subseteq \cdots \subseteq I^g \subseteq I$, where $n = |G|$, thus $I = I^g$. The prove for the case $I \subseteq I^g$ analogous as above. Hence I is $G-$stable.

Theorem 1 Let $R * G$ be a crossed product with G finite, and let R be an FPR, then the crossed product $R * G$ is an FPR if and only if the map $\phi : \mathcal{L}(R * G) \to \mathcal{L}(R); P \mapsto P \cap R$, is a one to one onto correspondence between the set of ideals of $R * G$ and the set of ideals of R.

Proof. $(\Rightarrow)$ It is easy to see that ϕ is a map. Let $I \lhd R$, by Lemma 1 and Lemma 2, I is $G-$stable, thus $I * G \lhd R * G$ such that $I = (I * G) \cap R$, hence ϕ is onto. Let $P_1, P_2 \in \mathcal{L}(R * G)$. Since we suppose that $R * G$ is an FPR, by Lemma 1, we may suppose that $P_1 \subset P_2$. If $\phi(P_1) = \phi(P_2)$, that is $P_1 \cap R = P_2 \cap R$, then by [1, Theorem 16.6(iii)], we have $P_1 = P_2$, contradiction. Hence ϕ is one to one.

$(\Leftarrow)$ Let P be an ideal of $R * G$ and let $IJ \subseteq P$ for some ideals I and J of $R * G$, thus $(I \cap R)(J \cap R) \subseteq P \cap R$. Since R is an FPR, we have $I \cap R \subseteq P \cap R$ or $J \cap R \subseteq P \cap R$. Because ϕ is one to one, for each ideal K of $R * G$, we have $K = (K \cap R) * G$. Then $I = (I \cap R) * G \subseteq (P \cap R) * G = P$ or $J = (J \cap R) * G \subseteq (P \cap R) * G = P$. Hence $R * G$ is an FPR.

Corollary 1 Let K be a field, G be a finite group, then the group ring $K[G]$ is an FPR if and only if $G = <1>$.

Definition 1 Let R be a ring, and let G be a group acting on R as automorphisms. R will be called $G - FPR$ if every $G-$stable ideal of R is $G-$prime.

The following two properties are just analogues of corresponding results of FPR.

Proposition 1 Let R be a ring, and let G be a group acting on R as automorphisms, then R is a $G - FPR$ if and only if the set of $G-$stable ideals of R is linearly ordered under inclusion and every $G-$stable ideal of R is idempotent.

Proposition 2 Let R be a ring, and let G be a group acting on R as automorphisms. If R is a $G - FPR$, then R is semiprime.

Theorem 2 Let $R * G$ be a crossed product with G finite, and let R be a ring, then the following are equivalent:

(i) the crossed product $R * G$ is an FPR;

(ii) (a) R is a $G - FPR$;

(b) the map $\phi : \mathcal{L}(R * G) \to G - \mathcal{L}(R); P \mapsto P \cap R$, is a one to one onto correspondence between the set of ideals of $R * G$ and the set of $G-$stable ideals of R.

Proof. $(i) \Rightarrow (ii)$ Let $A \triangleleft R$ be $G-$stable. If $BC \subseteq A$ for some $G-$stable ideals B, C of R, then $(B * G)(C * G) \subseteq A * G$, thus $B * G \subseteq A * G$ or $C * G \subseteq A * G$ since $R * G$ is an FPR. So $B \subseteq A$ or $C \subseteq A$. So R is a $G - FPR$ and (a) is proved. Obviously, ϕ is a map. Let $A \in G - \mathcal{L}(R)$, then $A * G \in \mathcal{L}(R * G)$ and $\phi(A * G) = (A * G) \cap R = A$. Hence ϕ is onto. Let $P_1, P_2 \in \mathcal{L}(R * G)$ such that $P_1 \neq P_2$, then by Lemma 1 we may suppose that $P_1 \subset P_2$. If $\phi(P_1) = \phi(P_2)$, that is $P_1 \cap R = P_2 \cap R$, then $P_1 = P_2$ by [1, Theorem 16.6(iii)], contradiction. Hence ϕ is one to one.

$(ii) \Rightarrow (i)$ Let $P \in \mathcal{L}(R * G)$, If $IJ \subseteq P$ for some ideals I and J of $R * G$. Then $(I \cap R)(J \cap R) \subseteq P \cap R$ and $I \cap R, J \cap R, P \cap R \in G - \mathcal{L}(R)$, by (a) of (ii), we have $I \cap R \subseteq P \cap R$ or $J \cap R \subseteq P \cap R$. Then by (b) of (ii), we have $I \subseteq P$ or $J \subseteq P$. So P is prime, hence $R * G$ is an FPR.

Example 1 There is a ring R which is a $G - FPR$ but not an FPR. Let $R = K \oplus K$, where K is a field. Obviously R is not an FPR. Let $g : R \to R, (x_1, x_2) \mapsto (x_2, x_1)$, then $g \in Aut(R)$. Let $G =< g >$ acting on R by $r^g = g(r)$, for all $r \in R$. It is easy to see that R have only two $G-$stable ideals: 0 and R, so R is a $G - FPR$.

Using the relationship between R*G and R^G, we can easily obtain the following result.

Proposition 3 Let R be a ring and let G be a finite group acting on R as automorphisms. Suppose that $|G|^{-1} \in R$. If the skew group ring $R * G$ is an FPR, then the fix ring R^G is also FPR.

In [3], a ring R is called an *almost fully prime ring* if each nonzero proper ideal of R is prime. It is clear that fully prime rings are almost fully prime rings, however [3, Example 2.4] gives a ring which is almost fully prime but not an FPR, and Example 1 also demonstrates this fact. Now, we denote $AFPR$ as the ring which is almost fully prime but not prime. The following are two results of $AFPR$:

Lemma 3[3] Let R be a ring whose set of ideals is not linearly ordered. Then R is an $AFPR$ if and only if

 1. R is a fully idempotent ring which has exactly two minimal ideals,

 2. each minimal ideal of R is contained in every nonminimal ideal of R, and

 3. the set of all nonminimal ideals of R is linearly ordered.

Lemma 4[3] Let R be a ring whose set of ideals is linearly ordered. Then R is an $AFPR$ if and only if it has a unique minimal ideal and every ideal of R except the minimal one is idempotent.

Remark Let R be an $AFPR$, and let G be a finite group acting on R as automorphisms.

(1) Let R be as in Lemma 3, analogous the proof of Lemma 2, we can sea that each nonminimal ideal of R is $G-$stable, but in Example 1, the minimal ideals of $R : I_1 = K \oplus 0, I_2 = 0 \oplus K$ are not $G-$stable.

(2) Let R be as in Lemma 3, then R is not $G-$prime if and only if the minimal ideals of R are $G-$stable.

(3) Let the crossed product $R * G$ be an $AFPR$ as in Lemma 3 and let P_1, P_2 be the minimal ideals of $R * G$, then $P_1 \cap R = 0$ if and only if $P_2 \cap R = 0$. Because if $P_1 \cap R = 0$, then 0 is $G-$prime by [1, Lemma 14.1]. So R is $G-$prime, thus by [1, Lemma 16.2], we have $P_2 \cap R = 0$, analogously for the other case.

Example 2 Let R and G be as in Example 1. Let $R * G$ be the skew group ring. Obviously $R * G$ is an FPR and R is an $AFPR$ with R being $G - FPR$.

Motivated by Example 2, we have:

Theorem 3 Let $R * G$ be a crossed product with G finite and $R * G$ fully prime. Then R is an $AFPR$ if and only if

(i) R has exactly two minimal ideals which are prime;

(ii) the map $\phi : \mathcal{L}(R * G) \to \overline{\mathcal{L}(R)}; P \mapsto P \cap R$, is a one to one onto correspondence between the set of ideals of $R * G$ and the set of ideals of R except the minimal ideals.

Proof. Because $R * G$ is an FPR, thus R is $G-$prime.

Suppose R is an $AFPR$. Firstly, we show that $\mathcal{L}(R)$ must not be linearly ordered under inclusion. If not, By Lemma 4, R has a unique minimal ideal I such that $I^2 = 0$. By Lemma 2 I is $G-$stable, so $(I * G)^2 = 0$, it is a contradiction with R is $G-$prime. Thus by Lemma 3 R has exactly two minimal ideals which are prime. Hence (i) holds. By Remark(2), it is easy to check that ϕ is an onto map. Let $P_1, P_2 \in \mathcal{L}(R * G)$ such that $P_1 \neq P_2$. We may suppose that $P_1 \subset P_2$ since $R * G$ is an FPR. Then by [1, Theorem

88

16.6(iii)] $\phi(P_1) \subset \phi(P_2)$, it follows that ϕ is one to one. Hence (ii) holds.

Conversely, let $0 \neq P \triangleleft R$ and $IJ \subseteq P$ for some ideals I, J of R such that $P \subseteq I, P \subseteq J$. If P is minimal, then by (i) P is prime. If not, then by (ii) P, I, J are $G-$stable, thus $(I * G)(J * G) \subseteq P * G$. So $I * G \subseteq P * G$ or $J * G \subseteq P * G$ since $R * G$ is an FPR. Then we have $J = (J * G) \cap R \subseteq (P * G) \cap R = P$ or $I = (I * G) \cap R \subseteq (P * G) \cap R = P$. Thus P is prime. By (i) we have R is not prime. Hence R is an $AFPR$.

If $R * G$ is not an FPR, but R is an $AFPR$, then we have

Theorem 4 Let $R * G$ be a crossed product with G finite. and let R be an $AFPR$ whose set of ideals is not linearly ordered under inclusion. Then the crossed product $R * G$ is an $AFPR$ if and only if

(i) the map $\phi_1 : \mathcal{L}(R * G) \to \mathcal{L}(R); P \mapsto P \cap R$, is a one to one onto correspondence between the set of ideals of $R * G$ and the set of ideals of R;

or (ii) (a) $R * G$ has exactly two minimal ideals P_1, P_2 which are prime;

(b) the map $\phi_2 : \overline{\mathcal{L}(R * G)} \to \overline{\mathcal{L}(R)}; P \mapsto P \cap R$, is a one to one onto correspondence between the set of ideals of $R * G$ except the minimal ideals and the set of ideals of R except the minimal ideals;

or (iii) (a) $R * G$ has a unique minimal ideal P_0 which is prime and nilpotent;

(b) the map $\phi_3 : \overline{\mathcal{L}(R * G)} \to \overline{\mathcal{L}(R)}; P \mapsto P \cap R$, is a one to one onto correspondence between the set of ideals of $R * G$ except P_1 and the set of ideals of R except the minimal ideals.

Proof. ($\Leftarrow$) Suppose that (i) holds. From ϕ_1 is one to one onto map, we have that $R * G$ satisfies the condition of Lemma 3. Hence $R * G$ is an $AFPR$. Suppose (ii) holds. Let $0 \neq P \triangleleft R * G$, and $IJ \subseteq P$ for some ideals I, J of $R * G$ such that $P \subseteq I, P \subseteq J$. If P is minimal, then by (a) of (ii) P is prime. If not, then $(I \cap R)(J \cap R) = (P \cap R) \neq 0$ by (b)of (ii), thus $I \cap R \subseteq P \cap R$ or $J \cap R \subseteq P \cap R$ since R is an $AFPR$. By (b) of (ii) we have $I = (I \cap R) * G \subseteq (P \cap R) * G = P$ or $J = (J \cap R) * G \subseteq (P \cap R) * G = P$ thus P is prime. By (a) of (ii), $R * G$ is not prime. Hence $R * G$ is an $AFPR$. Suppose (iii) holds. Similar to the above proof of case (ii), we also know that $R * G$ is an $AFPR$.

($\Rightarrow$) By Lemma 3, R has exactly two minimal ideals I_1, I_2 . There are two cases for the set $\mathcal{L}(R * G)$ of ideals of $R * G$ to consider.

Case 1. $\mathcal{L}(R * G)$ is not linearly ordered.

By Lemma 3 $R * G$ has exactly two minimal ideals P_0, P_1 which are prime. There are two cases for R to consider:

(1) R is not $G-$prime.

By Remark (2), each ideal $I \lhd R$ is $G-$stable, it follows that ϕ_1 is an onto map. Let $0 \neq P \in \mathcal{L}(R * G)$, we have $\phi_1(P) \neq 0$. Otherwise, 0 is $G-$prime, i.e. R is $G-$prime, contradiction. Suppose that there are $0 \neq P, P' \in \mathcal{L}(R * G)$ such that $P \neq P'$ and $\phi_1(P) = \phi_1(P')$, thus $\phi_1(P) = \phi_1(P') \neq 0$. If P, P' are exactly the minimal ideals of $R * G$, then $0 \neq P = (P \cap R) * G = (P' \cap R) * G = P'$, since P and P' are minimal, contradiction. If not, By Lemma 3 we may suppose that $P \subset P'$, then by [1, Theorem 16.6(iii)] we have $\phi_1(P) = P \cap R \subset P' \cap R = \phi_1(P')$, contradiction. So ϕ_1 is one to one. Hence (i) holds.

(2) R is $G-$prime.

Let $P \in \overline{\mathcal{L}(R * G)}$, then by [1, Lemma 14.2] $P \cap R$ is $G-$stable. By Remark (2) $P \cap R \in \overline{\mathcal{L}(R)}$, thus ϕ_2 is a map. Obviously, ϕ_2 is an onto map. Let $0 \neq P \in \overline{\mathcal{L}(R * G)}$, we have $\phi_2(P) \neq 0$. Otherwise, By [1, Lemma 16.2] P is minimal, contradiction. Let $0 \neq P, P' \in \overline{\mathcal{L}(R * G)}$ such that $P \neq P'$. By Lemma 3 we may suppose that $P \subset P'$, then by [1, Theorem 16.6(iii)] we have $\phi_1(P) = P \cap R \subset P' \cap R = \phi_1(P')$. So ϕ_1 is one to one. Hence (ii) holds.

Case 2. $\mathcal{L}(R * G)$ is linearly ordered.

By Lemma 4, $R * G$ has a unique minimal ideal P_0 which is prime and nilpotent. Firstly, we show that R must be $G-$prime. Otherwise, I_1, I_2 are $G-$stable by Remark (2), then $I_1 * G, I_2 * G \lhd R * G$, thus $I_1 * G \subset I_2 * G$ or $I_2 * G \subset I_1 * G$ since $\mathcal{L}(R * G)$ is linearly ordered. we suppose that $I_1 * G \subset I_2 * G$, so $I_1 = (I_1 * G) \cap R \subseteq (I_2 * G) \cap R = I_2$, contradiction. Let $P \in \overline{\mathcal{L}(R * G)}$, then by [1, Lemma 14.2] $P \cap R$ is $G-$stable. By Remark(2) $P \cap R \in \overline{\mathcal{L}(R)}$, thus ϕ_3 is a map. Obviously, ϕ_3 is an onto map. Let $0 \neq P \in \overline{\mathcal{L}(R * G)}$, we have $\phi_3(P) \neq 0$. Otherwise, By [1, Lemma 16.2] P is minimal, contradiction. Let $0 \neq P, P' \in \overline{\mathcal{L}(R * G)}$ such that $P \neq P'$, then we may suppose $P \subset P'$ since $\mathcal{L}(R * G)$ is linearly ordered. By [1, Theorem 16.6(iii)] we have $P \cap R \subset P' \cap R$. So ϕ_3 is one to one. Hence (iii) holds.

Theorem 5 Let $R * G$ be a crossed product with G finite, and let R be an $AFPR$ whose set of ideals is linearly ordered. Then the crossed product $R * G$ is an $AFPR$ if and only if the map $\phi : \mathcal{L}(R * G) \to \mathcal{L}(R); P \mapsto P \cap R$, is a one to one onto correspondence between the set of ideals of $R * G$ and the set of ideals of R.

Proof. Suppose $R * G$ is an $AFPR$. By Lemma 2, ϕ is an onto map. If $0 \neq P \in \mathcal{L}(R * G)$, we have $\phi(P) \neq 0$. Otherwise, 0 is $G-$prime. But by Lemma 2 and Lemma 4 we have $I^2 = 0$ with I is $G-$stable, where I is the minimal ideal of R, contradiction. Let $0 \neq P, P' \in \mathcal{L}(R * G)$ such that $P \neq P'$. Before proving $\phi(P) \neq \phi(P')$, we show that $\mathcal{L}(R * G)$ is linearly ordered under inclusion. If not, by Lemma 3 $R * G$ has two minimal ideals P_1, P_2 which are prime. Thus we have that either $P_1 \cap R = 0$ or $P_1 \cap R \neq 0$.

If $P_1 \cap R = 0$. Then 0 is $G-$prime, contradiction as above. Thus $P_1 \cap R \neq 0$. By Remark (3) $P_2 \cap R \neq 0$. If $P_1 \cap R \neq P_2 \cap R$. Then we may suppose that $P_1 \cap R \subset P_2 \cap R$ since $\mathcal{L}(R)$ is linearly ordered. By [1, Theorem 16.6(iii)] we have $P_1 \subset P_2$, contradiction. If $P_1 \cap R = P_2 \cap R$. Then $P_1 = (P_1 \cap R) * G = (P_2 \cap R) * G = P_2$ since P_1, P_2 are minimal, contradiction. Hence $\mathcal{L}(R * G)$ is linearly ordered under inclusion. Thus we may suppose $P \subset P'$. By [1, Theorem 16.6(iii)] we have $\phi(P) = P \cap R \subset P' \cap R = \phi(P')$. Hence ϕ is one to one.

The converse proof is obtained immediately from Lemma 4.

Definition 2 Let R be a ring, and let G be a group acting on R as automorphisms. R will be called $G - AFPR$ if every $G-$stable ideal except zero ideal is $G-$prime.

If we omit the condition which R is an $AFPR$ in Theorem 4, then we have

Theorem 6 Let R be a ring, and let $R * G$ be a crossed product with G finite. If the crossed product $R * G$ satisfies the condition that the set of ideals is not linearly ordered under inclusion. Then the crossed product $R * G$ is an $AFPR$ if and only if

(i) (a) R is a $G - AFPR$;

(b) the map $\phi_1 : \mathcal{L}(R * G) \to G - \mathcal{L}(R); P \mapsto P \cap R$, is a one to one onto correspondence between the set of ideals of $R * G$ and the set of $G-$stable ideals of R;

or (ii) (a) R is a $G - FPR$;

(b) $R * G$ has exactly two minimal ideals P_0, P_1 such that they are primes;

(c) the map $\phi_2 : \overline{\mathcal{L}(R * G)} \to G - \mathcal{L}(R); P \mapsto P \cap R$, is a one to one onto correspondence between the set of ideals of $R * G$ except $\{P_0, P_1\}$ and the set of $G-$stable ideals of R.

Proof. ($\Rightarrow$) By Lemma 3 $R * G$ has exactly two minimal ideals P_0, P_1. There are two cases for $P_0 \cap R$ to consider:

Case (1) $P_0 \cap R \neq 0$.

Then by Remark (3) $P_1 \cap R \neq 0$, it follows that $(P_0 \cap R)(P_1 \cap R) = 0$, so R is not $G-$prime. Let $0 \neq P \in G - \mathcal{L}(R)$, and let $IJ \subseteq P$ for some ideals $I, J \in G - \mathcal{L}(R)$, then $(I * G)(J * G) \subseteq (P * G)$, it follows that $I * G \subseteq P * G$ or $J * G \subseteq P * G$ since $R * G$ is an $AFPR$. Thus

$I = (I * G) \cap R \subseteq (P * G) \cap R = P$ or $J = (J * G) \cap R \subseteq (P * G) \cap R = P$, so P is a prime. Hence R is a $G - AFPR$. Obviously, ϕ_1 is an onto map. Next we show that ϕ_1 is one to one. Let $0 \neq P \in \overline{\mathcal{L}(R * G)}$. Then by [1, Theorem 16.6(iii)] we have $P \cap R \supset P_0 \cap R \neq 0$. Let $0 \neq P, P' \in \mathcal{L}(R * G)$ such that $P \neq P'$. If $P, P' \in \{P_1, P_2\}$. Then we have $P \cap R \neq P' \cap R$. Otherwise, we have $P = (P \cap R) * G = (P' \cap R) * G = P'$ since P, P' are minimal, contradiction. Suppose at lease one of P, P' is not in $\{P_1, P_2\}$, then by Lemma 4, we may suppose that $P \subset P'$, by [1, Theorem 16.6(iii)] we have $\phi_1(P) = P \cap R \neq \phi_1(P') = P' \cap R$. So ϕ_1 is one to one. Hence (i) hold.

Case (2) $P_0 \cap R = 0$.

Then by [1, Lemma 14.2] 0 is $G-$prime, it follows that R is $G-$prime. The same reason as in case (1), if $0 \neq P \in G - \mathcal{L}(R)$, we know that P is $G-$prime, hence R is a $G - FPR$. Obviously, ϕ_2 is an onto map. Let $0 \neq P \in \overline{\mathcal{L}(R * G)}$, we have $\phi_2(P) \neq 0$. Otherwise, By [1, Lemma 16.2] P is minimal since R is $G-$prime, contradiction. Suppose that there are $0 \neq P, P' \in \overline{\mathcal{L}(R * G)}$ such that $P \neq P'$. By Lemma 4 we may suppose that $P \subset P'$, then by [1, Theorem 16.6(iii)] we have $\phi_2(P) = P \cap R \subset P' \cap R = \phi_2(P')$. So ϕ_2 is one to one. Hence (ii) hold.

($\Leftarrow$) Let $0 \neq P \lhd R * G$ and $IJ \subseteq P$ for some ideals I, J of $R * G$. Then $(I \cap R)(J \cap R) \subseteq P \cap R$. Suppose that (i) holds. then by (b) of (i) we have $P \cap R \neq 0$. It follows that $I \cap R \subseteq P \cap R$ or $J \cap R \subseteq P \cap R$ by (a) of (i). By (b) of (i) we have $I = (I \cap R) * G$, $J = (J \cap R) * G$, $P = (P \cap R) * G$, thus $I \subseteq P$ or $J \subseteq P$, so P is prime. By (a) of (i), there are non$-$zero $G-$stable ideals I_1, I_2 of R such that $I_1 I_2 = 0$, thus $(I_1 * G)(I_2 * G) = 0$, so $R * G$ is not prime. Hence $R * G$ is an $AFPR$. Suppose that (ii) holds. If P is minimal, then P is prime by (b) of (ii). If not, then by [1, Lemma 16.2] we have $P \cap R \neq 0$. It follows that $I \cap R \subseteq P \cap R$ or $J \cap R \subseteq P \cap R$ by (a) of (ii). By (c) of (ii) we have $I = (I \cap R) * G$, $J = (J \cap R) * G$, $P = (P \cap R) * G$, thus $I \subseteq P$ or $J \subseteq P$. So P is prime. By (b) of (ii) $R * G$ is not prime. Hence $R * G$ is an $AFPR$.

If we omit the condition which R is an $AFPR$ in Theorem 5, then we have

Theorem 7 Let R be a ring, and let the crossed product $R * G$ with G finite such that whose set of ideals is linearly ordered. Then $R * G$ is an $AFPR$ if and only if

$\quad$ (i) (a) R is a $G - AFPR$;

$\qquad$ (b) the map $\phi_1 : \mathcal{L}(R * G) \to G - \mathcal{L}(R); P \mapsto P \cap R$, is a one to one onto correspondence between the set of ideals of $R * G$ and the set of

92

$G-$stable ideals of R;

   or (ii) (a) R is a $G - FPR$;

    (b) $R * G$ has a unique minimal ideal P_0 such that it is prime and $P_0^2 = 0$;

    (c) the map $\phi_2 : \overline{\mathcal{L}(R * G)} \to G - \mathcal{L}(R); P \mapsto P \cap R$, is a one to one onto correspondence between the set of ideals of $R * G$ except the only minimal ideal P_0 and the set of $G-$stable ideals of R .

Proof. $(\Rightarrow)$ By Lemma 4, $R * G$ has exactly one minimal ideal P_0 such that $P_0^2 = 0$. There are two case for $P_0 \cap R$ to consider.

Case (1) $P_0 \cap R \neq 0$.

Then we have $(P_0 \cap R)^2 = 0$, so R is not $G-$prime. Let $0 \neq P \in G - \mathcal{L}(R)$ and let $IJ \subseteq P$ for some $I, J \in G - \mathcal{L}(R)$, then $(I * G)(J * G) \subseteq P * G$, thus $J * G \subseteq P * G$ or $I * G \subseteq P * G$ since $R * G$ is an $AFPR$, so $J = (J * G) \cap R \subseteq (P * G) \cap R = P$ or $I = (I * G) \cap R \subseteq (P * G) \cap R = P$, i.e. P is $G-$prime. Hence R is a $G - AFPR$. Obviously, ϕ_1 is an onto map. Let $0 \neq P \in \overline{\mathcal{L}(R * G)}$. By [1, Theorem 16.6(iii)] we have $0 \neq P_0 \cap R \subset P \cap R$. Let $0 \neq P_1, P_2 \in \overline{\mathcal{L}(R * G)}$ such that $P_1 \neq P_2$, we may suppose that $P_1 \subset P_2$ since $\mathcal{L}(R * G)$ is linearly ordered. So by [1, Theorem 16.6(iii)] we have $\phi_1(P_1) = P_1 \cap R \neq P_2 \cap R = \phi_1(P_2)$, thus ϕ_1 is one to one. Hence (i) hold.

Case (2) $P_0 \cap R = 0$.

Then by [1, Lemma 14.2] 0 is $G-$prime, it follows that R is $G-$prime. Let $0 \neq P \in G - \mathcal{L}(R)$, as in case (1) we can show that P is $G-$prime, hence R is a $G - FPR$. Obviously, ϕ_2 is an onto map. Similarly as in case (1) we have ϕ_2 is one to one. Hence (ii) hold.

$(\Leftarrow)$ Suppose that (i) holds. By (a) there are nonzero $G-$stable ideals I, J of R such that $IJ = 0$, then $(I * G)(J * G) = 0$, hence $R * G$ is not prime. Let $0 \neq P \in \mathcal{L}(R * G)$ and $P_1 P_2 \subseteq P$ for some ideals P_1, P_2 of $R * G$. Then $(P_1 \cap R)(P_2 \cap R) \subseteq P \cap R$, but $P \cap R \neq 0$, otherwise, 0 is $G-$prime in R, it is a contradiction with R is a $G - AFPR$. Thus by (a) of (i) $P_1 \cap R \subseteq P \cap R$ or $P_2 \cap R \subseteq P \cap R$. By (b) we have that $P_1 = (P_1 \cap R) * G \subseteq (P_1 \cap R) * G = P$ or $P_2 = (P_2 \cap R) * G \subseteq (P \cap R) * G = P$, so P is prime. Hence $R * G$ is an $AFPR$. Suppose that (ii) holds. by (b) of (ii) $R * G$ is not prime. Let $0 \neq P \lhd R * G$. if $P = P_0$, by (b) of (ii) P is prime. If $P \neq P_0$, as above we can prove that P is prime, hence $R * G$ is an $AFPR$.

References

1. D. S. Passman, Infinite Crossed Products, Academic Press, (San Diego), 1980.

2. W. D. Blair and H. Tsutsui, Fully prime rings, Comm. Algebra. 22(1994) no.13. 5388-5400.

3. H. Tsutsui, Fully prime rings II, Comm. Algebra. 24(1996) no.9. 2981-2989.

4. S. Motgomery, Fixed Rings of Finite Automorphism Groups of Associative Rings, Lecture Notes in Math, 818, Springer, (Berlin), 1980.

5. F. W. Anderson and K. R. Fuller, Rings and Categories of Modules, New York Springer Verlag(1973).

On a left H-ring with Nakayama automorphism

Jiro, Kado

January 19, 2005

1 Nakayama isomorhphism

Let R and S be rings. We recall that a Morita duality between the category of the finitely generated left R-module $_R\mathcal{M}$ and the category of the finitely generated right S-module $\mathcal{M}_S$. If there exists contravariant functors $\mathcal{H}$: $_R\mathcal{M} \to \mathcal{M}_S$ $\mathcal{H}'$: $\mathcal{M}_S \to_R \mathcal{M}$ such that $\mathcal{H}'$ $\mathcal{H}$ and $\mathcal{H}$ $\mathcal{H}'$ are isomorphic to the identity functors of $_R\mathcal{M}$ $\mathcal{M}_S$, then it is called that $(\mathcal{H} : \mathcal{H}')$ is a Morita duality between $_R\mathcal{M}$ and $\mathcal{M}_S$. In this case $_R\mathcal{M}$ (or $\mathcal{M}_S$) is said to be dual to $\mathcal{M}_S$ (or $_R\mathcal{M}$). Especially, when $_R\mathcal{M}$ be dual to $\mathcal{M}_R$, R is said to be self-dual or to have self-duality. Let R be a left artinian ring. Put $S = End(E(_R(R/J(R))))$. It is well-known that $_R\mathcal{M}$ is dual to $\mathcal{M}_S$ if and only if $E(_R(R/J(R)))$ is finitely generated ([3],[9],[4]).

We turn our attention to those rings for which the $_RR_R$-dual $Hom(-,_R R_R)$ defines a duality between the category of finitely generated left and right modules over R. We call such a ring R to be *quasi-Frobenius* ring. Now we shall study the more deeper structure theorem about QF-rings.

For later use, we shall generalize the concept of 'Nakayama automorphism' to 'Nakayama isomorphism' for basic artinian rings.

Let R be a basic QF-ring and $\Theta = \{e_1, \ldots, e_n\}$ be a complete set of orthogonal primitive idempotents. For each $e_i \in \Theta$, there exists an unique $f_i \in \Theta$ such that $(e_iR : Rf_i)$ is an i-pair i.e. $f_iR/f_iJ \cong S(e_iR)$ $Re_i/Je_i \cong S(Rf_i)$.

Then

$$\begin{pmatrix} e_1 & e_2 & \cdots & e_n \\ f_1 & f_2 & \cdots & f_n \end{pmatrix}$$

is a permutation of $e_1, ..., e_n$. This permutation is called *Nakayama permutation* of $\{e_1, \ldots, e_n\}$ or of R. If there exists a ring automorphism ϕ of R satisfying $\phi(e_i) = f_i$ for all i, then ϕ is called a *Nakayama automorphism* of R. Haack has studied self-dualitty of Nakayama rings. Although he did not succeed, his result [[5] ,Theorem 3.1,] states that basic QF-Nakayama rings have Nakayama automorphisms. In Chapter 5, we shall present many examples which have a Nakayama automorphism. On the other hand, Koike has constructed several QF-rings which have no Nakayama automorphism([7]).

Let R be a basic left artinian ring such that $E(_R(R/J(R)))$ is finitely generated and $\Theta = \{e_1, \ldots, e_n\}$ be a complete set of orthogonal primitive idempotents of R. Since $G = E(_R(R/J(R))$ is finitely generate, $_R\mathcal{M}$ is Morita dual to $\mathcal{M}_{End(G)}$. In particular, if R is isomorphic to $T = End(G)$, then R has self-duality. This is a principal result for the study of self-duality. However, in spite of this result, it is not easy to find those artinian rings which have self-duality; even if we find an aritinian ring with duality, it seems to be difficult to verify whether it has self-duality or not. Finite dimensional algebra over a field, QF-rings and Nakayama rings are typical artinian rings which have self-duality. Therefore we shall define 'Nakayama isomorphism' as follows.

Put $G_i = E(_R(Re_i/J(Re_i)))$, then $G \cong \bigoplus_{i=1}^{n} G_i$. Therefore the endomorphism ring $T = End(G)$ is identified with the matrix ring:

$$\begin{pmatrix} [G_1, G_1] & \cdots & [G_1, G_n] \\ & \cdots & \\ [G_n, G_1] & \cdots & [G_n, G_n] \end{pmatrix}$$

Let f_i be the matrix such that (i, i)-position is the unity of $[G_i, G_i]$ and all other entries are zero maps. Then $\{f_1, \ldots, f_n\}$ is a complete set of orthogonal primitive idempotents of T. Here, if there exists a ring isomorphism ϕ from R to T such that $\phi(e_i) = f_i$ for all i, we call it a *Nakayama isomorphism* with respect to Θ. Of course, when R is a basic QF-ring, it is a just Nakayama automorphism of R.

Now we will discuss the problem whether special artinian rings (we call H-ring) have a Nakayama isomorphism or not.

Let R be a left H-ring with its complete set Θ of orthogonal primitive idempotents $\Theta = \{e_{11}, \ldots, e_{1n(1)}, \ldots, e_{m1}, \ldots, e_{mn(m)}\}$ satisfying

(1) each $e_{i1}R$ is an injective module

(2) $J(e_{i,k-1}R) \cong e_{ik}R$ for $k = 2, \ldots, n(i)$

(3) $e_{ik}R \not\cong e_{jt}R$ for $i \neq j$.

For each $e_{i1}R$, by the Fuller's Theorem ([2]), there exists an unique $Re_{\sigma(i)\rho(i)}$ such that

(1) $(e_{i1}R : Re_{\sigma(i)\rho(i)})$ is an i-pair

(2) $Re_{\sigma(i)\rho(i)}/S_{j-1}(Re_{\sigma(i)\rho(i)}) \cong E(T(Re_{ij}))$ for $i = 1, \ldots, mk = 1, \ldots, n(i)$
([13]).

In the above notations, put $g_i = e_{\sigma(i)\rho(i)}$, and g_{il}, denote the generator $g_i + S_{k-1}(Rg_i)$ of $Rg_i/S_{k-1}(Rg_i)$ for $i = 1, \ldots, m; k = 1, \ldots, n(i)$, and put
$$G = Rg_{11} \oplus \cdots \oplus Rg_{1n(1)} \oplus \cdots \oplus Rg_{m1} \oplus Rg_{mn(m)}$$

Since G is isomorphic to $E({}_R(R/J(R)))$ by (2) of above argument , G is finitely generated. So ${}_R\mathcal{M}$ is Morita dual to $\mathcal{M}_{T(R)}$ by the functor $Hom_R(-, {}_RG_T)$, where $T = T(R) = End(G)$. Therefore we call this ring $T = T(R) = End(G)$ the dual ring of R. In order to investigate the structure of $T = T(R)$, we express

$$T = \begin{pmatrix} [g_{11}, g_{11}] & \cdots & [g_{11}, g_{i1}] & \cdots & [g_{11}, g_{mn(m)}] \\ \vdots & & \vdots & & \vdots \\ [g_{i1}, g_{11}] & \cdots & [g_{i1}, g_{i1}] & \cdots & [g_{i1}, g_{mn(m)}] \\ \vdots & & \vdots & & \vdots \\ [g_{mn(m)}, g_{11}] & \cdots & [g_{mn(m)}, g_{i1}] & \cdots & [g_{mn(m)}, g_{mn(m)}] \end{pmatrix}$$

where $[g_{ij}, g_{kl}] = Hom(Rg_{ij}, Rg_{kl})$ for all i, j, k, l. Let h_{ij} be the matrix such that (ij, ij)-position is unity of $[g_{ij}, g_{ij}]$ and all other entries are zero maps. Then $\mathcal{F} = \{h_{i1}, \ldots, h_{1n(1)}, \ldots, h_{m1}, \ldots, h_{mn(m)}\}$ is a complete set of orthogonal primitive idempotents of T. Further T have the following nice properties.

Proposition 1.1 ([6],Prop3.3). *T is a basic left H-ring such that*

(1) *$h_{i1}T$ is injective for $i = 1, \ldots, m$*

(2) *$J(h_{i,k-1}T) \cong h_{ik}T$ for $i = 1, \ldots, m\ k = 2, \ldots, n(i)$.*

Moreover, if $(e_{i1}R : Re_{kt})$ is an i-pair , $(h_{i1}T : Th_{kt})$ is also an i-pair.

The next theorem is the first result which is proved by using the representative matrix rings. We note that a special case is proved in Section and Nakayama rings with a strictly increasing admissible sequence are those types.

Theorem 1.2 ([6], Th.5.1). *Let R be a basic left ring which is homogeneous type, i.e. $\sigma(i) = i$ for all $i = 1, \ldots, m$. Then R has a Nakayama isomorphism. Therefore these rings have self-duality.*

From now on, R is a general left H-ring. Let S be a two-sided ideal of R which is simple as a left ideal and right ideal. Now we shall study the structure of the dual ring of R/S. Before proving proposition, we shall prove several lemmas.

Lemma 1.3. *$S = S(e_{i1}R)$ some for i.*

Proof. Since $S \cong S(e_{ij}R)$ some for i and $S(e_{ij}R) \cong e_{i1}J(R)^{j-1} \subseteq e_{i1}R$, we see that $S \cong S(e_{i1})$. Since S is a two-sided ideal, we have $S = S(e_{i1})$.

From now on, we assume that $S = S(e_{i1}R), \sigma(i) \neq 1$ and $\rho(i) > 1$ for $g_i = g_{i1} = e_{\sigma(i)\rho(i)}$. Put $\overline{R} = R/S$. .

Lemma 1.4. *For any $e \in \Theta$ such that $e \neq e_{i1}$ and $e \neq g_i$, we have that $eRS = 0$ and $SRe = 0$. Consequently eR and Re become naturally $\overline{R}$-modules.*

Proof. Since $Re_{i1}/J(R)e_{i1} \cong S(Rg_i)$ and R is basic ,$eRS = eS = 0$ if $e \neq e_{i1}$. Since $g_iR/g_iJ(R) \cong S(e_{i1}R)$, $SRe = Se = 0$ if $e \neq g_i$.

Lemma 1.5. *For any $e \in \Theta$ such that $e \neq e_{i1}$ and $e \neq g_i$, we have that $eRS = 0$ and $SRe = 0$. Consequently eR and Re become naturally $\overline{R}$-modules.*

We note that in this paper, we ma assume that R is indecomposable as a ring. Put $\overline{R} = R/S$ and $\overline{r} = r + S$ for r in R.

98

Lemma 1.6. *(1) For $g_{k1} \neq g_{i1}$, Rg_{k1} is an injective $\overline{R}$-module and, moreover $Rg_{k1} \cong E(\overline{Re_{k1}}/J(\overline{Re_{k1}}))$ as a $\overline{R}$-module.*
(2) $J(Rg_{i1})$ is an injective left $\overline{R}$-module. Moreover, $J(Rg_{i1}) \cong E(\overline{Re_{i1}})$ as a $\overline{R}$-module.

Proof. (1) Since $S = S(e_{i1}R) = S(Rg_{i1})$, $SRg_{kl} = 0$. So SRg_{kl} is a left $\overline{R}$-module. Since SRg_{kl} is an injective R-module, SRg_{kl} is also injective as a $\overline{R}$-module. As $S(_R R_{kl})$ is also a simple left $\overline{R}$-module, we see $Rg_{k1} \cong E(\overline{Re_{k1}}/J(\overline{Re_{k1}}))$ as a $\overline{R}$-module.

(2) Since $SJ(Rg_{i1}) = SJ(R)Rg_{i1} = 0$, $J(Rg_{i1})$ is a left $\overline{R}$-module. If $J(Rg_{i1}) = 0$, then we see that $g_{i1} = e_{i1}$ and $n(i) = 1$. $J(R)e_{i1} = 0$ and $(e_{i1}R : Re_{i1})$ is i-pair, we also see that $e_{i1}J(R) = 0$. So $e_{i1}Re_{jk} = 0$ for $jk \neq i1$. Therefore it follows that $e_{i1}R$ is a direct summand of R as an ideal, which is contradicts the assumption that R is indecomposable as a ring. So $J(Rg_{i1}) \neq 0$.

To show that $J(Rg_{i1})$ is injective as a $\overline{R}$-module, let $\overline{I}$ be a left ideal of $\overline{R}$ and $\psi : \overline{I} \to J(Rg_{i1})$ a $\overline{R}$-homomorphism. Since $_R Rg_{i1}$ is injective, we have a homomorphism $\psi^* :_R \overline{R} \to_R Rg_{i1}$ such that ψ^* is an extension of ψ. Putting $x = \psi(\overline{1})$, we may show $Rx \subseteq J(Rg_{i1})$. If $Rx = Rg_{i1}$, then $S(e_{i1}R)x = S(e_{i1}R)\psi(\overline{1}) = \psi((S(e_{i1})) = 0$, whence $S(e_{i1}R)Rx = S(e_{i1}R)Rg_{i1} = 0$; this is a contradiction. Therefore $Rx \subseteq J(Rg_{i1})$.
Next, since $S(J(Rg_{i1}))$ is a simple left $\overline{R}$-module, we see $S(J(Rg_{i1})) \cong \overline{Re_{i1}}/J(\overline{Re_{i1}})$, and it follows that $J(Rg_{i1}) \cong E(\overline{Re_{i1}})$ as a $\overline{R}$-module. Our proof is complete.

Now we put

$$G' = Rg_{11} \oplus \cdots \oplus Rg_{1n(1)} \oplus \cdots \oplus Rg_{i-1,1} \oplus \cdots \oplus Rg_{i-1,n(i-1)}$$
$$\oplus J(Rg_{i1}) \oplus Rg_{i2} \oplus \cdots \oplus Rg_{in(i)} \oplus \cdots \oplus Rg_{m1} \oplus \cdots \oplus Rg_{mn(m)}.$$

Then, as we saw above, G' is a left $\overline{R}$-module and $G' \cong E(\overline{R}/J(\overline{R}))$. We put $T' = End(G')$ and show the following lemma.

Lemma 1.7. *There is a ring isomorphism between $T/S(h_{i1}T)$ and T'.*

Proof. We express T and T' as follows:

$$T = \begin{pmatrix} [11,11] & \cdots & [11,k1] & \cdots & [11,mn(m)] \\ \vdots & \ddots & \vdots & \cdots & \vdots \\ [i1,11] & \cdots & [i1,k1] & \cdots & [i1,mn(m)] \\ \vdots & & \vdots & & \vdots \\ [mn(m),11] & \cdots & [mn(m),k1] & \cdots & [mn(m),mn(m)] \end{pmatrix},$$

where $[ij,kl] = \mathrm{Hom}(Rg_{ij}, Rg_{kl})$.

$$T' = \begin{pmatrix} [11,11] & \cdots & [11,J(k1)] & \cdots & [11,mn(m)] \\ \vdots & \ddots & \vdots & \cdots & \vdots \\ [J(i1),11] & \cdots & [J(i1),J(k1)] & \cdots & [J(i1),mn(m)] \\ [i2,11] & \cdots & [i2,k1] & \cdots & [i2,mn(m)] \\ \vdots & & \vdots & & \vdots \\ [mn(m),11] & \cdots & [mn(m),k1] & \cdots & [mn(m),mn(m)] \end{pmatrix},$$

where $[ij,kl] = Hom(Rg_{ij}, Rg_{kl})$, $[J(i1),kl] = Hom(J(Rg_{ij}), Rg_{kl})$, and $[kl,J(i1)] = Hom(Rg_{kj}, J(Rg_{i1}))$.

$$S(Th_{i1}) = \begin{pmatrix} 0 & \cdots & 0 & 0 & 0 & \cdots & 0 \\ 0 & \cdots & 0 & 0 & 0 & \cdots & 0 \\ 0 & \cdots & 0 & X & 0 & \cdots & 0 \\ 0 & \cdots & 0 & 0 & 0 & \cdots & 0 \\ 0 & \cdots & 0 & 0 & 0 & \cdots & 0 \end{pmatrix},$$

where $X = \{\alpha : \alpha \in Hom_R(Rg_{i1}, Rg_{\sigma(i)\rho(i)}), Im(\alpha) \subseteq S(R_{\sigma(i)\rho(i)})\}$. Now, we define a mapping $\phi_{st,kl}$ from T to T' as follows:

Case 1. For $g_{kl} \neq g_{i1}$, we define $\phi_{i1,kl} : [Rg_{i1}, Rg_{kl}] \to [J(Rg_{i1}), Rg_{kl}]$ by $\phi_{i1,kl}(\alpha) = \alpha \mid_{J(Rg_{i1})}$ Then $\phi_{i1,kl}$ is homomorphism as an abelian group. To show $\phi_{i1,kl}$ is an isomorphism. Let $\beta \in [J(Rg_{i1}), Rg_{kl}]$. Consider the diagram

$$0 \to J(Rg_{i1}) \to Rg_{i1}$$
$$\downarrow \beta$$
$$Rg_{kl},$$

Since Rg_{kl} is injective, there exists α such that $\alpha \mid_{J(Rg_{i1})} = \beta$. Thus $\phi_{i1,kl}$ is an epimorphism. To show $\phi_{i1,kl}$ is monomorphism, assume $\phi_{i1,kl}(\alpha) = \alpha \mid_{J(Rg_{i1})} = 0$ and $\alpha \neq 0$.

Then $Ker(\alpha) = J(Rg_{i1})$. Since $Im(\alpha) \cong Rg_{i1}/J(Rg_{i1})$, we see that $Im(\alpha) = S(Rg_{i1})$. By assumption, $E(Rg_{i1}) \cong Rg_{i1}/J(Rg_{\alpha(i)\rho(i)})$, and hence $S(Rg_{i1}) \cong S(Rg_{\alpha(i)\rho(i)})$, and it follows $Rg_{i1} \cong Rg_{\alpha(i)\rho(i)}$. So $g_{i1} = g_{\alpha(i)\rho(i)}$, this is a contradiction. Then $\alpha = 0$ and hence $\phi_{i1,kl}$ is a monomorphism.

Case 2. For $g_{kl} \neq g_{i1}$, we define
$\phi_{kl,i1} : [Rg_{kl}, Rg_{i1}] \to [J(Rg_{kl}), Rg_{i1}]$ by $\phi_{kl,i1}(\alpha) = \alpha$.
This is well defined, since $Rg_{kl} \not\cong Rg_{i1}$. Clearly $\phi_{kl,i1}$ is an isomorphism as an abelian group.

Case 3. We define
$\phi_{i1,i1} : [Rg_{i1}, Rg_{i1}] \to [J(Rg_{i1}), J(Rg_{i1})]$ by $\phi_{i1,i1}(\alpha) = \alpha \mid_{J(Rg_{i1})}$.
We also see as above that $\phi_{i1,i1}$ is an isomorphism.

Case 4. For $g_{kl} = g_{\sigma(i)\rho(i)}$, we define
$\phi_{i1,sigma(i)\rho(i)} : [Rg_{i1}, Rg_{\sigma(i)\rho(i)}] \to [J(Rg_{i1}), Rg_{\sigma(i)\rho(i)}]$ by $\phi_{i1,\sigma(i)\rho(i)}(\alpha) = \alpha \mid_{J(Rg_{i1})}$.
We can see that $\phi_{i1,\sigma(i)\rho(i)}$ is an epimorphism as above and $Ker(\phi_{i1,\sigma(i)\rho(i)}) = \{\alpha \in [Rg_{i1}, Rg_{\sigma(i)\rho(i)}] : Im(\alpha) \subseteq S(Rg_{\sigma(i)\rho(i)})\}$. So $\phi_{i1,\sigma(i)\rho(i)}$ induces an isomorphism : $[Rg_{i1}, Rg_{\sigma(i)\rho(i)}]/[J(Rg_{i1}), Rg_{\sigma(i)\rho(i)}] \cong [J(Rg_{i1}, Rg_{\sigma(i)\rho(i)})]$
Note that $X = Ker(\phi_{i1,\sigma(i)\rho(i)}) \neq 0$.

Case 5. For other g_{kl}, g_{st}, we define
$\phi_{kl,st} : [Rg_{kl}, Rg_{st}] \to [Rg_{kl}, Rg_{st}]$ by $\phi_{kl,st} = $ identity map.

Here consider the componentwise map:

$$\phi = \begin{pmatrix} \phi_{11,11} & \cdots & \phi_{11,kl} & \cdots & \phi_{11,mn(m)} \\ \vdots & \ddots & \vdots & \cdots & \vdots \\ \phi_{i1,11} & \cdots & \phi_{i1,kl} & \cdots & \phi_{[i1,mn(m)} \\ \vdots & & \vdots & & \vdots \\ \phi_{mn(m),11} & \cdots & \phi_{mn(m),kl} & \cdots & \phi_{mn(m),mn(m)} \end{pmatrix} : T \to T'$$

For $\phi_{pq,kl}$ and $\phi_{kl,st}$, we see that $\phi_{pq,kl}\phi_{kl,st} = \phi_{pq,st}$. Therefore ϕ is a ring epimorphism and

$$Ker(\phi) = \begin{pmatrix} 0 & \cdots & 0 & 0 & 0 & \cdots & 0 \\ 0 & \cdots & 0 & 0 & 0 & \cdots & 0 \\ 0 & \cdots & 0 & X & 0 & \cdots & 0 \\ 0 & \cdots & 0 & 0 & 0 & \cdots & 0 \\ 0 & \cdots & 0 & 0 & 0 & \cdots & 0 \end{pmatrix},$$

where $X = Ker(\phi)$ Therefore ϕ induces an isomorphism:

$$\phi' : T/S(h_{i1}T) \to T'.$$

Theorem 1.8. *If R has a Nakayama isomorphism, then $\overline{R} = R/S$ has also a Nakayama isomorphism.*

Proof. Next we shall show that $\overline{R}$ also has a Nakayama isomorphism. Recall $\{h_{ij} : i = 1, \cdots, m, j = 1, \ldots, n(m)\}$ be the matrix units of T. By the assumption, there exists an isomorphism $\Phi : R \to T$ such that $\Phi(e_{ij}) = h_{ij}$ for all i, j. We see that $\Phi(e_{i1}) = g_{i1}$, since the identity of the $(i1, i1)$-component of T is g_{i1}. Since Φ transfers the (ik, jt)-component of R to the (ik, jt)-component of T, we have a key result that Φ trasfers

$$\begin{pmatrix} 0 & \cdots & 0 & 0 & 0 & \cdots & 0 \\ 0 & \cdots & 0 & 0 & 0 & \cdots & 0 \\ 0 & \cdots & 0 & S(e_{i1}R) & 0 & \cdots & 0 \\ 0 & \cdots & 0 & 0 & 0 & \cdots & 0 \\ 0 & \cdots & 0 & 0 & 0 & \cdots & 0 \end{pmatrix}$$

onto

$$\begin{pmatrix} 0 & \cdots & 0 & 0 & 0 & \cdots & 0 \\ 0 & \cdots & 0 & 0 & 0 & \cdots & 0 \\ 0 & \cdots & 0 & X & 0 & \cdots & 0 \\ 0 & \cdots & 0 & 0 & 0 & \cdots & 0 \\ 0 & \cdots & 0 & 0 & 0 & \cdots & 0 \end{pmatrix},$$

Therefore Φ induces an isomorphism $\Phi' : \overline{R} \to T/Ker(\phi)$ and $\phi'\Phi' : \overline{R} \to T'$ is the desired Nakayama isomorphism.

Next theorem is obtained.

Theorem 1.9 ([6],Prop5.4). *Let R be a basic left H-ring. If R has a Nakayama isomorphism, then $R/S(R)$ has also a Nakayama isomorphsim.*

Proof. We will prove in case R is QF-ring. Assume $(e_i R; Re_{\sigma(i)})$ is i-pair. So $g_i = e_{\sigma(i)}$. We put $\overline{R} = R/S(R)$. Let be

$$\begin{pmatrix} e_1 e_2 \ldots e_n \\ f_1 f_2 \ldots f_n \end{pmatrix}$$

a Nakayama permutation. Each Jg_i is an injective $\overline{R} = R/S(R)$- module by the same proof of (2) of Lemma 3.6. Then $G' = Jg_1 \oplus Jg_2 \oplus \cdots \oplus Jg_n$ is a minimal injective cogenrator $\overline{R}$-module. Now we can construct a ring epimorphism from the dual ring $T(R)$ to the dual ring $T(\overline{R})$.

We define

$\phi_{i,k} : [Rg_i, Rg_k] \to [Jg_i, Jg_k]$ by $\phi_{i,k}(\alpha) = \alpha \mid_{Jg_i}$ for all i, k. By the same proof of Lemma 3.7, $\phi_{i,k}$ is an epimorphism and if $k \neq \sigma(i)$, $\phi_{i,k}$ is monomorphism and if $k = \sigma(i)$, $X_i = Ker(\phi_{i,k}) = \{\alpha : \alpha \in Hom_R(Rg_i, Rg_{\sigma(i)}), Im(\alpha) \subseteq S(R_{\sigma(i)})\}$.

We express $T(R)$ and $T(\overline{R})$ as follows:

$$T(R) = \begin{pmatrix} [1,1] & \ldots & [1,k] & \ldots & [1,m] \\ \vdots & \ddots & \vdots & \ldots & \vdots \\ [i,1] & \ldots & [i,k] & \ldots & [i,m] \\ \vdots & & \vdots & & \vdots \\ [m,1] & \ldots & [m,k] & \ldots & [m,m] \end{pmatrix},$$

where $[i,k] = \mathrm{Hom}(Rg_i, Rg_k)$.

$$T(\overline{R}) = \begin{pmatrix} [J(1),J(1)] & \ldots & [J(1),J(k)] & \ldots & [J(1),J(m)] \\ \vdots & \ddots & \vdots & \ldots & \vdots \\ [J(i),J(1)] & \ldots & [J(i),J(k)] & \ldots & [J(i),J(m)] \\ \vdots & & \vdots & & \vdots \\ [J(m),J(1)] & \ldots & [J(m),J(k)] & \ldots & [J(m),J(m)] \end{pmatrix},$$

where $[J(i), J(k)] = Hom(Jg_i, Jg_k)$.

Here consider the componentwise map:

$$\phi = \begin{pmatrix} \phi_{1,1} & \cdots & \phi_{1,k} & \cdots & \phi_{1,m} \\ \vdots & \ddots & \vdots & \cdots & \vdots \\ \phi_{i,1} & \cdots & \phi_{i,k} & \cdots & \phi_{[i,m} \\ \vdots & & \vdots & & \vdots \\ \phi_{m,1} & \cdots & \phi_{m,k} & \cdots & \phi_{m,m} \end{pmatrix} : T(R) \to T(\overline{R}).$$

We put

$$X_{ik} = \begin{cases} X_i & \text{if} k = \sigma(i) \\ 0 & \text{if} k \neq \sigma(i). \end{cases} \tag{1}$$

, then

$$ker(\phi) = \begin{pmatrix} X_{1,1} & \cdots & X_{1,k} & \cdots & X_{1,m} \\ \vdots & \ddots & \vdots & \cdots & \vdots \\ X_{i,1} & \cdots & X_{i,k} & \cdots & X_{i,m} \\ \vdots & & \vdots & & \vdots \\ X_{m,1} & \cdots & X_{m,k} & \cdots & X_{m,m} \end{pmatrix}.$$

We put

$$S_{ik} = \begin{cases} S(e_i Re_{\sigma(i)}) & \text{if} k = \sigma(i) \\ 0 & \text{if} k \neq \sigma(i). \end{cases} \tag{2}$$

, then

$$S(R) = \begin{pmatrix} S_{1,1} & \cdots & S_{1,k} & \cdots & S_{1,m} \\ \vdots & \ddots & \vdots & \cdots & \vdots \\ S_{i,1} & \cdots & S_{i,k} & \cdots & S_{i,m} \\ \vdots & & \vdots & & \vdots \\ S_{m,1} & \cdots & S_{m,k} & \cdots & S_{m,m} \end{pmatrix}.$$

Recall $\{h_i : i = 1, \cdots, m\}$ be the matrix units of T. By the assumption, there exists an isomorphism $\Phi : R \to T$ such that $\Phi(e_i) = h_i$ for all $i,$. We see that Φ transfer $S(R)$ to $ker(\phi)$ componentwisely. Therefore Φ induces an isomorphism $\Phi' : \overline{R} \to T/ker(\phi)$. This is a Nakayama isomorphism. Our proof is comlete.

References

[1] F. W. Anderson and K. R. Fuller: *Rings and categories of modules (second edition)* Graduate Texts in Math. **13**, Springer-Verlag, Heidelberg/New York/Berlin (1991)

[2] K.R.Fuller *On indecomposable injectives over artinian rings*, Pacific J.Math. **29**, 1969,115-135

[3] G. Azumaya: *A duality theory for injective modules*, Amer. J. Math. **81** 1959, 249-278

[4] B.J.Muller: *On Morita duality*, Canad.J.Math , **21**, 1969, 1338-1347

[5] J.K.Haack: *Self-duality and serial rings*, J.Algebara **59** ,1979, 345-363

[6] J. Kado and K. Oshiro: *Self-Duality and Harada Rings*, J. Algebra **211** (1999), 384-408

[7] K. Koike, *Eamples of QF rings without Nakayama automorphism and H-rings without self-duality* J. Algebra **241** (2001), 731-744

[8] H. Kupisch,*A characterization of Nakayama Rings*, Comm. in Algebra **23** (2) (1995), 739-741

[9] K. Morita, *Duality for modules and its applications to the theory of rings with minimum condition*, Sci. Rep. Tokyo Kyoiku Daigaku **6** (1958), 89-142

[10] T. Nakayama, *Note on uniserial and generalized uniserial rings*, Proc. Imp. Acad. Tokyo **16** (1940), 285-289

[11] T. Nakayama,*On Frobenius algebra II*, Ann of Math. **42** (1941), 1-21

[12] K. Oshiro, *Lifting modules, extending modules and their applications to QF-rings*, Hokkaido Math. J. **13** (1984), 310-338

[13] K. Oshiro, *lifting modules, extending modules and their applications to generalized uniserial rings*, Hokkaido Math. J. **13** (1984), 310-338

[14] K. Oshiro, *On Harada ring I*, Math. J. Okayama Univ. **31** (1989), 161-178

[15] K. Oshiro, *On Harada ring II*, Math. J. Okayama Univ. **31** (1989), 179-188

[16] K. Oshiro and K. Shigenaga, *On H-rings with homogeneous socles*, Math. J. Okayama Univ. **31** (1989), 189-196

[17] K. Oshiro, *On Harada ring III*, Math. J. Okayama Univ. 32 (1990), 111-118

ISOMORPHISM CLASSES OF ALGEBRAS WITH RADICAL CUBE ZERO

I. KIKUMASA[1] AND H. YOSHIMURA[2]

Department of Mathematics, Faculty of Science
Yamaguchi University, Yamaguchi 753-8512, Japan
[1] *E-mail: kikumasa@yamaguchi-u.ac.jp*
[2] *E-mail: yoshi@yamaguchi-u.ac.jp*

We present the canonical forms of finite dimensional local quasi-Frobenius (QF) algebras Λ over a field k such that the radical cubed is zero and Λ modulo the radical is a product of copies of k and determine the isomorphism classes of those algebras Λ under some condition.

Introduction

In [2], we studied 'commutative' local QF algebras Λ over a field k satisfying the condition that

$$(*) \quad J^3 = 0 \quad \text{and} \quad \Lambda/J \text{ is a product of copies of } k$$

where J is the radical of Λ. In particular we determined the isomorphism classes of those k-algebras under some conditions on k. In this paper we consider this problem generally in not necessarily commutative case.

In Section 1 we provide preliminary results on the congruence of matrices for the classification of local QF k-algebras with $(*)$. In Section 2 we present the canonical forms of local QF k-algebras with $(*)$ and in Section 3 we determine the isomorphism classes of those k-algebras in a low dimensional case.

Throughout this paper, k is a field with $k^* = k - \{0\}$ the multiplicative group, all k-algebras mean 'not necessarily commutative' finite dimensional algebras over k and isomorphisms between k-algebras mean k-algebra isomorphisms. We denote by ch k the characteristic of k and by P' the transpose of a matrix P over k. For positive integers m and n, we denote by $M_{m,n}(k)$ and $M_n(k)$ the set of $m \times n$ matrices over k and the set of $n \times n$ matrices over k, respectively.

1. Matrix Congruence

In [2] we showed that the set of isomorphism classes of local (resp. QF) commutative k-algebras of dimension $n + 2$ with the condition $(*)$ in the introduction corresponds with the set of equivalence classes of nonzero (resp. nonsingular) symmetric $n \times n$ matrices over k with respect to some equivalence relation, which is related to the congruence of matrices. In general, as will be shown in the next section (Proposition 2.1), the set of isomorphism classes of 'not necessarily commutative' those k-algebras corresponds with the set of equivalence classes of 'all' nonzero $n \times n$ matrices over k.

Thus, in this section we provide preliminary results on the classification of nonzero matrices by congruence.

Definition 1.1. Let $X, Y \in M_n(k)$. Then X is said to be *congruent to Y* if there exists a $P \in GL_n(k)$ such that $X = PYP'$.

Definition 1.2. We denote the elementary matrices as follows.

- $P(i, j) =$ the matrix obtained by exchanging the i-th and the j-th rows of the identity matrix.

- $P(i; c) =$ the matrix obtained by multiplying the i-th row of the identity matrix by a nonzero element c of k.

- $P(i, j; c) =$ the matrix obtained by adding c multiple of the j-th row of the identity matrix to the i-th row.

Lemma 1.1. *Any nonzero 2×2 matrix $A = (a_{ij})$ over k is congruent to either an upper triangular matrix $T = (b_{ij})$ with $b_{11} \neq 0$ or a matrix*
$$U = \begin{pmatrix} 0 & 1 \\ -1 & 0 \end{pmatrix}.$$

Proof. Case 1. *A is congruent to a matrix with a nonzero diagonal entry.* In this case, by (congruence with) $P(1, 2)$ if necessary, we may assume that $a_{11} \neq 0$. Then, by $P(2, 1; -a_{11}^{-1} a_{21})$, A becomes an upper triangular matrix of the form T.

Case 2. *Otherwise.* In this case, we may assume that $a_{11} = a_{22} = 0$. Also, note that $a_{12} = -a_{21}$, because if otherwise, then we set
$$P = \begin{pmatrix} a_{12} & -a_{21} \\ 1 & 1 \end{pmatrix}$$

to obtain PAP' whose $(2, 2)$-entry is $a_{12} + a_{21} \neq 0$, which is a contradiction. Thus, $a_{12} = -a_{21}$, from which we have $P(2, a_{12}^{-1}) A P(2, a_{12}^{-1})' = U$. $\qquad\square$

The next lemma is the key to the classification of local k-algebras with the condition $(*)$.

Lemma 1.2. (1) *Assume that* $ch\,k \neq 2$. *Then, any nonzero* $n \times n$ *matrix over* k *is congruent to one of the following matrices.*

(a) *An upper triangular matrix of the form*

$$
T = \begin{pmatrix}
a_{11} & \cdots & \cdots & \cdots & \cdots & a_{1n} \\
 & \ddots & & & & \vdots \\
 & & a_{pp} & \cdots & \cdots & a_{pn} \\
 & & & 0 & \cdots & 0 \\
 & 0 & & & \ddots & \vdots \\
 & & & & & 0
\end{pmatrix}, \ \text{where each } a_{ii} \neq 0.
$$

(b) *A matrix of the form*

$$
U = \begin{pmatrix}
U_1 & & & & & \\
 & \ddots & & & 0 & \\
 & & U_t & & & \\
 & & & 0 & & \\
 & 0 & & & \ddots & \\
 & & & & & 0
\end{pmatrix}, \ \text{where each } U_i = \begin{pmatrix} 0 & 1 \\ -1 & 0 \end{pmatrix}.
$$

(2) *Assume that* $ch\,k = 2$. *Then, any nonzero* $n \times n$ *matrix over* k *is congruent to one of the following matrices.*

(a) *An upper triangular matrix of the form* T *above.*
(b) *A matrix of the form* U *above.*
(c) *A matrix of the form*

$$
W = \begin{pmatrix}
a_{11} & \cdots & \cdots & \cdots & \cdots & \cdots & a_{1n} \\
 & \ddots & & & & & \vdots \\
 & & a_{pp} & \cdots & \cdots & \cdots & a_{pn} \\
 & & U_1 & & & & \\
 & & & \ddots & & 0 & \\
 & & & & U_t & & \\
 & 0 & & & 0 & & \\
 & & & & & \ddots & \\
 & & & & & & 0
\end{pmatrix}, \ \text{where each } a_{ii} \neq 0.
$$

Proof. (1) **Claim.** Any 3×3 matrix of the form

$$A = \begin{pmatrix} a & b & c \\ 0 & 0 & d \\ 0 & -d & 0 \end{pmatrix}, \quad \text{where } a \neq 0, d \neq 0$$

is congruent to an upper triangular matrix.

Indeed, if $b = c = 0$, then we see that

$$PAP' = \begin{pmatrix} a & 2a & 2a \\ 0 & a & 2a \\ 0 & 0 & a \end{pmatrix}, \quad \text{where } P = \begin{pmatrix} 1 & 1 & 0 \\ 1 & 1 & ad^{-1} \\ 1 & 0 & ad^{-1} \end{pmatrix}.$$

On the other hand, assume that either $b \neq 0$ or $c \neq 0$. Then by $P(2,3)$ if necessary and by $P(1; s)$ for some $s \in k$, we may assume that $c = 1$; furthermore, by $P(2,3; -b)$ that $b = 0$. We then see that

$$PAP' = \begin{pmatrix} a & 3a & 3a \\ 0 & 2a & 4a \\ 0 & 0 & 2a \end{pmatrix}, \quad \text{where } P = \begin{pmatrix} 1 & d^{-1} & 0 \\ 1 & d^{-1} & a \\ 1 & -d^{-1} & a \end{pmatrix}.$$

This completes the proof of the claim.

Now, to prove the lemma, we shall show the following assertion by induction on n.

For any nonzero $n \times n$ matrix $A = (a_{ij})$ over k, A is congruent to an upper triangular matrix of the form T if A is congruent to a matrix with a nonzero diagonal entry; A is congruent to a matrix of the form U if otherwise.

The case $n = 2$ follows from the proof of Lemma 1.1. Assume that $n > 2$ and the assertion holds for $n' < n$.

Case (I). *A is congruent to a matrix with a nonzero diagonal entry.* In this case, we may assume by $P(1, i)$ for some i that $a_{11} \neq 0$ and by $P(i, 1; -a_{11}^{-1}a_{i1})$ for $i = 2, \ldots, n$ that $a_{21} = \cdots = a_{n1} = 0$, i.e.,

$$A = \begin{pmatrix} a & * & \cdots & * \\ 0 & & & \\ \vdots & & A_1 & \\ 0 & & & \end{pmatrix}$$

where $0 \neq a \in k$ and $A_1 \in M_{n-1}(k)$. If $A_1 = O$, then obviously A is of the form T. Assume that $A_1 \neq O$. Then by induction hypothesis, there exists

$P_1 \in GL_{n-1}(k)$ such that $P_1 A_1 P_1'$ is either of the form T or of the form U. Now, set

$$Q_1 = \begin{pmatrix} 1 & 0 & \cdots & 0 \\ 0 & & & \\ \vdots & & P_1 & \\ 0 & & & \end{pmatrix}.$$

Case (i). $P_1 A_1 P_1'$ *is of the form T*. In this case, we see that $Q_1 A Q_1'$ is of the form T.

Case (ii). $P_1 A_1 P_1'$ *is of the form U*. In this case, we have

$$Q_1 A Q_1' = \begin{pmatrix} X_1 & Y_1 \\ O & V_1 \end{pmatrix}, \quad \text{where } X_1 = \begin{pmatrix} a & * & * \\ 0 & 0 & 1 \\ 0 & -1 & 0 \end{pmatrix} \in M_3(k),$$

$Y_1 \in M_{3,n-3}(k)$, $V_1 \in M_{n-3}(k)$ is of the form U or a zero matrix and O is a zero matrix. By the claim, there exists $P_2 \in GL_3(k)$ such that $P_2 X_1 P_2'$ is a nonsingular upper triangular matrix. Set

$$Q_2 = \begin{pmatrix} P_2 & O \\ O & I_{n-3} \end{pmatrix}$$

where I_{n-3} is the identity matrix of $M_{n-3}(k)$. If $n = 3$ or $V_1 = O$, then we see that $(Q_2 Q_1) A (Q_2 Q_1)'$ is of the form T. If otherwise, then

$$(Q_2 Q_1) A (Q_2 Q_1)' = \begin{pmatrix} * & * & \cdots & \cdots \\ 0 & * & \cdots & \cdots \\ & & X_2 & Y_2 \\ & O & & \\ & & O & V_2 \end{pmatrix}, \quad \text{where } X_2 = \begin{pmatrix} a' & * & * \\ 0 & 0 & 1 \\ 0 & -1 & 0 \end{pmatrix} \in M_3(k)$$

with $0 \neq a' \in k$, $Y_2 \in M_{3,n-5}(k)$, $V_2 \in M_{n-5}(k)$ is of the form U or a zero matrix. Applying the same argument above to the $(n-2) \times (n-2)$ matrix

$$\begin{pmatrix} X_2 & Y_2 \\ O & V_2 \end{pmatrix},$$

we have $Q_3 \in GL_n(k)$ such that

$$(Q_3 Q_2 Q_1) A (Q_3 Q_2 Q_1)' = \begin{pmatrix} * & * & * & * & \cdots & \cdots \\ 0 & * & * & * & \cdots & \cdots \\ 0 & 0 & * & * & \cdots & \cdots \\ 0 & 0 & 0 & * & \cdots & \cdots \\ & & & & X_3 & Y_3 \\ & & O & & & \\ & & & & O & V_3 \end{pmatrix}.$$

We continue in this manner to see that A is congruent to an upper triangular matrix of the form T.

Case (II). *A is not congruent to any matrix with a nonzero diagonal entry.* In this case, by $P(i,j)$ for some i, j and by the proof of Lemma 1.1, we may assume that

$$A = \begin{pmatrix} 0 & 1 & a_{13} & \cdots & a_{1n} \\ -1 & 0 & a_{23} & \cdots & a_{2n} \\ a_{31} & a_{32} & & & \\ \vdots & \vdots & & A_1 & \\ a_{n1} & a_{n2} & & & \end{pmatrix}$$

where $A_1 \in M_{n-2}(k)$. By $P(i,1;-a_{i2})$ and $P(i,2;a_{i1})$ for $i = 3,\ldots,n$, a_{i1} and a_{i2} can be taken to be all zero. If $a_{1j} \neq 0$ for some $j \geq 3$, then by $P(1,j;1)$, A becomes a matrix whose $(1,1)$ entry is $a_{1j} \neq 0$, which contradicts the hypothesis of A. Thus, $a_{1j} = 0$ for all $j \geq 3$. Similarly, $a_{2j} = 0$ for all $j \geq 3$. Consequently, we may assume that

$$A = \begin{pmatrix} 0 & 1 & 0 & \cdots & 0 \\ -1 & 0 & 0 & \cdots & 0 \\ 0 & 0 & & & \\ \vdots & \vdots & & A_1 & \\ 0 & 0 & & & \end{pmatrix}.$$

Note by the hypothesis of A that A_1 is also not congruent to any matrix with a nonzero diagonal entry. If $A_1 = 0$, then obviously A is of the form U. If otherwise, then by induction hypothesis there exists $P_1 \in GL_{n-2}(k)$ such that $P_1 A_1 P_1'$ is of the form U. Therefore, we set

$$Q = \begin{pmatrix} I_2 & O \\ O & P_1 \end{pmatrix}$$

to see that QAQ' is of the form U, which completes the proof of (1).

(2) This follows from a similar proof of (1) except the claim. $\qquad\square$

Remark 1.1. Lemma 1.2(1) does not hold for a field k of ch $k = 2$, because it can be shown that a matrix

$$A = \begin{pmatrix} 1 & 0 & 1 \\ 0 & 0 & 1 \\ 0 & 1 & 0 \end{pmatrix}$$

over $k = \mathbb{Z}_2$ is congruent to neither an upper triangular matrix of the form T nor a matrix of the form U. Indeed, it is obvious that A is not congruent to a matrix of the form U. On the other hand, suppose that PAP' is an upper triangular matrix for some $P = (p_{ij}) \in GL_3(k)$. Since A, and hence PAP', is a nonsingular matrix, the diagonal entries of PAP' are all nonzero, i.e.,

$$p_{11}(p_{11} + p_{13}) = p_{21}(p_{21} + p_{23}) = p_{31}(p_{31} + p_{33}) = 1,$$

from which we have $p_{13} = p_{23} = p_{33} = 0$. This contradicts the nonsingularity of P.

2. k-algebras of Type $(1, n, 1)$

Most of results for commutative algebras in $[^2,$ Section 1] can be modified for not necessarily commutative algebras. Thus in this section we give a brief outline of the results.

Throughout this section, let n be a fixed positive integer unless otherwise stated.

Definition 2.1. Let Λ be a k-algebra with $J = Rad(\Lambda)$ the radical. Then we say that Λ is *of type* $(1, n, 1)$ if $\dim \Lambda/J = 1$, $\dim J/J^2 = n$, $\dim J^2 = 1$ and $J^3 = 0$.

Definition 2.2. Let Λ be a k-algebra of type $(1, n, 1)$ with $J = Rad(\Lambda)$ and let

$$\{u_1 + J^2, \ldots, u_n + J^2\} \quad \text{and} \quad \{u\}$$

be k-bases of J/J^2 and J^2, respectively. Then, for each $i, j = 1, \ldots, n$, there exists an $a_{ij} \in k$ such that

$$u_i u_j = a_{ij} u.$$

We say that the $n \times n$ matrix $A = (a_{ij})$ over k is a *representative matrix* of Λ (with respect to $\{u_1, \ldots, u_n\}$ and $\{u\}$). Note that A is a nonzero matrix.

Remark 2.1. A k-algebra Λ of type $(1, n, 1)$ with a representative matrix A is commutative if and only if A is a symmetric matrix.

The following two lemmas hold even for not necessarily commutative k-algebras of type $(1, n, 1)$, from which most of results on the classification of those commutative algebras in $[^2,$ Section 1] can be extended in general.

Lemma 2.1. (cf. [2Lemma 1.3]) *Let Λ_i $(i = 1, 2)$ be a k-algebra of type $(1, n, 1)$ with A_i a representative matrix. Then the following conditions are equivalent:*

(1) $\Lambda_1 \cong \Lambda_2$;

(2) $PA_2P' = aA_1$ *for some* $P \in GL_n(k)$ *and* $a \in k^*$.

Lemma 2.2. (cf. [2Lemma 1.4]) *Let Λ be a k-algebra of type $(1, n, 1)$ with A a representative matrix and let $Soc\,({}_\Lambda\Lambda)$ and $Soc\,(\Lambda_\Lambda)$ be the left and the right socle of Λ, respectively. Then it holds that*

$$\dim_k Soc\,({}_\Lambda\Lambda) \ = \ \dim_k Soc\,(\Lambda_\Lambda) \ = \ n + 1 - \operatorname{rank} A.$$

In particular, Λ is QF if and only if $A \in GL_n(k)$.

Proof. Assume that $A = (a_{ij})$ is a representative matrix of Λ with respect to $\{u_1, \ldots, u_n\}$ and $\{u\}$. Set $J = Rad(\Lambda)$, $S_l = Soc({}_\Lambda\Lambda)$ and $S_r = Soc(\Lambda_\Lambda)$. Then we have the following.

$$J^2 \subset S_l \ = \ \{x \in J \mid u_ix = 0 \ (i = 1, \ldots, n)\},$$
$$J^2 \subset S_r \ = \ \{x \in J \mid xu_i = 0 \ (i = 1, \ldots, n)\}.$$

As in the proof of [2Lemma 1.4], we see that

$$S_l/J^2 \ \cong \ \{\boldsymbol{a} \in k^{(n)} \mid X\boldsymbol{a} = \boldsymbol{0}\},$$
$$S_r/J^2 \ \cong \ \{\boldsymbol{a} \in k^{(n)} \mid \boldsymbol{a}'X = \boldsymbol{0}\}$$

as k-spaces, from which

$$\dim_k S_l \ = \ \dim_k S_r \ = \ \dim_k S_l/J^2 + 1 \ = \ (n - \operatorname{rank} X) + 1.$$

The last assertion follows from the fact that Λ is QF if and only if $S_l \cong {}_\Lambda(\Lambda/J)$ and $S_r \cong (\Lambda/J)_\Lambda$ (e.g. [1Theorem 31.3]). $\qquad\square$

Let $k\langle x_1, \ldots, x_n \rangle$ be the free algebra over k in the n indeterminates $x_1, \ldots, x_n$ commuting any element of k.

Definition 2.3. Let $A = (a_{ij})$ be a nonzero $n \times n$ matrix over k and choose a nonzero entry a_{pq} of A. Then we set

$$\Lambda_A = k\langle x_1, \ldots, x_n \rangle / I$$

where

$$I = \langle x_ix_j - a_{ij}a_{pq}^{-1}x_px_q, x_p^2x_q \mid 1 \leq i, j \leq n, (i, j) \neq (p, q) \rangle.$$

It is easy to see that

$$Rad(\Lambda_A) = \langle x_1, \ldots, x_n \rangle / I, \ \Lambda_A / Rad(\Lambda_A) \cong k \ \text{ and } \ Rad(\Lambda_A)^3 = 0$$

and that

$$\{(x_1 + I) + Rad(\Lambda_A)^2, \ldots, (x_n + I) + Rad(\Lambda_A)^2\}$$

and

$$\{a_{pq}^{-1} x_p x_q + I\}$$

are k-bases of $Rad(\Lambda_A)/Rad(\Lambda_A)^2$ and $Rad(\Lambda_A)^2$, respectively. Thus we have the following.

Lemma 2.3. (cf. [2 Proposition 1.6]) *For any nonzero $n \times n$ matrix A over k, Λ_A is a k-algebra of type $(1, n, 1)$ with A a representative matrix.*

Now, we define an equivalence relation $\sim$ on $M_n^*(k) := M_n(k) - \{O\}$ by the condition (2) of Lemma 2.1, i.e., for $A_1, A_2 \in M_n^*(k)$,

$$A_1 \sim A_2$$

if $PA_2P' = aA_1$ for some $P \in GL_n(k)$ and $a \in k^*$. We then note that A_1 being congruent to A_2 implies $A_1 \sim A_2$, but the converse does not holds. Let

$$M_n^*(k)/\sim \quad \text{and} \quad GL_n(k)/\sim$$

be the set of equivalence classes of $M_n^*(k)$ and $GL_n(k)$ with respect to $\sim$, respectively and let

$$\mathcal{D}_n \quad \text{and} \quad \mathcal{C}_n$$

be the set of isomorphism classes of k-algebras of type $(1, n, 1)$ and QF k-algebras of type $(1, n, 1)$, respectively. Then by Lemmas 2.1, 2.2 and 2.3 we see the following.

Proposition 2.1. *Let k be a field and let n be a positive integer. Then the map from the equivalence class of A to the equivalence class of Λ_A defines a one to one correspondence from $M_n^*(k)/\sim$ (resp. $GL_n(k)/\sim$) onto $\mathcal{D}_n$ (resp. $\mathcal{C}_n$).*

By virtue of Proposition 2.1, to determine $\mathcal{D}_n$ or $\mathcal{C}_n$, we need to do equivalence classes of $M_n^*(k)$ or $GL_n(k)$ with respect to the relation $\sim$. Indeed, according to Proposition 2.1 and Lemma 1.2, we obtain the following 'canonical forms' of k-algebras of type $(1, n, 1)$.

Theorem 2.1. (cf. [2Theorem 1.8]) *Let k be a field and let n be a positive integer.*

(1) *Assume that* $\mathrm{ch}\, k \neq 2$. *Then any k-algebra Λ of type $(1, n, 1)$ is isomorphic to one of the following algebras.*

(a) $\Lambda_T = k\langle x_1, \ldots, x_n\rangle / \langle x_i x_j - a_{ij} a_{pp}^{-1} x_p^2, x_p^3 \mid (i, j) \neq (p, p)\rangle$

where $T = (a_{ij})$ is an upper triangular $n \times n$ matrix over k in Lemma 1.2(1).

(b) $\Gamma_U = k\langle x_1, \ldots, x_n\rangle / \langle x_i x_j - b_{ij} x_1 x_2, x_1^2 x_2 \mid (i, j) \neq (1, 2)\rangle$

where $U = (b_{ij})$ is an $n \times n$ matrix over k in Lemma 1.2(1).

(2) *Assume that* $\mathrm{ch}\, k = 2$. *Then any k-algebra Λ of type $(1, n, 1)$ is isomorphic to one of the algebras Λ_T, Γ_U in (1)(a)(b) and*

(c) $\Lambda_W = k\langle x_1, \ldots, x_n\rangle / \langle x_i x_j - a_{ij} a_{pp}^{-1} x_p^2, x_p^3 \mid (i, j) \neq (p, p)\rangle$

where $W = (a_{ij})$ is an $n \times n$ matrix over k in Lemma 1.2(2).

By Theorem 2.1 and Lemma 2.2 we have the following.

Corollary 2.1. (cf. [2Corollary 1.9]) *We may replace 'any k-algebra' in Theorem 2.1 with 'any QF k-algebra' if the matrices T, U and W in the theorem are provided with the nonsingularity.*

Remark 2.2. (1) In any case $\mathrm{ch}\, k = 2$ or $\neq 2$, the algebras Λ_T and Γ_U are not isomorphic. Indeed, if otherwise, then $PUP' = aT$ for some $P \in GL_n(k)$ and $a \in k^*$. But, we see that the $(1,1)$-entry of PUP' is zero, while the one of aT is $aa_{11} \neq 0$, a contradiction.

(2) Let Λ be a k-algebra of type $(1, n, 1)$ with a representative matrix $T = (a_{ij})$ which is upper triangular. Then by congruence with appropriate matrices we may assume that $a_{nn} \neq 0$. We set

$$P = \begin{pmatrix} b_1 & & \\ & \ddots & 0 \\ & & b_{n-1} \\ 0 & & & 1 \end{pmatrix}, \quad \text{where } b_i = \begin{cases} a_{nn} a_{in}^{-1} & (\text{if } a_{in} \neq 0) \\ 1 & (\text{if otherwise}) \end{cases}$$

to obtain

$$a_{nn}^{-1} PTP' = \begin{pmatrix} * & \cdots & * & e_1 \\ & \ddots & \vdots & \vdots \\ & & * & e_{n-1} \\ 0 & & & 1 \end{pmatrix}, \quad \text{where each } e_i \in \{0, 1\}.$$

It then follows from Lemma 2.1 that a representative matrix T of Λ may be replaced by a matrix of the form above.

3. k-algebras of Type $(1, 2, 1)$

In this last section we determine the isomorphism classes of all k-algebras of type $(1, 2, 1)$.

Definition 3.1. For $a \in k$ and $e \in \{0, 1\}$, we set the following.

- $\Lambda_{(a,e)} = k\langle x, y\rangle/\langle x^2 - ay^2, xy - ey^2, yx, y^3\rangle$.
- $\Gamma = k\langle x, y\rangle/\langle x^2, y^2, xy + yx\rangle$.

Remark 3.1. (1) $\Lambda_{(a,e)}$ and Γ are k-algebras of type $(1, 2, 1)$ for which

$$A_{(a,e)} = \begin{pmatrix} a & e \\ 0 & 1 \end{pmatrix} \quad \text{and} \quad B = \begin{pmatrix} 0 & 1 \\ -1 & 0 \end{pmatrix}$$

are representative matrices, respectively.

(2) The following holds:

(i) $\Lambda_{(a,e)}$ is commutative if and only if $e = 0$.

(ii) $\Lambda_{(a,e)}$ is QF if and only if $a \neq 0$.

Lemma 3.1. (1) *Any k-algebra of type $(1, 2, 1)$ is isomorphic to one of the algebras*

$$\Lambda_{(0,0)}, \quad \Lambda_{(0,1)}, \quad \Lambda_{(a,0)}, \quad \Lambda_{(a,1)} \ (a \in k^*) \quad \text{and} \quad \Gamma.$$

Also, these five algebras are not isomorphic to each other.

(2) *Let a, $b \in k^*$. Then,*

(i) $\Lambda_{(a,0)} \cong \Lambda_{(b,0)}$ *if and only if $ab^{-1} \in (k^*)^2 := \{c^2 \mid c \in k^*\}$.*

(ii) $\Lambda_{(a,1)} \cong \Lambda_{(b,1)}$ *if and only if $a = b$.*

Proof. (1) The first assertion follows from Theorem 2.1 and Remark 2.2(2), while the second follows from Remark 2.2(1) and Remark 3.1(2).

(2) (i) follows from [2, Lemma 4.2 (2)]. (ii) Assume that $\Lambda_{(a,1)} \cong \Lambda_{(b,1)}$. Then there exist $P = (p_{ij}) \in GL_2(k)$ and $c \in k^*$ such that

$$(*) \quad PA_{(a,1)}P' = cA_{(b,1)}.$$

Comparing the $(1, 2)$ and $(2, 1)$-entries of both sides of $(*)$, we see that

$$\begin{cases} ap_{11}p_{21} + p_{11}p_{22} + p_{12}p_{22} = c \\ ap_{11}p_{21} + p_{12}p_{21} + p_{12}p_{22} = 0. \end{cases}$$

We subtract the equations to obtain $c = p_{11}p_{22} - p_{12}p_{21} = \det P$, while we take the determinants of both sides of $(*)$ to obtain $a(\det P)^2 = bc^2$. Thus, $ac^2 = bc^2$, from which we have $a = b$. $\qquad\square$

Let $\{a_i \mid i \in I\}$ be a complete set of representatives of the group k^* modulo the subgroup $(k^*)^2$. By Remark 3.1 and Lemma 3.1 we obtain the following proposition, which completely determines the set $\mathcal{D}_2$ (resp. $\mathcal{C}_2$) of isomorphism classes of (resp. QF) k-algebras of type $(1, 2, 1)$.

Proposition 3.1. *For any field k, the following holds:*

- $\mathcal{C}_2 = \{\Lambda_{(a,1)} \mid a \in k^*\} \cup \{\Gamma\} \cup \{\Lambda_{(a_i,0)} \mid i \in I\}$.
- $\mathcal{D}_2 = \mathcal{C}_2 \cup \{\Lambda_{(0,e)} \mid e = 0, 1\}$.

In particular, we have:

- $|\mathcal{C}_2| = |k| + |k^*/(k^*)^2|$.
- $|\mathcal{D}_2| = |k| + |k^*/(k^*)^2| + 2$.

Corollary 3.1. *Let k be a field. Then there exist infinitely many QF k-algebras, up to isomorphism, of type $(1, 2, 1)$ if and only if k is an infinite field.*

Example 3.1. (1) If k is an infinite field, then $|\mathcal{C}_2| = |k|$.
(2) If k is a finite field, then

$$|\mathcal{C}_2| = \begin{cases} |k| + 2 & (\text{if } \operatorname{ch} k \neq 2) \\ |k| + 1 & (\text{if } \operatorname{ch} k = 2). \end{cases}$$

Acknowledgments

The authors would like to thank the organizers and the staffs of the 4th China-Korea-Japan International Symposium for their hospitality. The authors would also like to thank the referee for useful suggestions.

References

1. F.W. Anderson and K.R. Fuller, Rings and Categories of Modules, 2nd ed., GTM 13, Springer-Verlag, 1992.
2. I. Kikumasa and H. Yoshimura, Commutative algebras with radical cube zero, *Comm. in Algebra* **31** (2003), 1837–1858.

ON LIFTING PROPERTIES OF MODULES

YOSUKE KURATOMI

Kitakyushu National College of Technology,
5-20-1 Shii, Kokuraminami, Kitakyushu, Fukuoka, 802-0985, JAPAN
E-mail: kuratomi@kct.ac.jp

A module M is said to be lifting, if it satisfies the following lifting property : For any submodule X of M, there exists a direct summand of M which is a co-essential submodule of X in M. This property is a notable property of (semi)perfect ring which was introduced by H.Bass in 1960. Since then, many researchers has been studying this property. The purpose of this peper is to consider some results of lifting modules.

1. Preliminaries

A R-module M is said to be *extending* (CS) if it satisfies the following extending property: For any submodule X of M, there exists a direct summand of M which contains X as an essential submodule, that is, for any submodule X of M, there exists a closure of X in M which is a direct summand of M. Dually, M is said to be a *lifting* module, if it satisfies the dual property: For any submodule X of M, there exists a direct summand of M which is a co-essential submodule of X, that is, for any submodule X of M, there exists a co-closure of X in M which is a direct summand of M. (cf., [24]).

The extending property is a notable property of (quasi-)injective and (quasi-)continuous modules. It was Utumi who first paid attention to this property. Utumi [30] introduced continuous rings by using the extending property. In 1974, continuous rings were generalized as (quasi-)continuous modules by Jeremy [13]. Since then, it was not until Harada's work on extending property for simple submodules of modules with completely indecomposable decompositions that the study of these modules progressed (cf.[7]-[11]). This method in the study by Harada urged the research of these module.

On the other hand, in 1960, Bass [2] introduced (semi)perfect rings. The lifting property is a notable property of (semi)perfect ring. In 1983, by using this property, Oshiro [24] introduced (quasi-)semiperfect modules as these generalization. Moreover, he applied the study of lifting and extending modules to one of QF-rings and Nakayama rings, and characterized a Harada ring that is a new Artinian ring like the nucleus of both QF-rings and also Nakayama rings (cf.[25]-[27]). For this reason, extending and lifting property of modules take roots inside of ring theory, and so these property have been studied by many researchers since the early 1980s. However, for these modules, many fundamental problems remain as open problems. In the study of extending modules, the existence of injective hull is useful for determining the structure of extending modules. However, modules do not always have a projective cover. For the reason, the study of lifting modules has not been made more than that of extending modules.

In Section 2, we study the relation between classical artinian rings and these properties. In Section 3, we introduce a new concept of relative projectivity that is dual to the generalized injectivity and investigate some characteristics of this projectivity. In Section 4, using the results of Section 3, we give a characteristic for a finite direct sum of lifting modules to be lifting for the given decomposition. In Section 5, we prove that any lifting module over right perfect rings has an indecomposable decomposition and has the internal exchange property.

Throughout this paper all rings are associative and R will always denote a ring with unity. Modules are unital right R-modules unless indicated otherwise. Let M be a module. A submodule S of M is said to be *small* in M (denoted by $S \ll M$) if $M \neq K + S$ for any proper submodule K of M. Let N and L be submodules of M. Let N and K be submodules of M with $K \subseteq N$. K is said to be a *co-essential* submodule of N in M if $N/K \ll M/K$ and we write $K \subseteq_c N$ in M in this case. Let X be a submodule of M. X is called a *co-closed* submodule in M if X has no proper co-essential submodules in M. X' is called a *co-closure* of X in M if X' is a co-closed submodule of M with $X' \subseteq_c X$ in M (cf.[5], [24]). In general, submodules of a module M do not always have a co-closure in M. For example, $2\mathbb{Z}$ has no co-closure in $\mathbb{Z}_{\mathbb{Z}}$. $K <_\oplus N$ means that K is a direct summand of N. A module M is said to have the (*finite*) *exchange property* if, for any (finite) index set I, whenever $M \oplus N = \oplus_I A_i$ for modules N and A_i, then $M \oplus N = M \oplus (\oplus_I B_i)$ for submodules $B_i \subseteq A_i$. A module M is said to have the (*finite*) *internal exchange property* if, for any (finite) direct sum decomposition $M = \oplus_I M_i$ and any $X <_\oplus M$, there exist $N_i \subseteq M_i$ ($i \in I$) such that $M = X \oplus (\oplus_I N_i)$. A module M is called *lifting* if, for any submodule X of M, there exists a direct summand X^* of M such that

120

$X^* \subseteq_c X$ in M. Let $\{M_i | i \in I\}$ be a family of modules and let $M = \oplus_I M_i$. M is said to be a *lifting module for* the decomposition $M = \oplus_I M_i$ if, for any submodule X of M, there exist $X^* \subseteq M$ and $\overline{M_i} \subseteq M_i$ $(i \in I)$ such that $X^* \subseteq_c X$ and $M = X^* \oplus (\oplus_I \overline{M_i})$.

For background, basic results and applications of extending and lifting properties, the reader is referred to the texts of Harada [9], Mohamed and Müller [20], Dung *et al.* [4], Wisbauer [32] and the recent survey of Oshiro [28].

2. Classical Artinian Rings

A ring R is said to be *QF* if it is left and right artinian and right self injective. A ring R is said to be *Nakayama* if it is right and left serial, that is, for any primitive idempotent e in R, the submodules of eR_R and submodules of $_RRe$ form a chain by inclusion. These rings are left-right symmetric. An R-module M is said to be a *small* module if it is small in its injective hull. Dually, M is said to be a *cosmall* module if, for any projective module P and any epimorphism $f : P \to M$, $\ker f$ is an essential submodule of P. A ring R is said to be *right (left) Harada* if it is right (left) artinian with the following condition : Any non-small right (left) R-module contains a non-zero injective submodule. A ring R is said to be *right (left) co-Harada* if it satisfies ACC on right (left) artinian with the following condition : Any non-cosmall right (left) R-module contains a non-zero projective direct summand. In [27], Oshiro proved the following : A ring R is right Harada if and only if R is a left co-Harada. However right Harada rings need not be right co-Harada.

These artinian rings are closely related to extending and lifting modules as follows:

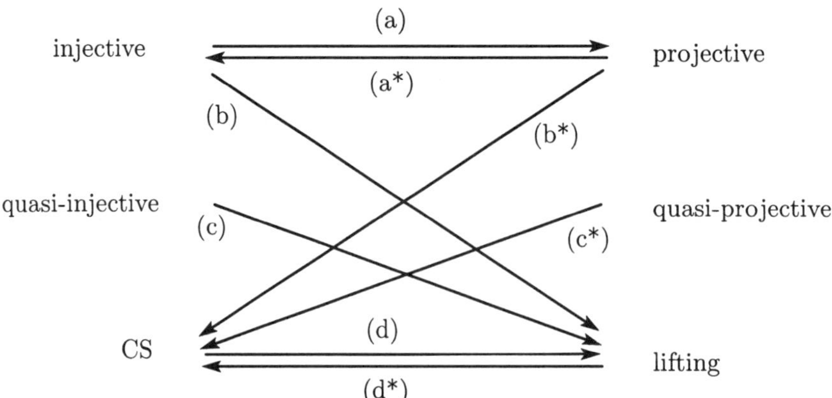

As is well known, R is QF $\Leftrightarrow$ (a) $\Leftrightarrow$ (a^*). It is shown in [26] that R is Nakayama $\Leftrightarrow$ (d) $\Leftrightarrow$ R is a right perfect ring with (d^*) $\Leftrightarrow$ (c) $\Leftrightarrow$ (c^*). It is shown in [27] that R is right Harada $\Leftrightarrow$ R is left co-Harada $\Leftrightarrow$ (b) and R is left Harada $\Leftrightarrow$ R is right co-Harada $\Leftrightarrow$ (b^*).

Hence QF-rings and Nakayama rings are right and left Harada rings.

3. Generalized Projectivity

In 2002, Hanada-Kuratomi-Oshiro [6] introduced a new injectivity that is called *generalized injective* (this is called *ojective* by Mohamed and Müller [21]) and studied the problem 'When is a direct sum of extending modules extending?' Mohamed and Müller [22] considered a dual notion *generalized projective* (*dual-cojective*) as follows:

Definition 3.1. Let A and B be modules. A is said to be *generalized $B-projective$* ($B-dual$ *ojective*) if, for any homomorphism $f : A \to X$ and any epimorphism $g : B \to X$, there exist decompositions $A = A_1 \oplus A_2$, $B = B_1 \oplus B_2$, a homomorphism $h_1 : A_1 \to B_1$ and an epimorphism $h_2 : B_2 \to A_2$ such that $g \circ h_1 = f|_{A_1}$ and $f \circ h_2 = g|_{B_2}$.

Remark (1) Any B-projective module is generalized B-projective.

(2) Let A and B be indecomposable modules. Then A is generalized B-projective if and only if, for any homomorphism $f : A \to X$ and any epimorphism $g : B \to X$, (i) if Im $f \neq X$, then f is liftable to $A \to B$ or (ii) if Im $f = X$, then either f is liftable to $A \to B$ or there exists a homomorphism h such that $f \circ h = g$. Hence, the concept of generalized projectivity is the same as that of almost projectivity (cf. [1], [10]).

We do not know whether generalized projectivity closed under direct summands or not. However we can give some characterizations of generalized projectivity as follows.

Proposition 3.1. *Let A is generalized B-projective. Then the following holds.*

(1) *For any direct summand B' of B, A is generalized B'-projective* ([22]).

(2) *If A satisfies the finite internal exchange property, then A' is generalized B-projective for any direct summand A' of A* ([17]).

(3) *If any submodule of A has a co-closure in A (e.g. A is a module over right perfect rings), then A' is generalized B-projective for any direct summand A' of A* ([17]).

A ring R is *right perfect* if every right R-module has a projective cover. Now we consider the following condition: (*) Any submodule of M has a co-closure in M. By [24, Theorem 1.3], we note that every module M over right perfect rings satisfies the condition (*). Hence Proposition 3.1 (3) implies the following: Let A and B be modules over a right perfect ring. If A is generalized B-projective then A' is generalized B-projective for any direct summand A' of A.

A module A is said to be *small B-projective* if, for any epimorphism $g : B \to X$ and any homomorphism $f : A \to X$ with $\mathrm{Im} f \ll X$, there exists a homomorphism $h : A \to B$ such that $g \circ h = f$ (cf. [15]). If B is a lifting module, then a generalized B-projective module is small B-projective (cf. [17]).

A lifting module M is said to be *discrete* (*semiperfect*) if M satisfies the following condition (D):

(D) If $X \subseteq M$ such that M/X is isomorphic to a direct summand of M, then X is a direct summand.

A lifting module M is said to be *quasi-discrete* (*quasi-semiperfect*) if M satisfies the following condition (D'):

(D') If M_1 and M_2 are direct summands of M such that $M = M_1 + M_2$, then $M_1 \cap M_2$ is a direct summand of M.

Note that projective $\Rightarrow$ quasi-projective $\not\Rightarrow$ discrete $\Rightarrow$ quasi-discrete $\Rightarrow$ lifting (cf. [24], [20]). If any submodule has co-closure then quasi-projective modules are discrete.

For quasi-discrete modules, the following holds (cf. [18]).

Proposition 3.2. (1) *Let N be a quasi-discrete module (or a lifting module over right perfect rings) and let $M = M_1 \oplus \cdots \oplus M_n$ be lifting for $M = M_1 \oplus \cdots \oplus M_n$. If M_i is generalized N-projective $(i = 1, \cdots, n)$, then M is generalized N-projective.*

(2) *Let M be a quasi-discrete module (or a lifting module over right perfect ring) and $N = N_1 \oplus \cdots \oplus N_m$ be lifting for $N = N_1 \oplus \cdots \oplus N_m$. If N_i and M are relative generalized projective $(i = 1, \cdots, m)$, then M is generalized N-projective.*

We do not know whether the proposition above holds for lifting modules or not.

4. Direct sums of lifting modules

Lifting and extending modules take roots inside of ring theory, so these modules have been studied by many researchers. However the following fundermental problems remain as open problems:

Problem A When is a direct sum $\oplus_I M_\alpha$ of extending modules $\{M_\alpha\}_I$ extending ?

Problem B When is a direct sum $\oplus_I M_\alpha$ of lifting modules $\{M_\alpha\}_I$ lifting ?

In special cases, these problems have been studied by several authors, e.g., Baba-Harada[1], Dung[3], Harada-Oshiro[11], Harmanci-Smith[12], Kado-Kuratomi-Oshiro[14], Keskin[16]. However, in general, these problems are unsolved even in the case that the index set I is finite. In [6], we studied Problem A and obtained the following:

Theorem Let M_1 and M_2 be extending modules and put $M = M_1 \oplus M_2$. Then M is extending for $M = M_1 \oplus M_2$ if and only if M_i is generalized M_j-injective $(i \neq j)$.

In this section, we consider the dual problem by using generalized relative projectivity. The following is one of main results in this paper.

Theorem 4.1. *(cf.*[17], *Theorem 3.1]) Let M_1 and M_2 be lifting modules and put $M = M_1 \oplus M_2$. Then M is lifting for $M = M_1 \oplus M_2$ if and only if M_i' is generalized M_j-projective for any $M_i' <_\oplus M_i$ $(i \neq j)$.*

When the induction is applied to these results the following are obtained.

Theorem 4.2. *Let $M_1, \cdots, M_n$ be lifting modules and put $M = M_1 \oplus \cdots \oplus M_n$. Then the following are equivalent:*

(1) M is lifting for $M = M_1 \oplus \cdots \oplus M_n$;

(2) A and B are relative generalized projective for any $A <_\oplus M(I)$ and $B <_\oplus M(J)$, where I and J are any two disjoint nonempty subsets of $\{1, 2, \cdots, n\}$, $M(I) = \oplus_{i \in I} M_i$ and $M(J) = \oplus_{j \in J} M_j$;

(3) M_i' and T are relative generalized projective for any $M_i' <_\oplus M_i$ and any $T <_\oplus (\oplus_{j \neq i} M_j)$, where $i \in \{1, \cdots, n\}$.

Theorem 4.3. *Let $M_1, \cdots, M_n$ be lifting modules and put $M = M_1 \oplus \cdots \oplus M_n$. Assume that M has condition $(*)$, that is, any submodule of M has a co-closure in M. Then the following conditions are equivalent:*

(1) M is lifting with the finite internal exchange property;

(2) M is lifting for $M = M_1 \oplus \cdots \oplus M_n$;

(3) $M(I)$ and $M(J)$ are relative generalized projective for any two nonempty disjoint subsets I and J of $\{1, 2, \cdots, n\}$, where $M(I) = \oplus_{i \in I} M_i$ and $M(J) = \oplus_{j \in J} M_j$;

124

(4) M_i and $\oplus_{j \neq i} M_j$ are relative generalized projective for all $i \in \{1, \cdots, n\}$.

Theorem 4.4. Let $M_1, \cdots, M_n$ be lifting modules with the finite internal exchange property and put $M = M_1 \oplus \cdots \oplus M_n$. Then the following conditions are equivalent:

(1) M is lifting with the finite internal exchange property;

(2) M is lifting for $M = M_1 \oplus \cdots \oplus M_n$;

(3) $M(I)$ and $M(J)$ are relative generalized projective for any two nonempty disjoint subsets I and J of $\{1, 2, \cdots, n\}$, where $M(I) = \oplus_{i \in I} M_i$ and $M(J) = \oplus_{j \in J} M_j$;

(4) M_i and $\oplus_{j \neq i} M_j$ are relative generalized projective for all $i \in \{1, \cdots, n\}$.

As an immediate consequence of Theorem 4.2 and Theorem 4.4, we obtain the following.

Corollary 4.1. Let $M_1, \cdots, M_n$ be lifting modules (with the finite internal exchange property) and put $M = M_1 \oplus \cdots \oplus M_n$. If M_i and M_j are relative projective $(i \neq j)$, then M is lifting (with the finite internal exchange property).

Remark Let N and L be submodules of M. N is called a *supplement* of L if it is minimal with the property $M = N + L$, equivalently, $M = N + L$ and $N \cap L \ll N$. M is said to be *amply supplemented* if, for any submodules A, B of M with $M = A + B$ there exists a supplement P of A such that $P \subseteq B$. In the case M is amply supplemented, Corollary 4.1 has already been proved by Keskin [16].

By Proposition 3.2, in the case of quasi-discrete modules over any ring or lifting modules over a right perfect ring, generalized projectivity close under direct sums. So we see the following.

Theorem 4.5. Let $M_1, \cdots, M_n$ be quasi-discrete modules over any ring or lifting modules over a right perfect ring and put $M = M_1 \oplus \cdots \oplus M_n$. Then the following conditions are equivalent:

(1) M is lifting with the (finite) internal exchange property;

(2) M is lifting for $M = M_1 \oplus \cdots \oplus M_n$;

(3) M_i is generalized M_j-projective for all $i \neq j$.

A module H is called hollow if H is indecomposable lifting. Since any hollow module satisfies the condition (D'), we see the following.

Corollary 4.2. *Let $H_1, \cdots , H_n$ be hollow modules and put $M = H_1 \oplus \cdots \oplus H_n$. Then the following conditions are equivalent:*
(1) M is lifting with the (finite) internal exchange property;
(2) M is lifting for $M = H_1 \oplus \cdots \oplus H_n$;
(3) H_i is generalized H_j-projective for all $i \neq j$.

The following is due to Baba-Harada [1].

Theorem 4.6. *Let $\{H_i\}_I$ be a family of hollow modules with local endomorphism rings and put $M = \oplus_I H_i$. Then the following conditions are equivalent:*
(1) M is lifting;
(2) H_i is generalized H_j-projective $(i \neq j)$ and $\{H_i\}_I$ is $lsTn$;

5. Lifting modules over right perfect rings

In 1984, Okado [23] has studied the decomposition of extending modules over right noetherian rings and obtained the following.

Theorem 5.1. *(cf. [23]) A ring R is right noetherian if and only if every extending R-module is expressed as a direct sum of indecomposable (uniform) modules.*

As a dual problem, we consider the following: Which ring R has the property that every lifting R-module has an indecomposable decomposition ? In this section we consider this problem. The following give a characterization of right perfect ring.

Proposition 5.1. *(cf. [2]) A ring R is right perfect if and only if every projective right R-module is lifting.*

$\Sigma \oplus_{\lambda \in \Lambda} X_\lambda \subseteq X$ is called a *local summand* of X, if $\Sigma \oplus_{\lambda \in F} X_\lambda < \oplus X$ for every finite subset $F \subseteq \Lambda$.

The following lemma due to Oshiro[24] is useful. For Okado's result above, the first lemma was used.

Lemma 5.1. *If every local summand of M is a direct summand, then M has an indecomposable decomposition.*

Lemma 5.2. *Every local summand of projective lifting modules is a direct summand.*

The following is a main result in this section.

Theorem 5.2. ([[19]]) *If R is a right perfect ring, then every local summand of lifting modules is a direct summand. Hence any lifting module over right perfect rings has an indecomposable decomposition.*

The following is essentially due to Harada [9].

Theorem 5.3. *Let $M = \oplus_I M_\alpha$, where each M_α has a local endomorphism ring. Then the following conditions are equivalent:*

(1) M has the internal exchange property (in the direct sum $M = \oplus_I M_\alpha$);

(2) M has the (finite) exchange property;

(3) Every local summand of M is a direct summand.

By the proof of [31, Proposition 1], we see

Lemma 5.3. *Let H be a hollow module. If $H \oplus H$ has the internal exchange property, then H has a local endomorphism ring.*

By Lemma 5.3, we see the following

Theorem 5.4. *Let R be a right perfect ring and let H be a hollow module. Then $End(H)$ is a local ring.*

The following is immediate from Theorem 5.2, 5.3 and 5.4.

Theorem 5.5. *Any lifting module over right perfect rings has the exchange property.*

The following is immediate from Proposition 5.1 and Theorem 5.5 (cf.[29], [33]).

Corollary 5.1. *Any projective module over right perfect rings has the exchange property.*

By Theorem 4.6 and 5.5, we see the following.

Theorem 5.6.

Let R be a right perfect ring and let $M = \oplus_I H_i$, where each H_i is hollow. Then the following conditions are equivalent:

(1) M is lifting with the (internal) exchange property;

(2) M is lifting for $M = \oplus_I H_i$;

(3) M is lifting;

(4) (a) H_i is generalized H_j-projective $(i \neq j)$

    (b) $\{H_i\}_I$ is lsTn;

(5) (a) H_i is generalized H_j-projective $(i \neq j)$

    (b) Every local summand of M is a direct summand.

We do not know whether any lifting module has the internal exchange property or not.

References

1. Y. Baba and M. Harada, On almost M-projectives and almost M-injectives. *Tsukuba J. Math.* **14**, 53–69 (1990).
2. H. Bass, Finitistic dimension and a homological generarization of semiprimary rings. *Trans. Amer. Math. Soc.* **95**, 466–488 (1960).
3. N. V. Dung, On indecomposable decomposition of CS-modules II. *J. Pure and Applied Algebra* **119**, 139–153 (1997).
4. N. V. Dung, D.V. Huynh, P.F Smith and R. Wisbauer, Extending modules. Pitman Research Notes in Mathematics Series 313, Longman Group Limited, 1994; 224pp.
5. L. Ganesan and N. Vanaja, Modules for which every submodule has a unique coclosure. *Comm. Algebra* **30**, 2355–2377 (2002).
6. K. Hanada, Y. Kuratomi and K. Oshiro, On direct sums of extending modules and internal exchange property. *Journal of Algebra* **250**, 115–133 (2002).
7. M. Harada, On modules with lifting properties. *Osaka J. Math.* **19**, 189–201 (1982).
8. M. Harada, On modules with extending properties *Osaka J. Math.* **19**, 203–215 (1982).
9. M. Harada, Factor categories with applications to direct decomposition of modules. LN Pure Appl. Math. 88, Dekker, New York, 1983; ???pp.
10. M. Harada and A. Tozaki, Almost M-projectives and Nakayama rings. *J. Algebra* **122**, 447–474 (1989).
11. M. Harada and K. Oshiro, On extending property of direct sums of uniform module. *Osaka J. Math.* **18**, 767–785 (1981).
12. A. Harmanci and P. F. Smith, Finite direct sums of CS-modules. *Houston Journal of Mathematics* **19**, 523–532 (1993).
13. L. Jeremy, Sur les modules et anneaux quasi-continus. *Canad. Math. Bull.* **17**, 217–228 (1974).
14. J. Kado, Y. Kuratomi, K. Oshiro, CS-property of direct sums of uniform modules. *International Symposium on Ring Theory, Trends in Math*, 149–159 (2001).
15. D. Keskin, Finite Direct Sums of (D1)-modules. *Turkish Journal of Math.* **22**, 85–91 (1998).
16. D. Keskin, On lifting modules. *Comm. Algebra* **28**, 3427–3440 (2000).
17. Y. Kuratomi, On direct sums of lifting modules and internal exchange property. *to appear in Comm. Algebra*.

18. Y. Kuratomi, On direct sums of quasi-discrete modules and those of lifting module over right perfect rings. *preprint.*

19. Y. Kuratomi and C. Chang, Lifting modules over right perfect rings. *preprint.*

20. S.H. Mohamed and B.J. Müller, Continuous and Discrete Modules. London Math. Soc., LN 147, Cambridge Univ. Press, 1990; 126pp.

21. S.H. Mohamed and B.J. Müller, Ojective modules. *Comm. Algebra* **30**, 1817–1827 (2002).

22. S.H. Mohamed and B.J. Müller, Co-ojective modules. *preprint.*

23. M. Okado, On The Decomposition of Extending Modules. *Math. Japonica* **29**, 939–941 (1984).

24. K. Oshiro, Semiperfect modules and quasi-semiperfect modules. *Osaka J. Math.* **20**, 337–372 (1983).

25. K. Oshiro, Lifting modules, extending modules and their applications to QF-rings. *Hokkaido Math. J.* **13**, 310–338 (1984).

26. K. Oshiro, Lifting modules, extending modules and their applications to generalized uniserial rings. *Hokkaido Math. J.* **13**, 339–346 (1984).

27. K. Oshiro, On Harada rings I,II,III. *Math. J. Okayama Univ.* **31**, 161–178, 179–188 (1989), **32**, 111–118 (1990).

28. K. Oshiro, Theories of Harada in Artinian Rings and applications to classical Artinian Rings. *International Symposium on Ring Theory, Trends in Math,* 279–301 (2001).

29. J. Stock, On rings whose projective modules have the exchange property. *J. Algebra* **103**, 437–453 (1986).

30. Y. Utumi, On continuous regular rings. *Canad. Math. Bull.* **4**, 63–69 (1961).

31. R. B. Warfield, A Krull-Schmidt theorem for infinite sums of modules. *Proc. Amer. Math. Soc.* **22**, 460–465 (1969).

32. R. Wisbauer, Foundations of Module and Ring Theory. Gordon and Breach Science Publications, 1991; 606pp.

33. K. Yamagata, On projective modules with the exchange property. *Sci. Rep. Tokyo Kyoiku Daigaku Sec. A* **12**, 149–158 (1974).

ON REGULAR RINGS WITH THE PROPERTY (DF)

MAMORU KUTAMI

Department of Mathematics, Faculty of Science,
Yamaguchi University, Yamaguchi 753-8512, JAPAN
E-mail: kutami@yamaguchi-u.ac.jp

A regular ring R is said to satisfy the property (DF) if the class of directly finite projective R-modules is closed under finite direct sums. The notion of the property (DF) was first given by the author in 1985, from considerations for directly finiteness of projective modules over directly finite regular rings with the comparability axiom, and it was proved in 1996 that unit-regular rings with s-comparability have the property (DF). In this paper, we study regular rings with the property (DF).

1. Preliminaries

The notion of the property (DF) was born in 1985 from the study of directly finite projective modules over directly finite regular rings with the comparability axiom [8]. In 1996, we showed that unit-regular rings with s-comparability have the property (DF), and using this result effectively, we could study directly finite projective modules over these rings [10]. In this paper, we shall study regular rings with the property (DF).

In Section 2, on the basis of above considerations, more generally we treat regular rings with s-comparability, and we show that these rings have the property (DF) (Theorem 2.2). But, we notice that there exists an example of a typical regular ring which does not have the property (DF). Therefore we have a problem: Which regular rings have the property (DF)?

In Section 3, we treat the above problem for regular rings with weak comparability, and we give a new condition (C) for studying the property (DF) of regular rings with weak comparability. We show that every stably finite regular ring satisfies the condition (C) if and only if it is a simple unit-regular ring with s-comparability for some positive integer s (Theorem 3.1), from which we see that every stably finite regular ring with the condition (C) has the property (DF) (Corollary 3.1). Meanwhile, Ara et al.[5] proved that every simple regular ring with weak comparability has the property (DF) if and only if it satisfies s-comparability (see Theorem 3.2). It is unknown that there exists a simple regular ring which does not satisfy weak comparability. Therefore, at the present time, we may consider that every simple regular ring has the property (DF) if and only if it satisfies s-comparability for some positive integer s. From the results in Sections 2 and 3, the property (DF) for regular rings seems to be closely related to s-comparability. Thus we have a question: Does the property (DF) for regular rings characterize s-comparability?

In Section 4, we shall show that the answer for the above question is negative, by giving new constructions of unit-regular rings (not always satisfying s-comparability) which have the property (DF).

Throughout this paper, a ring is an associative ring with identity and modules are unitary right modules.

We recall a Notation and well-known Definitions.

Notation 1.1. Let R be a ring. For two R-modules M and N, we use $M \leq N$ (resp. $M < N$; $M \leq_\oplus N$) to mean that M is a submodule (resp. a proper submodule; a direct summand) of N, and $M \lesssim N$ (resp. $M \lesssim_\oplus N$) means that M is isomorphic to a submodule (resp. a direct summand) of N. For a cardinal number k and an R-module M, kM denotes the direct sum of k-copies of M.

Definition 1.1. A ring R is said to be *regular* if for each $x \in R$ there exists $y \in R$ such that $xyx = x$, and R is said to be *unit-regular* if for each $x \in R$ there exists a unit (i.e., an invertible) element $u \in R$ such that $xux = x$. A module M is *directly finite* provided that M is not isomorphic to a proper direct summand of itself. If M is not directly finite, then M is said to be *directly infinite*. A ring R is said to be directly finite if the R-module R_R is directly finite, and R is said to be *stably finite* if the matrix ring $M_n(R)$ is directly finite for all positive integers n.

Now, we recall the definition of the property (DF).

Definition 1.2. ([9]). A ring R is said to *have the property* (DF) provided that $P \oplus Q$ is directly finite for any directly finite projective R-modules P and Q.

All basic results concerning regular rings can be found in Goodearl's book [6].

2. Regular Rings with S-comparability

We recall the definition of s-comparability for regular rings.

Definition 2.1. Let R be a regular ring and s be a positive integer. Then R is said to *satisfy s-comparability* provided that for any $x, y \in R$, either $xR \precsim s(yR)$ or $yR \precsim s(xR)$. In particular, 1-comparability is said to be *the comparability axiom*.

It is well-known that a regular ring with s-comparability is a prime ring by the similar proof of one of [6, Proposition 8.5], and that a regular ring with s-comparability for some $s > 1$ always satisfies 2-comparability ([3, Theorem 2.8]). Here, we give typical Examples of regular rings with s-comparability.

Example 2.1. ([6, Examples 8.1, 8.7 and 18.19]). (1) There exists a simple unit-regular ring with the comparability axiom. For example, choose a field F and set $R_n = M_{2^n}(F)$ for all $n = 0,1,2,\ldots$. For each n, define a map $\varphi_n:$ $R_n \to R_{n+1}$ according to the rule

$$\varphi_n(x) = \begin{pmatrix} x & 0 \\ 0 & x \end{pmatrix},$$

and let $R = \varinjlim R_n$. Then R is a simple unit-regular ring with the comparability axiom.

(2) There exists a simple unit-regular ring with 2-comparability which does not satisfy the comparability axiom. For example, choose a field F and set $R_n = M_{3^n}(F) \times M_{3^n}(F)$ for $n = 0,1,2,\ldots$. For each n, define a map $\varphi_n: R_n \to R_{n+1}$ according to the rule

$$\varphi_n(x, y) = \left(\begin{pmatrix} x & 0 & 0 \\ 0 & x & 0 \\ 0 & 0 & y \end{pmatrix}, \begin{pmatrix} x & 0 & 0 \\ 0 & y & 0 \\ 0 & 0 & y \end{pmatrix} \right),$$

and let $R = \varinjlim R_n$. Then R is a simple unit-regular ring with 2-comparability which does not satisfy the comparability axiom.

To study the forms of directly finite projective modules over regular rings with s-comparability, we give the following definition and conditions.

Definition 2.2. For an R-module A, its *trace ideal* is $tr(A) = \sum f(A)$ where f ranges over all R-homomorphisms from A to R.

Remark 2.1. Let R be a regular ring, and let A, B be finitely generated projective R-modules. Then (1) $tr(A) \leq tr(B)$ if and only if $A \precsim kB$ for some positive integer k ([3, p.25]). (2) In particular, when R is a regular ring with s-comparability, either $tr(A) \leq tr(B)$ or $tr(B) \leq tr(A)$ by [3, Proposition 2.1].

Lemma 2.1. ($[^3$, Proposition 2.5] and $[^{11}$, Lemma 1.5]). *Let R be a regular ring with s-comparability, and let A, B be finitely generated projective R-modules. Then*

(1) *If $tr(A) < tr(B)$, then $\aleph_0 A \lesssim B$.*
(2) *If B is nonzero directly finite and $\aleph_0 A \lesssim B$, then $tr(A) < tr(B)$.*

Let R be a regular ring with s-comparability, and P be a (non-finitely) countably generated projective R-module with a cyclic directly finite decomposition $P = \oplus_{i=1}^{\infty} P_i$ which satisfies (*):

(*) There exists no nonzero cyclic projective R-module T such that $T \lesssim 2P_i$ for all $i \in I'$, where I' is an infinite subset of $\{1, 2, \ldots\}$.

Then we may assume from (*) and Lemma 2.1(1) that $tr(P_1) \geq tr(P_2) \geq \ldots$ for $P = \oplus_{i=1}^{\infty} P_i$, by arranging the index set $\{1, 2, \ldots\}$. For this decomposition $P = \oplus_{i=1}^{\infty} P_i$, we can consider the following conditions (A) and (B):

(A) There exists a positive integer m such that $tr(P_m) = tr(P_{m+1}) = \cdots$.

(B) There exists a sequence $1 = n_1 < n_2 < \cdots$ of positive integers such that $tr(P_{n_1}) > tr(P_{n_2}) > \cdots$.

Using the above conditions (*), (A) and (B), we can give forms of directly finite projective modules over regular rings with s-comparability, as follows.

Theorem 2.1. ($[^{11}$, Theorem 2.11]). *Let R be a regular ring with s-comparability, and P be a projective R-module with a cyclic decomposition $P = \oplus_{i \in I} P_i$. Then P is directly finite if and only if $P = \oplus_{i \in I} P_i$ satisfies (1) or (2) or (3) as follows:*

(1) *P is finitely generated and P_i's are directly finite.*
(2) *P is (non-finitely) countably generated, P_i's are directly finite and $P = \oplus_{i=1}^{\infty} P_i$ satisfies (*) and (A) such that $\oplus_{i=m}^{\infty} P_i \lesssim tP_m$ for some positive integer t.*
(3) *P is (non-finitely) countably generated, P_i's are directly finite and $P = \oplus_{i=1}^{\infty} P_i$ satisfies (*) and (B).*

Using Theorem 2.1, we have the following result.

Theorem 2.2. ($[^{11}$, Theorem 2.12]). *Every regular ring with s-comparability always has the property (DF).*

Hence, we shall look for other regular rings with the property (DF). But, unfortunately there exists an abelian right self-injective regular ring which does not have the property (DF), as follows.

Example 2.2. ([9, Example] and [15, Theorem 2.5]). Choose a field F, and set $R_{2^n} = \prod_{i=1}^{2^n} F_i$, where $F_i = F$ for each i. Map each $R_{2^{n-1}} \to R_{2^n}$, given by the rule $x \to (x, x)$, and set $R = \varinjlim R_{2^n}$. Let $Q(R)$ be the maximal right quotient ring of R. Then $Q(R)$ is an abelian right self-injective regular ring which does not have the property (DF).

From Example 2.2, we see that typical regular rings do not have the property (DF) in general. Therefore we have a problem: Which regular rings have the property (DF)?

3. Weak Comparability

We shall treat the above problem (in Section 2) for regular rings with weak comparability.

Definition 3.1. ([17]). A regular ring R satisfies *weak comparability* if for each nonzero $x \in R$, there exists a positive integer n such that $n(yR) \lesssim R$ implies $yR \lesssim xR$ for all $y \in R$, where the n depends on x.

Remark 3.1. The notion of weak comparability was first introduced by O'Meara [17], to prove that simple directly finite regular rings with weak comparability must be unit-regular [6, Open Problem 3]. Thereafter properties for regular rings with weak comparability have been studying in many papers (see [1], [2], [4], [12], [13], [16] etc.).

Lemma 3.1. ([17, Proposition 2]). *A regular ring with weak comparability must either have bounded index of nilpotence or be a prime ring. Also every regular ring of bounded index of nilpotence satisfies weak comparability.*

Lemma 3.2. ([17, Corollary 2]). *Let R be a directly finite simple regular ring with s-comparability. Then R is a unit-regular ring with weak comparability.*

We give some typical Examples for regular rings with weak comparability, as follows.

Example 3.1. (1) There exists a simple unit-regular ring with weak comparability which does not have bounded index of nilpotence. For example, choose fields $F_1, F_2, \ldots$, set $R_n = M_{n!}(F_n)$ for all n, and set $R = \prod_{n=1}^{\infty} R_n$. Let M be a maximal two-sided ideal of R which contains $\oplus R_n$. Then R/M is a simple unit-regular ring with weak comparability which does not have bounded index of nilpotence, from [6, Example 10.7] and Lemma 3.2.

(2) ([17, Example 2]). There exists a non-simple prime unit-regular ring with weak comparability. For example, choose a simple non-artinian unit-regular ring S which satisfies the comparability axiom (see Example 2.1(1)). Let F be the center of S, and so it is a field. Since S is non-artinian, we can choose an infinite sequence $e_1, e_2, \ldots, e_n, \ldots$ of nonzero orthogonal idempotents of S. Let $f_n = e_1 + e_2 + \cdots + e_n$ for all n, and set $J = \cup_{n=1}^{\infty} f_n S f_n$. Let $R = F + J$. Then R is a non-simple prime unit-regular ring with weak comparability.

From Lemma 3.1 and Example 2.2, we see that unit-regular rings with weak comparability do not have the property (DF) in general. Hence we shall give a new condition (C) to study the property (DF) for regular rings with weak comparability, as follows. Here, we notice that the condition (C) is seemed to be a natural, slight strengthing of weak comparability.

Definition 3.2. A regular ring R is said to *satisfy the condition* (C) provided that for each nonzero $x \in R$, there exists a positive integer n such that $R \not\lesssim n(yR)$ $(y \in R)$ implies $yR \lesssim xR$, where the n depends on x.

Lemma 3.3. ([16, Lemma 2.1]). *Let R be a stably finite regular ring with the condition (C). Then R satisfies weak comparability.*

Proposition 3.1. ([16, Proposition 2.3]). *If R is a simple regular ring with s-comparability, then R satisfies the condition (C).*

Now, we give a characterization of the condition (C) for a stably finite regular ring, as follows.

Theorem 3.1. ([16, Theorem 2.4]). *Let R be a stably finite regular ring. Then R satisfies the condition (C) if and only if R is a simple unit-regular ring with s-comparability for some positive integer s.*

By Theorem 3.1 and Theorem 2.2, we have the following.

Corollary 3.1. *Every stably finite regular ring with the condition (C) has the property (DF).*

Meanwhile, Ara, Pardo and Perera gave a characterization of the property (DF) for simple regular rings with weak comparability, as follows.

Theorem 3.2. ($[^5$, Theorem 4.4] and $[^2$, Theorem 4.3]). *Every simple regular ring with weak comparability has the property* (DF) *if and only if it satisfies s-comparability for some positive integer s.*

We notice that there exists a simple unit-regular ring with weak comparability which does not satisfy s-comparability, as follows.

Example 3.2. ($[^{17}$, Example 1]). For each positive integer n, let $p_n = n^2 + 4n + 1, w_n = (p_1 + 2)(p_2 + 2) \cdots (p_{n-1} + 2)$, and let R_n be the direct product of three copies of the ring of $w_n \times w_n$ matrices over a fixed F. Let R be the direct limit of the sequence $R_1 \to R_2 \to \cdots$ where the ring maps $R_n \to R_{n+1}$ are given by the rule $(A, B, C) \longrightarrow$

$$
\left(\begin{pmatrix} A & & & & \\ & \ddots & & & \\ & & A & & \\ & & & B & \\ & & & & C \end{pmatrix}, \begin{pmatrix} A & & & & \\ & B & & & \\ & & \ddots & & \\ & & & B & \\ & & & & C \end{pmatrix}, \begin{pmatrix} A & & & & \\ & B & & & \\ & & C & & \\ & & & \ddots & \\ & & & & C \end{pmatrix} \right)
$$

and where the indicated repetitions occur p_n times. Then R is a simple unit-regular ring with weak comparability which does not satisfy s-comparability.

By Theorem 3.2 and Example 3.2, we also see that there exists a simple unit-regular ring which does not have the property (DF), comparing with Example 2.2. By the way, it is unknown that there exists a simple regular ring which does not satisfy weak comparability. Therefore, at the present time, we may consider that every simple regular ring has the property (DF) if and only if it satisfies s-comparability for some positive integer s.

4. New Constructions

From the results in Sections 2 and 3, the property (DF) for regular rings is seemed to be closely related to s-comparability. Thus we have a question: gDoes the property (DF) for regular rings characterize s-comparability?h In this section, we shall show that the answer for this question is negative. For this purpose, we give new constructions of unit-regular rings (not always satisfying s-comparability) which have the property (DF), by treating some factor rings of direct products of unit-regular rings.

We first give a characterization of the property (DF) for unit-regular rings, and for this purpose we need some Lemmas.

Lemma 4.1. ($[^{10}$, Lemma 1]). *Let R be a unit-regular ring, and let P be a projective R-module with a cyclic decomposition $P = \oplus_{i \in I} P_i$. Then the following conditions (1) through to (3) are equivalent:*

(1) *P is directly infinite.*

(2) *There exists a nonzero principal right ideal X of R such that $X \lesssim \oplus_{i \in I-F} P_i$ for all finite subsets F of I.*

(3) *There exists a nonzero principal right ideal X of R such that $\aleph_0 X \lesssim_\oplus P$.*

Lemma 4.2. ($[^5$, Proposition 4.2]). *Let R be a ring. Then R has the property (DF) if and only if $P \oplus Q$ is directly finite for any directly finite countably generated projective R-modules P and Q.*

Now, we can give a characterization of the property (DF) for unit-regular rings using Lemmas 4.1 and 4.2, as follows.

Theorem 4.1. ($[^{14}$, Proposition 3]). *Let R be a unit-regular ring. Then the following conditions are equivalent:*

(1) *R has the property (DF).*

(2) *For each nonzero principal right ideal X of R and each decompositions $X = A_1 \oplus B_1, A_j = A_{2j} \oplus B_{2j}$ and $B_j = A_{2j+1} \oplus B_{2j+1}$ for each $j = 1, 2, \ldots$, there exists a nonzero principal right ideal Y of R such that $Y \lesssim \oplus_{k=n}^{\infty} A_k$ for all positive integers n or $Y \lesssim \oplus_{k=n}^{\infty} B_k$ for all positive integers n.*

Using Theorem 4.1 effectively, we have the following Theorem which makes for plenty of unit-regular rings with the property (DF).

Theorem 4.2. ($[^{14}$, Theorem 8]). *Let I be a set, and let $\{R_i\}_{i \in I}$ be a family of unit-regular rings. Then the ring $(\prod_{i \in I} R_i)/(\oplus R_i)$ always has the property (DF).*

From Theorem 4.2, we see that for any unit-regular ring R, the ring $(\prod_{i=1}^{\infty} R)/(\oplus R)$ has the property (DF) but it does not satisfy s-comparability, since it is a non-prime ring.

Remark 4.1. For any regular ring R, it is well-known that $(\prod_{i=1}^{\infty} R)/(\oplus R)$ is a right and left $\aleph_0$-injective regular ring (see $[^7]$ or $[^6$, p386]).

Notation 4.1. Let I be a set. We use $|I|$ to denote the cardinal number of I. For each element $x = (x_i) \in \prod_{i \in I} R_i$, we set $supp(x) = \{i \in I \mid x_i \neq 0\}$, where the R_i's are rings. We denote $\oplus^\beta R_i = \{x \in \prod_{i \in I} R_i \mid |supp(x)| \leq \beta\}$ for each infinite cardinal number β.

Here, we consider a more generalization of Theorem 4.2, and we can give the following Theorem.

Theorem 4.3. ([14, Theorem 10]). *Let $\{R_i\}_{i \in I}$ be a family of unit-regular rings with $|I| = \alpha$ $(> \aleph_0)$, and let β be a cardinal number with $\aleph_0 \leq \beta < \alpha$. Then the following conditions are equivalent:*

(1) *$(\prod_{i \in I} R_i)/(\oplus^\beta R_i)$ has the property (DF).*
(2) *$|\{i \in I | R_i$ does not have the property (DF)$\}| \leq \beta$.*

By Theorem 4.3, we have the following Corollary.

Corollary 4.1. *Let R be a unit-regular rings with the property (DF). Then $(\prod_{i \in I} R)/(\oplus^\beta R)$ has the property (DF), where $\aleph_0 \leq \beta < |I|$.*

We also obtain the following Theorem.

Theorem 4.4. ([14, Notes (1)]). *Let $\{R_i\}_{i \in I}$ be a family of unit-regular rings. Then the ring $\prod_{i \in I} R_i$ has the property (DF) if and only if so does R_i for all $i \in I$.*

Finally, using Theorems 4.2 and 4.3, we can show that the property (DF) for unit-regular rings is not inherited by factor rings and subrings in general, as follows.

Example 4.1. Let R be an abelian regular ring which does not have the property (DF) (see Example 2.2). We set $T = (\prod_{i \in I} R)/(\oplus R)$, where $\aleph_0 < |I|$. Then T is a unit-regular ring with the property (DF) by Theorem 4.2, and a factor ring $(\prod_{i \in I} R)/(\oplus^{\aleph_0} R)$ of T does not have the property (DF) by Theorem 4.3. Also, note that $R \cong (\oplus R + 1 \cdot R)/(\oplus R) \leq (\prod_{i \in I} R)/(\oplus R) = T$.

References

1. P. Ara and K.R. Goodearl, The almost isomorphism relation for simple regular rings. *Publ. Mat. UAB* **36**, 369–388 (1992).
2. P. Ara, K.R. Goodearl, E. Pardo and D.V. Tyukavkin, K–theoretically simple von Neumann regular rings. *J. Algebra* **174**, 659–677 (1995).
3. P. Ara, K.C. O'Meara and D.V. Tyukavkin, Cancellation of projective modules over regular rings with comparability. *J. Pure Appl. Algebra* **107**, 19–38 (1996).

4. P. Ara and E. Pardo, Refinement monoids with weak comparability and applications to regular rings and C^*–algebras. *Proc. Amer. Math. Soc.* **124(3)**, 715–720 (1996).

5. P. Ara, E. Pardo and F. Perera, The structure of countably generated projective modules over regular rings. *J. Algebra* **226**, 161–190 (2000).

6. K.R. Goodearl, Von Neumann regular rings. 2nd Edn.; Pitman: London, 1979; Krieger: Malabar, Florida, 1991; 412pp.

7. D. Handelman, Homomorphisms of C^* algebras to finite AW^* algebras. *Michigan Math. J.* **28**, 229–240 (1981).

8. M. Kutami, On projective modules over directly finite regular rings satisfying the comparability axiom. *Osaka J. Math.* **22**, 815–819 (1985).

9. M. Kutami, Projective modules over regular rings of bounded index. *Math. J. Okayama Univ.* **30**, 53–62 (1988).

10. M. Kutami, On unit–regular rings satisfying s–comparability. *Osaka J. Math.* **33**, 983–995 (1996).

11. M. Kutami, On regular rings with s–comparability. *Comm. Algebra* **27(6)**, 2917–2933 (1999).

12. M. Kutami, Regular rings with comparability and some related properties. *Comm. Algebra* **30(7)**, 3337–3349 (2002).

13. M. Kutami, On von Neumann regular rings with weak comparability. *J. Algebra* **265**, 285–298 (2003).

14. M. Kutami, A construction of unit-regular rings which satisfy (DF). *Comm. Algebra* **32(4)**, 1509–1517 (2004).

15. M. Kutami and I. Inoue, The property (DF) for regular rings whose primitive factor rings are artinian. *Math. J. Okayama Univ.* **35**, 169–179 (1993).

16. M. Kutami and H. Tsunashima, Unit–regular rings satisfying weak comparability. *Comm. Algebra* **29(3)**, 1131–1140 (2001).

17. K.C. O'Meara, Simple regular rings satisfying weak comparability. *J. Algebra* **141**, 162–186 (1991).

SEMILATTICE GRADED WEAK HOPF ALGEBRA AND ITS QUANTUM DOUBLE*

FANG LI   AND   HAIJUN CAO

Department of Mathematics
Zhejiang University
Hangzhou, Zhejiang 310028, China
E-mail: fangli@zju.edu.cn   hjcao99@163.com

In this paper, over a field k, for a so-called semilattice graded weak Hopf algebra H, we show that it is a weak Hopf sub-algebra of crossed product of $k\bar{G}$ over the summand of indecomposable components of all idempotents of $G(H)$ in case H is pointed and give the structure theorem of the quantum double $D(H)$ of H through bicrossed products and quantum doubles in case H is commutative.

Because of the important role of Hopf algebra in the theory of quantum group and related mathematical physics, the meaning of some weaker concepts of Hopf algebra is understood and paid close attention more and more along with the deepening of researches. A well-known example is a weak Hopf algebra, which is introduced in [L1] for studying the non-invertible solution of Yang-Baxter Equation based on this class of bialgebras (in [L1] and [L5]), and there is a tight relation between weak Hopf algebra and regular monoid, for example, a semigroup algebra is a weak Hopf algebra if and only if the semigroup is a regular moniod. Obviously, it is necessary to find more non-trivial weak Hopf algebras. In this paper, we construct a so-called semilattice graded weak Hopf algebra. An example of semilattice graded weak Hopf algebra is just Clifford moniod algebra.

Firstly, we introduce some useful concepts.

H is called a *pre-bialgebra* if H is an algebra and also a coalgebra with comultiplication Δ which is an algebra morphism but usually without $\Delta(1) = 1 \otimes 1$. A bialgebra H over k is called a *weak Hopf algebra*[L1] if there

---

*Tthis work is supported by the natural science foundation of zhejiang province of china (no.102028) and partially by the cultivation fund of the key scientific and technical innovation project, ministry of education of china (no. 704004)

exists $T \in Hom_k(H, H)$ (the convolution algebra) satisfying $id * T * id = id$ and $T * id * T = T$, where T is called a *weak antipode* of H. A weak Hopf algebra H is called (1) a *perfect weak Hopf algebra* [L3] if its weak antipode T is an anti-bialgebra morphism satisfying $(id * T)(H) \subseteq C(H)$ (the center of H); (2) a *coperfect weak Hopf algebra* [L2] if its weak antipode is an anti-bialgebra morphism satisfying $\sum_{(x)} x'T(x'') \otimes x''' = \sum_{(x)} x''T(x''') \otimes x'$ for any $x \in H$; (3) a *biperfect weak Hopf algebra* if it is perfect and also coperfect. A semigroup with identity is called a *monoid*.

A semigroup S is called a *Clifford semigroup* [Pe] if it is a regular semigroup and all of its idempotents lie in its center $C(S)$. An equivalent definition is that a Clifford semigroup S is a semilattice of groups, which means that the set of maximal subgroups $\{G_\alpha : \alpha \in Y\}$ of S can be indexed by elements of a semilattice (i.e. a commutative semigroup of idempotents) Y such that $S = \cup_{\alpha \in Y} G_\alpha$ and $G_\alpha G_\beta \subseteq G_{\alpha\beta}$ for each $\alpha, \beta \in Y$. For each $\alpha, \beta \in Y$ with $\alpha\beta = \beta$ there exists a homomorphism $\varphi_{\alpha,\beta} : G_\alpha \to G_\beta$. The homomorphisms are such that $\varphi_{\alpha,\alpha}$ is the identity map on G_α, and if $\alpha\beta = \beta, \beta\gamma = \gamma$, then $\varphi_{\beta,\gamma}\varphi_{\alpha,\beta} = \varphi_{\alpha,\gamma}$. For any $\alpha, \beta \in Y$ and $a \in G_\alpha, b \in G_\beta$, the multiplication in S is given by $ab = \varphi_{\alpha,\alpha\beta}(a)\varphi_{\beta,\alpha\beta}(b)$. In a semilattice Y, a partial order $\leq$ is defined satisfying $\alpha \leq \beta$ if $\alpha\beta = \alpha$ for $\alpha, \beta \in Y$, which is called the *natural partial order* in Y.

It is easy to see for every Clifford monoid S, the semigroup algebra kS is a weak Hopf algebra and $kS = \bigoplus_{\alpha \in Y} kG_\alpha$ is a semilattice grading sum. As its natural generalization, we will define the following concept, which supply a way to obtain a new class of weak Hopf algebras through some given Hopf algebras.

A weak Hopf algebra H with weak antipode T is called a *semilattice graded weak Hopf algebra* if $H = \bigoplus_{\alpha \in Y} H_\alpha$ is a semilattice grading sum where H_α are Hopf sub-algebras of H with antipodes $T|_{H_\alpha}$ for all $\alpha \in Y$ and there are homomorphisms of Hopf algebras $\varphi_{\alpha,\beta}$ from H_α to H_β if $\alpha\beta = \beta$, such that for $a \in H_\alpha$ and $b \in H_\beta$, the multiplication $a * b$ in H can be given by $a * b = \varphi_{\alpha,\alpha\beta}(a)\varphi_{\beta,\alpha\beta}(b)$.

Thus the set of group-like elements of H is the Clifford monoid $G(H) = [Y; G(H_\alpha), \varphi_{\alpha,\beta}|_{G(H_\alpha)}]$.

1. Decomposition

It is well known that each coalgebra C is (uniquely) a direct sum of indecomposable subcoalgebras; moreover when C is cocommutative, the indecomposable components are irreducible. In 1995 Montgomery[Mo1] gave an alternate proof of this result and applied these results to show that for any pointed Hopf algebra H, there is a normal subgroup N of the group $G(H)$ of group-like elements such that H is a crossed product of $k(G/N)$ and the indecomposable component of the identity element of H. In this section , we will generalize this result to a pointed semilattice graded weak Hopf algebra, but here, we need H with weak antipode T an anti-algebra bijection. We firstly need some preparation works.

Let $\mathcal{C}$ be the set of simple sub-coalgebras of a coalgebra C. The quiver Γ_C is given as follows: (V) the vertices of Γ_C are the elements of $\mathcal{C}$; and (E) there exists an edge $S_1 \to S_2$ for $S_i \in \mathcal{C} \Leftrightarrow S_1 \wedge S_2 \neq S_1 + S_2$; C is called link-indecomposable (L.I.) if Γ_C is connected (as an undirect graph)[Mo1]. We will also say that S_1 and S_2 are linked if $S_1 \to S_2$ or $S_2 \to S_1$, and that S_1 and S_2 are connected (denoted as $S_1 \sim S_2$) if they are in the same connected component of Γ_C. And a subcoalgebra D of C is called link-indecomposable component (LIC) if it is maximal with respect to Γ_D is connected.

Just as in [Mo1], when C is pointed and for any $x, y \in G(C)$, we write $x \to y$ instead of $S_1 \to S_2$, where $S_1 = kx, S_2 = ky$. We call an element $c \in C$ is (x,y)-primitive if $\Delta(c) = x \otimes c + c \otimes y$. Obviously $k(x - y)$ are (x,y)-primitive, an (x,y)-primitive element c is non-trivial if $c \notin k(x - y)$. So for a pointed coalgebra C, $x \to y$ if and only if there exists a non-trivial (x,y)-primitive element.

In [Mo1], the author used normal subgroups and their quotient groups to construct the decomposition of a group algebra. Now, we hope to give its generalization to semilattice graded weak Hopf algebras through the so-called normal inverse sub-semigroups, that is Theorem 1.5.

Definition 1.1[Pe] Let S be an inverse semigroup with a semilattice E of idempotent elements. Define an inverse sub-semigroup N of S to be normal if it is full (i.e. $E \subset N$) and conjugative (i.e. $xNx^{-1} \subset N$ for all $x \in S$).

Obviously E is a normal sub-inverse semigroup of S.

Lemma 1.2[Pe] Let $S = [Y; G_\alpha, \varphi_{\alpha,\beta}]$ be a Clifford semigroup and N be the normal inverse sub-semigroup of S, then N is a Clifford subsemigroup with the form $[Y; N_\alpha, \psi_{\alpha,\beta}]$ where every N_α is a normal subgroup of G_α, $\psi_{\alpha,\beta} = \varphi_{\alpha,\beta}|_{G_\alpha}$ and $\varphi_{\alpha,\beta}(N_\alpha) \subseteq N_\beta$ if $\alpha \geq \beta$.

Lemma 1.3 Let $S = [Y; G_\alpha, \varphi_{\alpha,\beta}]$ be a Clifford semigroup and $N = [Y; N_\alpha, \psi_{\alpha,\beta}]$ a normal inverse sub-semigroup of S, then $\bar{S} = [Y; G_\alpha/N_\alpha, \phi_{\alpha,\beta}]$ is also a Clifford semigroup, where $\phi_{\alpha,\beta} : G_\alpha/N_\alpha \to G_\beta/N_\beta$ satisfying $\phi_{\alpha,\beta}(xN_\alpha) = \varphi_{\alpha,\beta}(x)N_\beta$ for $\alpha \geq \beta$.

Proof: It is a direct and easy proof based on the definition of Clifford semigroup.

Lemma 1.4[Mo1] (1) If C and D are pointed coalgebras, then $C \otimes D$ is pointed and $G(C \otimes D) = G(C) \otimes G(D)$.

(2) If $f : C \to D$ is a surjection of coalgebras and C is pointed, then D is pointed and $G(D) = f(G(C))$.

In fact if C and D are pointed indecomposable, then $C \otimes D$ is also indecomposable, but it is false that images of pointed indecomposable Hopf algebras are indecomposable.

Denote $C_{(g)}$ be the indecomposable component of C over an element g.

Theorem 1.5 Let $H \cong \bigoplus_{\alpha \in Y} H_\alpha$ be a pointed semilattice graded weak Hopf algebra with weak antipode T, which is an anti-algebra isomorphism, $G = G(H) = [Y; G_\alpha, \varphi_{\alpha,\beta}]$ for $G_\alpha = G(H_\alpha)$. Let $H_{(x)}$ denote the indecomposable component containing x. Then

(1) $H_{(x)} H_{(y)} \subseteq H_{(xy)}$ and $T(H_{(x)}) \subseteq H_{(T(x))}$. In particular, $H_{(e_\alpha)}$ is a Hopf sub-algebra of H for every idempotent e_α in G;

(2) For $B = \oplus_{\alpha \in Y} H_{(e_\alpha)}$, $N = G(B) = [Y; N_\alpha, \psi_{\alpha,\beta}]$ is a normal sub-inverse semigroup of G;

(3) G_β acts on $H_{(e_\alpha)}$ by $x_\beta \cdot h_\alpha = x_\beta h_\alpha x_\beta^{-1}$ for all $x_\beta \in G_\beta$ and $h \in H_{(e_\alpha)}$;

(4) $H \cong \bigoplus_{\alpha \in Y}(H_{(e_\alpha)} \#_{\sigma_{\alpha\alpha}} k(G_\alpha/N_\alpha))$, with cocycle $\sigma_{\alpha\alpha} : G_\alpha/N_\alpha \times G_\alpha/N_\alpha \to N_\alpha$;

(5) H is a weak Hopf sub-algebra of $B \#_\sigma k\bar{G}$ with $\bar{G} = [Y, G_\alpha/N_\alpha, \phi_\alpha]$ as in Lemma 1.3, where $\sigma = \sum_{\alpha,\beta \in Y} \sigma_{\alpha\beta}$ is the cocycle with $\sigma_{\alpha\beta} : G_\alpha/N_\alpha \times G_\beta/N_\beta \to N_{\alpha\beta}$.

Note that $B \#_\sigma k\bar{G}$ is usually not a weak Hopf algebra except for the fact that H is a Hopf algebra and $H = B \#_\sigma k\bar{G}$.

Proof: (1) If $x, y \in G$, then Lemma 1.4 implies that $H_{(x)} \otimes H_{(y)}$ is pointed indecomposable. Also multiplication $H_{(x)} \otimes H_{(y)} \to H_{(x)} H_{(y)}$ is a coalgebra surjection, and thus by Lemma 1.4, $H_{(x)} H_{(y)}$ is pointed with $G(H_{(x)} H_{(y)}) = \{zw | z \in G(H_{(x)}), w \in G(H_{(y)})\}$. Moreover, a similar argument to the one after Lemma 1.4 shows that $G(H_{(x)} H_{(y)})$ is connected. Thus $H_{(x)} H_{(y)}$ is link-indecomposable; since it contains xy, it must be contained in $H_{(xy)}$. It follows that $(H_{(e_\alpha)})^2 \subseteq H_{(e_\alpha)}$ and so $H_{(e_\alpha)}$ is a bialgebra for each $e_\alpha \in E(G)$. It remains to show that $T(H_{(x)}) \subseteq H_{(T(x))}$. Now T is bijective and thus $T : H^{cop} \to H$ is a coalgebra isomorphism, here H^{cop} is H with the opposite coalgebra structure. Thus $H_{(x)}$ indecomposable implies that $T(H_{(x)}^{cop})$ is indecomposable. Since $T(x) \in T(H_{(x)}^{cop})$, it follows that $T(H_{(x)}^{cop})$ is the indecomposable component containing $T(x)$. Thus

$T(H_{(x)}) \subseteq H_{(T(x))}$. So, $T(H_{(e_\alpha)}) \subseteq H_{(T(e_\alpha))} = H_{(e_\alpha)}$, by the fact that $H_{(e_\alpha)}$ is a sub-bialgebra of the Hopf algebra H_α, then $H_{(e_\alpha)}$ is a Hopf sub-algebra of H for every $\alpha \in Y$.

(2) Now $N = \{x \in G | x \sim e_\alpha, \exists e_\alpha \in E(G)\}$ (here $x \sim e_\alpha$ means that kx and ke_α are in the same connected component, see the definition in Section 1) since $B = \bigoplus_{\alpha \in Y} H_{(e_\alpha)}$ is the direct sum of indecomposable components containing e_α. Also N is a sub-semigroup of G, this can be proved directly. We have known $x \to y$ implies $xz \to yz$ as in [M1]. Thus if $x, y \in N$, there will be $e, f \in E(G)$ such that $e \sim x$ and $f \sim y$, then $xy \sim ey \sim ef$. Since $E(G)$ is a semilatiice, then $ef \in E(G)$ and $xy \in N$. Similarly if $e \sim x$ and $z \in G$, then $zez^{-1} \sim zxz^{-1}$, but $zez^{-1} \in E(G)$, so N is conjugative. Together with $E(G) \subset N$, we say that N is a normal sub-semigroup of G. Then we can write $N = [Y; N_\alpha, \psi_{\alpha,\beta}]$ with N_α is a normal subgroup of G_α, $\psi_{\alpha,\beta} = \varphi_{\alpha,\beta}|_{N_\alpha}$, and $N_\alpha \psi_{\alpha,\beta} \subset N_\beta$ if $\alpha \geq \beta$.

(3) For each $x_\alpha \in G_\alpha$, the map $\tau_{x_\alpha} : H_{(e_\alpha)} \to H_{(e_\alpha)}$ given by $h_\alpha \mapsto x_\alpha h_\alpha$ is a coalgebra automorphism of H_{e_α}. Thus $\tau_{x_\alpha}(H_{(e_\alpha)}) = x_\alpha H_{(e_\alpha)}$ is the indecomposable component of H containing x_α, and so $x_\alpha H_{(e_\alpha)} = H_{(x_\alpha)}$. Similarly $H_{(x_\alpha)} = H_{(e_\alpha)} x_\alpha$. Consequently $x_\alpha H_{(e_\alpha)} x_\alpha^{-1} = H_{(e_\alpha)}$.

(4)(5) Obviously $G(H_{(e_\alpha)}) = N_\alpha$ as defined in (2), because $G(H_{(e_\alpha)}) = \{x \in G | x \sim e_\alpha\}$. Hence, let $T_\alpha = \{t(\bar{x}_\alpha)\}$ be the set of distinct coset representatives of N_α in G_α, $H_\alpha = \bigoplus_{t(\bar{x}_\alpha) \in T_\alpha} H_{(t(\bar{x}_\alpha))} = \bigoplus_{t(\bar{x}_\alpha) \in T_\alpha} H_{(e_\alpha)} t(\bar{x}_\alpha)$. If we define an action, a cocycle σ as: $x_\beta \cdot h_\alpha = x_\beta h_\alpha x_\beta^{-1}, \sigma_{\alpha\beta}(\bar{x}_\alpha, \bar{y}_\beta) = t(\bar{x}_\alpha) t(\bar{y}_\beta) t(\overline{x_\alpha y_\beta})^{-1}$ where $\sigma_{\alpha\beta}(\bar{x}_\alpha, \bar{y}_\beta) \in N_{\alpha\beta}$.

Then, for any $h_\alpha, k_\alpha \in H_{(e_\alpha)}$,

$$(h_\alpha t(\bar{x}_\alpha))(k_\alpha t(\bar{y}_\alpha)) = h_\alpha t(\bar{x}_\alpha) t(\bar{x}_\alpha)^{-1} t(\bar{x}_\alpha) k_\alpha t(\bar{y}_\alpha)$$

$$= h_\alpha t(\bar{x}_\alpha) k_\alpha t(\bar{x}_\alpha)^{-1} t(\bar{x}_\alpha) t(\bar{y}_\alpha)$$

$$= h_\alpha t(\bar{x}_\alpha) k_\alpha t(\bar{x}_\alpha)^{-1} t(\bar{x}_\alpha) t(\bar{y}_\alpha) t(\overline{x_\alpha y_\alpha})^{-1} t(\overline{x_\alpha y_\alpha})$$

$$= h_\alpha (t(\bar{x}_\alpha) \cdot k_\alpha) \sigma_{\alpha\alpha}(\bar{x}_\alpha, \bar{y}_\alpha) t(\overline{x_\alpha y_\alpha}).$$

Thus $H_\alpha = \bigoplus_{t(\bar{x}_\alpha) \in T_\alpha} H_{(t(\bar{x}_\alpha))} \cong H_{(e_\alpha)} \#_{\sigma_{\alpha\alpha}} k(G_\alpha/N_\alpha)$, a crossed product. Moreover,

$$(h_\alpha t(\bar{x}_\beta))(k_\gamma t(\bar{y}_\delta)) = h_\alpha t(\bar{x}_\beta) t(\bar{x}_\beta)^{-1} t(\bar{x}_\beta) k_\gamma t(\bar{y}_\delta)$$

$$= h_\alpha t(\bar{x}_\beta) k_\gamma t(\bar{x}_\beta)^{-1} t(\bar{x}_\beta) t(\bar{y}_\delta)$$

$$= h_\alpha t(\bar{x}_\beta) k_\gamma t(\bar{x}_\beta)^{-1} t(\bar{x}_\beta) t(\bar{y}_\delta) t(\overline{x_\beta y_\delta})^{-1} t(\overline{x_\beta y_\delta})$$

$$= h_\alpha (t(\bar{x}_\beta) \cdot k_\gamma) \sigma_{\beta\delta}(\bar{x}_\beta, \bar{y}_\delta) t(\overline{x_\beta y_\delta})$$

is the multiplication of the crossed product $B \#_\sigma k(\bar{G})$ with cocycle $\sigma = \sum_{\alpha,\beta \in Y} \sigma_{\alpha\beta}$. Therefore

$$H = \bigoplus_{\alpha \in Y} H_\alpha = \bigoplus_{\alpha \in Y} \bigoplus_{t(\bar{x}_\alpha) \in T_\alpha} H_{(t(\bar{x}_\alpha))} \cong \bigoplus_{\alpha \in Y} (H_{(e_\alpha)} \#_{\sigma_{\alpha\alpha}} k(G_\alpha/N_\alpha))$$

$$\subset \bigoplus_{\alpha,\beta \in Y} H_{(e_\alpha)} \#_{\sigma_{\alpha\beta}} k(G_\beta/N_\beta) = (\bigoplus_{\alpha \in Y} H_{(e_\alpha)}) \#_\sigma (\bigoplus_{\alpha \in Y} k(G_\alpha/N_\alpha))$$

$$\cong B \#_\sigma k(\bar{G}).$$

Example 1.6 We return to $vsl_q(2)$, which we see in [L5] is decomposable. In this case the direct sum of indecomposable components containing idempotents is:

$$B = H_{(1)} \bigoplus H_{(K\overline{K})}$$

where $H_{(1)} = k1$ and $H_{(K\overline{K})} = k < \overline{K}F, \overline{K}E, KF, KE, \overline{K}K, K^2, \overline{K}^2 >$ with the same relations as Example 2. It is easy to find that $B = H_{(1)} \bigoplus H_{(K\overline{K})}$, $H_{(1)}$ and $H_{(K\overline{K})}$ are all stable under the action of T, so they are all weak Hopf sub-algebras of $vsl_q(2)$. Thus

$$vls_q(2) \cong (H_{(1)} \# k1) \bigoplus (H_{(K\overline{K})} \# k(Z_2 \bigoplus Z_2)) \subset B \#_\sigma k(\overline{G})$$

where $\overline{G} = [Y; G_\alpha/N_\alpha, \phi_{\alpha,\beta}]$, then $k(\overline{G}) \cong k1 \bigoplus k(Z_2 \bigoplus Z_2)$. The action, cocycle σ and multiplication are all defined as in Theorem 1.5.

2. Structure Of Quantum Double

In [L2], a new type of quasi-bicrossed products are constructed by means of weak Hopf skew-pairs of weak Hopf algebras as a generalization of Hopf pairs introduced by Takeuchi. As a special case, the quantum double of a finite dimensional biperfect (noncocommutative) weak Hopf algebra is built. Therefore, it will be interesting to research the structure and representation of the quantum double of a biperfect weak Hopf algebra H with semilattice grading structure $H = \bigoplus_{\alpha \in Y} H_\alpha$. In this section, we suppose a semilattice graded weak Hopf algebra $H = \bigoplus_{\alpha \in Y} H_\alpha$ is **commutative** with finite dimension. Then, it is easy to prove that H is biperfect .

Let B_α be a basis of H_α for every $\alpha \in Y$, then $B = \bigcup_{\alpha \in Y} B_\alpha$ is a basis of H. In this section, we always suppose that H satisfies $\sum_{(a)} T(a''')a' \otimes a'' = \sum_{(a)} T(a'')a''' \otimes a'$ for any $a \in H$ and $1_{H_\alpha} a_\beta \in B_{\alpha\beta}$ for any $a_\beta \in B_\beta$. Obviously, this condition is satisfied when H is a Clifford monoid algebra.

For any $a \in B$, let ϕ_a be the dual morphism of a in H^*, that is, $\phi_a(x) = \begin{cases} 1 \text{ if } a = x \\ 0 \text{ if } a \neq x. \end{cases}$ According to [L6], the quantum double $D(H)$ can be constructed from H with $(f\infty a)(g\infty b) = \sum_{(a)} fg(T^{-1}(a''')?a')\infty a''b$ for $f, g \in H^{op*}$, $a, b \in H$, where $g(T^{-1}(a''')?a')$ means the morphism: $x \longmapsto g(T^{-1}(a''')xa')$ for $x \in H$. As a k-linear space, $D(H) = H^{op*}\infty H$ possesses a basis $\{\phi_a \infty x : a, x \in B\}$ and the identity $1_{D(H)} = \varepsilon_H \infty 1_H$.

Thus as k-linear spaces, we have $D(H) = H^* \otimes H = H^* \otimes (\bigoplus_{\alpha \in Y} H_\alpha)$. Moreover, $H^* = (\bigoplus_{\alpha \in Y} H_\alpha)^* = \prod_{\alpha \in Y} H_\alpha^*$.

The set Y is finite since H is of finite dimension. So, $\prod_{\alpha \in Y} H_\alpha^* = \bigoplus_{\alpha \in Y} H_\alpha^*$. Thus, we get $D(H) = \bigoplus_{\alpha,\beta \in Y}(H_\alpha^* \otimes H_\beta)$ as linear spaces. Denote $D(H_\alpha, H_\beta) = H_\alpha^{op*} \otimes H_\beta$ and $Q_H(H_\alpha) = H_\alpha^{op*} \otimes H = \bigoplus_{\beta \in Y} D(H_\alpha, H_\beta)$. Then, as linear spaces,

$$D(H) = \bigoplus_{\alpha \in Y} Q_H(H_\alpha) = \bigoplus_{\alpha,\beta \in Y} D(H_\alpha, H_\beta). \tag{1}$$

According to (1), for each $\alpha \in Y$, H_α^* is embedded into H^* such that any $\varphi \in H_\alpha^*$ is mapped to $\overline{\varphi} \in H^*$ satisfying $\overline{\varphi}(u + v) = \varphi(u)$ for any element $u + v$ of $H = \oplus_{\beta \in Y} H_\beta$ where $u \in H_\alpha$ and $v \in \bigoplus_{\beta \neq \alpha} H_\beta$.

For α_1, α_2, β_1, $\beta_2 \in Y$, we consider the multiplication between $D(H_{\alpha_1}, H_{\beta_1})$ and $D(H_{\alpha_2}, H_{\beta_2})$ according to their embedding in H^*. For $x \in B_{\beta_1}$, $y \in B_{\beta_2}$, $a \in B_{\alpha_1}$ and its duality ϕ_a in $H_{\alpha_1}^{op*}$, $b \in B_{\alpha_2}$ and its duality ϕ_b in $H_{\alpha_2}^{op*}$, we have

$$(\phi_a \infty x)(\phi_b \infty y) = (\overline{\phi_a} \infty x)(\overline{\phi_b} \infty y) = \sum_{(x)} \overline{\phi_a}\,\overline{\phi_b}(T^{-1}(x''')?x')\infty x''y$$

$$= \sum_{(x)} \overline{\phi_a}\,\overline{\phi_b}(T^{-1}(x''')x'?)\infty x''y$$

$$= \sum_{(x)} \overline{\phi_a}\,\overline{\phi_b}(T^{-1}(x'')x'''?)\infty x'y$$

$$= \sum_{(x)} \overline{\phi_a}\,\overline{\phi_b}(\varepsilon_{\beta_1}(x'')1_{H_{\beta_1}}?)\infty x'y$$

$$= \sum_{(x)} \overline{\phi_a}\,\overline{\phi_b}(1_{H_{\beta_1}}?)\infty \varepsilon_{\beta_1}(x'')x'y$$

$$= \overline{\phi_a}\,\overline{\phi_b}(1_{H_{\beta_1}}?)\infty xy$$

where $\overline{\phi_a}\,\overline{\phi_b}(1_{H_{\beta_1}}?) \triangleq \begin{cases} 0 & \text{if } 1_{H_{\beta_1}}a \neq b \\ \overline{\phi_a} = \phi_a & \text{if } 1_{H_{\beta_1}}a = b \end{cases}$, since $1_{H_{\beta_1}}a \in B_{\beta_1\alpha_1}$.

Hence, $\overline{\phi_a}\,\overline{\phi_b}(1_{H_{\beta_1}}?)$ is always in $H_{\alpha_1}^{op*}$. And, $xy \in H_{\beta_1\beta_2}$. Therefore, $(\phi_a \infty x)(\phi_b \infty y) \in H_{\alpha_1}^{op*}\infty H_{\beta_1\beta_2}$. So we get

$$(H_{\alpha_1}^{op*}\infty H_{\beta_1})(H_{\alpha_2}^{op*}\infty H_{\beta_2}) \subseteq H_{\alpha_1}^{op*}\infty H_{\beta_1\beta_2},$$

146

that is,

$$D(H_{\alpha_1}, H_{\beta_1})D(H_{\alpha_2}, H_{\beta_2}) \subseteq D(H_{\alpha_1}, H_{\beta_1\beta_2}); \qquad (2)$$

and, if and only if $\alpha_2 \not\leq \alpha_1$ or $\alpha_2 \not\leq \beta_1$, the following holds:

$$D(H_{\alpha_1}, H_{\beta_1})D(H_{\alpha_2}, H_{\beta_2}) = 0 \qquad (3)$$

since in this case, $1_{H_{\beta_1}} a \notin H_{\alpha_2}$, then always $1_{H_{\beta_1}} a \neq b$.

In (2), let $\alpha_1 = \alpha_2 = \alpha$, then

$$D(H_\alpha, H_{\beta_1})D(H_\alpha, H_{\beta_2}) \subseteq D(H_\alpha, H_{\beta_1\beta_2}). \qquad (4)$$

We call $D(H_\alpha, H_\beta)$ the *bicrossed product* of two Hopf algebras H_α and H_β which are included in $H = \bigoplus_{\alpha \in Y} H_\alpha$; $Q_H(H_\alpha)$ the *bicrossed product* of H and its Hopf sub-algebra H_α. Denote $D(H_\alpha, H_\beta) = H_\alpha^{op*} \infty H_\beta$; $Q_H(H_\alpha) = H_\alpha^{op*} \infty H$.

A sub-ring K of a ring R is called a null sub-ring if there is an $n \in \mathbb{N}$ such that $K^n = 0$.

Firstly, we need the following lemmas on $D(H_\alpha, H_\beta)$ and $Q_H(H_\alpha)$:

Lemma 2.1 For all α, $\beta \in L$, $D(H_\alpha, H_\beta)$ are coalgebras and subrings of $D(H)$. For any $\alpha \not\leq \beta$, $D(H_\alpha, H_\beta)$ is a null subring. For any $\alpha \leq \beta$, $D(H_\alpha, H_\beta)$ is a pre-bialgebra under the same multiplication.

Proof: For $f \in H_\alpha^{op*}$ and $x \in H_\beta$,

(i) Define $\Delta : D(H_\alpha, H_\beta) \longrightarrow D(H_\alpha, H_\beta) \otimes D(H_\alpha, H_\beta)$ satisfying $\Delta(f \infty x) = \sum_{(f)} (f' \infty x') \otimes (f'' \infty x'')$, where $\Delta(f) = \sum_{(f)} f' \otimes f''$ according to the comultiplication of H_α^{op*}.

(ii) Define $\varepsilon : D(H_\alpha, H_\beta) \longrightarrow k$ satisfying $\varepsilon(f \infty x) = \varepsilon_{H_\alpha^{op*}}(f)\varepsilon_{H_\beta}(x)$.

Obviously, $(\Delta \otimes 1)\Delta = (1 \otimes \Delta)\Delta$.

For any $f \in H_\alpha^{op*}$, $x \in H_\beta$,

$(\varepsilon \otimes 1)\Delta(f \infty x) = (\varepsilon \otimes 1) \sum_{(f),(x)} (f' \infty x') \otimes (f'' \infty x'')$

$= \sum_{(f),(x)} \varepsilon(f' \infty x')(f'' \infty x'') = \sum_{(f),(x)} \varepsilon_{H_\alpha^{op*}}(f')\varepsilon_{H_\beta}(x')f'' \infty x''$

$= \sum_{(f),(x)} \varepsilon_{H_\alpha^{op*}}(f')f'' \infty \varepsilon_{H_\beta}(x')x'' = f \infty x;$

Similarly, $(1 \otimes \varepsilon)\Delta(f \infty x) = f \infty x$. Hence, $(\varepsilon \otimes 1)\Delta = (1 \otimes \varepsilon)\Delta = id$.

Therefore, $D(H_\alpha, H_\beta)$ becomes a coalgebra on Δ and ε.

The multiplication of $D(H_\alpha, H_\beta)$ is given as that of $D(H)$, that is, for α, $\beta \in Y$, a, $b \in B_\alpha$ and x, $y \in B_\beta$, $(\phi_a \infty x)(\phi_b \infty y) = \overline{\phi_a}\, \overline{\phi_b}(1_{H_\beta}?) \infty xy =$

$$\begin{cases} 0 & \text{if } 1_{H_\beta} a \neq b \\ \overline{\phi_a}\, \overline{\phi_b}(1_{H_\beta}?) \infty xy = \overline{\phi_a} \infty xy = \phi_a \infty xy & \text{if } 1_{H_\beta} a = b \end{cases}$$

In (4), let $\beta_1 = \beta_2 = \beta$, then $D(H_\alpha, H_\beta)D(H_\alpha, H_\beta) \subseteq D(H_\alpha, H_\beta)$ since $\beta\beta = \beta$. Then, every $D(H_\alpha, H_\beta)$ is a sub-ring of $D(H)$. By (3), if and only if $\alpha \not\leq \beta$, $D(H_\alpha, H_\beta)D(H_\alpha, H_\beta) = 0$. Hence, in this case, $D(H_\alpha, H_\beta)$ is a null sub-ring.

Now, suppose that $\alpha \leq \beta$. Since for any $b \in B_\alpha$, $1_{H_\beta}b \in B_{\alpha\beta}$ as we have defined, hence $\sum_{a \in B_\alpha} \phi_a(1_{H_\beta}b) = 1$, $D(H_\alpha, H_\beta)$ possesses the identity $\sum_{a \in S_\alpha} \phi_a \infty 1_{H_\beta}$. Therefore $D(H_\alpha, H_\beta)$ itself is an algebra.

For any $a, b \in B_\alpha$, $x, y \in B_\beta$,

$$\Delta((\phi_a \infty x)(\phi_b \infty y)) = \Delta(\overline{\phi_a}\,\overline{\phi_b}(1_{H_\beta}?)\infty xy)$$

$$= \begin{cases} 0 & \text{if } 1_{H_\beta}a \neq b \\ \Delta(\phi_a \infty xy) & \text{if } 1_{H_\beta}a = b \end{cases}$$

$$= \begin{cases} 0 & \text{if } 1_{H_\beta}a \neq b \\ \sum_{(\phi_a),(xy)}(\phi'_a \infty x'y') \otimes (\phi''_a \infty x''y'') & \text{if } 1_{H_\alpha}a = b. \end{cases}$$

$$\Delta(\phi_a \infty x)\Delta(\phi_b \infty y)$$

$$= (\sum_{(\phi_a),(x)}(\phi'_a \infty x') \otimes (\phi''_a \infty x''))(\sum_{(\phi_b),(y)}(\phi'_b \infty y') \otimes (\phi''_b \infty y''))$$

$$= \sum_{(\phi_a)(\phi_b),(x)(y)}(\phi'_a \infty x')(\phi'_b \infty y') \otimes (\phi''_a \infty x'')(\phi''_b \infty y'')$$

$$= \sum_{(\phi_a)(\phi_b),(x)(y)}(\overline{\phi'_a}\,\overline{\phi'_b}(1_{H_\beta}?)\infty x'y') \otimes (\overline{\phi''_a}\,\overline{\phi''_b}(1_{H_\beta}?)\infty x''y'').$$

For any $u, v \in H_\beta^*$, $s, t \in B_\alpha$,

$$\Delta((\phi_a \infty x)(\phi_b \infty y))(s \otimes u \otimes t \otimes v)$$

$$= \begin{cases} 0 & \text{if } 1_{H_\beta}a \neq b \\ \sum_{(\phi_a),(x)} \phi'_a(s)u(x'y')\phi''_a(t)v(x''y'') & \text{if } 1_{H_\beta}a = b \end{cases}$$

$$= \begin{cases} 0 & \text{if } 1_{H_\beta}a \neq b \\ \phi_a(st)u(x'y')v(x''y'') & \text{if } 1_{H_\beta}a = b. \end{cases}$$

$$\Delta(\phi_a \infty x)\Delta(\phi_b \infty y)(s \otimes u \otimes t \otimes v)$$

$$= \sum_{(\phi_a)(\phi_b),(x)(y)} \overline{\phi'_a}(s)\overline{\phi'_b}(1_{H_\beta}s)u(x'y')\overline{\phi''_a}(t)\overline{\phi''_b}(1_{H_\beta}t)v(x''y'').$$

And $\Delta(\phi_a \infty x)\Delta(\phi_b \infty y)(s \otimes u \otimes t \otimes v) = \phi_a(st)\phi_b(1_{H_\beta}st)u(x'y')v(x''y'')$

$$= \begin{cases} 0 & \text{if } 1_{H_\beta}a \neq b \\ \phi_a(st)u(x'y')v(x''y'') & \text{if } 1_{H_\beta}a = b. \end{cases}$$

Thus,

$$\Delta((\phi_a \infty x)(\phi_b \infty y)) = \Delta(\phi_a \infty x)\Delta(\phi_b \infty y).$$

In H_α^{op*}, for any $s, t \in B_\alpha$, $\Delta(\sum_{a \in S_\alpha} \phi_a)(s \otimes t) = \sum_{a \in B_\alpha} \phi_a(st) = \sum_{a \in G_\alpha} \delta_{a,st}$ may not equal to 1 , but $\Delta(\sum_{a \in S_\alpha} \phi_a) = (\sum_{b,c \in S_\alpha} \phi_b \otimes \phi_c)$. Thus, for $1_{D(H_\alpha, H_\beta)} = \sum_{a \in H_\alpha} \phi_a \infty 1_{H_\beta}$, $\Delta(1_{D(H_\alpha, H_\beta)})$ may not equal to $1_{D(H_\alpha, H_\beta)} \otimes 1_{D(H_\alpha, H_\beta)}$. Therefore Δ is an algebra morphism but not preserve the identity. It is easy to get $\varepsilon((\phi_a \infty x)(\phi_b \infty y)) = \varepsilon(\phi_a \infty x)\varepsilon(\phi_b \infty y)$ and $\varepsilon(1_{D(H_\alpha, H_\beta)}) = 1$. Therefore, ε is an algebra morphism.

Hence, $D(H_\alpha, H_\beta)$ becomes a pre-bialgebra.

Lemma 2.2 For any $\alpha \in Y$, $Q_H(H_\alpha)$ is a right ideal of $D(H)$ and itself is a coalgebra with comultiplication Δ satisfying $\Delta((\phi_a \infty x)(\phi_b \infty y)) = \Delta(\phi_a \infty x)\Delta(\phi_b \infty y)$ for any $x, y \in B$ and $a, b \in B_\alpha$. Moreover, $Q_H(H_\alpha) = N_H(H_\alpha) \oplus B_H(H_\alpha)$ where $N_H(H_\alpha) = \sum_{\beta \in Y, \beta \not\geq \alpha} D(H_\alpha, H_\beta)$

is a null right ideal of $D(H)$ and is a subcoalgebra and ideal of $Q_H(H_\alpha)$, $B_H(H_\alpha) = \sum_{\beta \in Y, \beta \geq \alpha} D(H_\alpha, H_\beta)$ is a sub-pre-bialgebra of $Q_H(H_\alpha)$ with $N_H(H_\alpha)B_H(H_\alpha) = 0$ and $B_H(H_\alpha)N_H(H_\alpha) \subseteq N_H(H_\alpha)$.

Proof : From (4) and (1), we get $Q_H(H_\alpha)D(H) \subseteq Q_H(H_\alpha)$ for any $\alpha \in Y$, which means that $Q_H(H_\alpha)$ is a right ideal of $D(H)$, and thus $D(H)$ can be decomposed into a direct sum of these ideals.

For $f \in H_\alpha^{op*}$ and $x \in B$,

(1) Define $Q_H(H_\alpha) \longrightarrow Q_H(H_\alpha) \otimes Q_H(H_\alpha)$ satisfying $\Delta(f \infty x) = \sum_{(f),(x)}(f' \infty x') \otimes (f'' \infty x'')$, where $\Delta(f) = \sum_{(f)} f' \otimes f''$ according to the comultiplication of H_α^{op*}.

(2) Define $\varepsilon : Q_H(H_\alpha) \longrightarrow k$ satisfying $\varepsilon(f \infty x) = \varepsilon_{H_\alpha^{op*}}(f)\varepsilon_H(x)$.

As for $D(H_\alpha, H_\beta)$ in Lemma 2.1, $Q_H(H_\alpha)$ is a coalgebra on Δ and ε. By the definition of Δ, $N_H(H_\alpha)$ and $B_H(H_\alpha)$ are both subcoalgebras of $Q_H(H_\alpha)$.

Let $\beta \not\geq \alpha$ and $\gamma \in Y$. From (3), $D(H_\alpha, H_\beta)D(H_\alpha, H_\gamma) = 0$. It means that $N_H(H_\alpha)Q_H(H_\alpha) = 0$. Specially, $N_H(H_\alpha)N_H(H_\alpha) = 0$ (i.e. $N_H(H_\alpha)$ is null) and $N_H(H_\alpha)B_H(H_\alpha) = 0$. For any $D(H_\gamma, H_\xi)$ in $D(H)$, $D(H_\alpha, H_\beta)D(H_\gamma, H_\xi) \subseteq D(H_\alpha, H_{\beta\xi})$. But, $\alpha \not\leq \beta$. So, $\alpha \not\leq \beta\xi$. Then, $D(H_\alpha, H_{\beta\xi}) \subseteq N_H(H_\alpha)$. Thus, $N_H(H_\alpha)$ is a right ideal of $D(H)$. If $\alpha \leq \gamma$, then $\alpha \not\leq \gamma\beta$ since $\alpha \not\leq \beta$, thus $D(H_\alpha, H_\gamma)D(H_\alpha, H_\beta) \subseteq D(H_\alpha, H_{\gamma\beta}) \subseteq N_H(H_\alpha)$. It follows that $B_H(H_\alpha)N_H(H_\alpha) \subseteq N_H(H_\alpha)$ and $N_H(H_\alpha)$ is an ideal of $Q_H(H_\alpha)$.

It is easy to see that $B_H(H_\alpha)$ possesses the identity $1_{B_H(H_\alpha)} = \sum_{a \in B_\alpha} \phi_a \infty 1_H$ and $B_H(H_\alpha)B_H(H_\alpha) \subseteq B_H(H_\alpha)$. So, $B_H(H_\alpha)$ is an algebra and a sub-ring of $Q_H(H_\alpha)$.

As in Lemma 2.1, we also have $\Delta((\phi_a \infty x)(\phi_b \infty y)) = \Delta(\phi_a \infty x)\Delta(\phi_b \infty y)$, $\varepsilon((\phi_a \infty x)(\phi_b \infty y)) = \varepsilon(\phi_a \infty x)\varepsilon(\phi_b \infty y)$, $\Delta(1_{B_H(H_\alpha)}) \neq 1_{B_H(H_\alpha)} \otimes 1_{B_H(H_\alpha)}$ and $\varepsilon(1_{B_H(H_\alpha)}) = 1$. Therefore, we know that $B_H(H_\alpha)$ is a pre-bialgebra.

A ring R is a *semilattice sum of subrings*[We] R_α, $\alpha \in \Omega$, if Ω is a semilattice, $R = \sum_{\alpha \in \Omega} R_\alpha$ and $R_\alpha R_\beta \subseteq R_{\alpha\beta}$; R is a *supplementary semilattice sum of subrings* R_α, $\alpha \in \Omega$, if R is a semilattice sum of subrings R_α, $\alpha \in \Omega$, and if for every $\alpha \in \Omega$, $R_\alpha \cap \sum_{\beta \neq \alpha} R_\beta = \{0\}$; i.e. if the sum is direct.

From the discussion above, we get the following main result:

Theorem 2.3 (STRUCTURE THEOREM) For a finite dimensional commutative semilattice graded weak Hopf algebra $H = \bigoplus_{\alpha \in Y} H_\alpha$ with weak antipode T, where Y is a semilattice and H_α a Hopf sub-algebra of H with antipode $T|_{H_\alpha}$ for each $\alpha \in Y$ and $\sum_{(a)} T(a''')a' \otimes a'' = \sum_{(a)} T(a'')a''' \otimes a'$ for any $a \in H$, suppose there exists a basis B_α of H_α for every $\alpha \in Y$, such that $B = \bigcup_{\alpha \in Y} B_\alpha$ a basis of H satisfying $1_{H_\alpha} a_\beta \in B_{\alpha\beta}$ for any $a_\beta \in B_\beta$. Then the quantum double $D(H)$ is a direct sum of right ideals $Q_H(H_\alpha)$, $\alpha \in Y$, where

(1) every $Q_H(H_\alpha)$ is a supplementary semilattice sum of subrings $D(H_\alpha, H_\beta)$ for $\beta \in Y$ and is an coalgebra with comultiplication Δ satisfying $\Delta((\phi_a \infty x)(\phi_b \infty y)) = \Delta(\phi_a \infty x)\Delta(\phi_b \infty y)$ for any x, $y \in B$ and a, $b \in B_\alpha$;

(2) $Q_H(H_\alpha) = N_H(H_\alpha) \oplus B_H(H_\alpha)$ where $N_H(H_\alpha) = \sum_{\beta \in Y, \beta \ngeq \alpha} D(H_\alpha, H_\beta)$ is a null right ideal of $D(H)$ and is a subcoalgebra and ideal of $Q_H(H_\alpha)$, $B_H(H_\alpha) = \sum_{\beta \in Y, \beta \geq \alpha} D(H_\alpha, H_\beta)$ is a sub-pre-bialgebra of $Q_H(H_\alpha)$ with $N_H(H_\alpha)B_H(H_\alpha) = 0$ and $B_H(H_\alpha)N_H(H_\alpha) \subseteq N_H(H_\alpha)$;

(3) $D(H_\alpha, H_\beta)$ are subcoalgebras of $Q_H(H_\alpha)$. If $\alpha \nleq \beta$, $D(H_\alpha, H_\beta)$ is a null sub-ring. If $\alpha \leq \beta$, $D(H_\alpha, H_\beta)$ is a pre-bialgebra. If $\alpha = \beta$, and hence $D(H_\alpha, H_\alpha) = D(H_\alpha)$, which means that every quantum double $D(H_\alpha)$ is a direct sum component of $D(H)$.

At last, as an application, we discuss the semi-simplicity of quantum doubles. In [Wi], it is shown that for a finite group G, the quantum double $D(G)$ is semisimple as an algebra if and only if the characteristic p of k does not divide the order $|G|$ of G. Here, we will consider the similar question for a semilattice grading weak Hopf algebra as above. In other hand, $D(H)$ is regular if and only if it is semisimple. So, in the sequel, we will only study the semisimplicity of $D(H)$.

Suppose $D(H)$ is an semisimple algebra for H satisfying the conditions in Theorem 2.3. A k-algebra is semisimple if it is a semismple right module over itself, and any sub-module of a semisimple module is semisimple. Then, $D(H)$ is semisimple as a right $D(H)$-module. From Theorem 2.3, every $Q_H(H_\alpha)$ is a right ideal of $D(H)$, then is a right $D(H)$-submodule of the right $D(H)$-module $D(H)$. Hence, $Q_H(H_\alpha)$ must be semisimple as a right $D(H)$-module.

Suppose $N_H(H_\alpha) \neq 0$ for an arbitrary fixed $\alpha \in Y$. Then there exists $\beta \in Y$ such that $\alpha \nleq \beta$. From the semi-simplicity of $Q_H(H_\alpha)$, we know that $N_S(H_\alpha)$ is also semisimple as a right $D(H)$-module since it is a right sub-module of $Q_H(H_\alpha)$. Then, $N_H(H_\alpha)$ can be decomposed as a direct sum of some simple right $D(H)$-submodules.

Let $\theta = \prod_{\lambda \in Y} \lambda$. Then $\theta \leq \lambda$ for all $\lambda \in Y$ and $\alpha \nleq \theta$ (otherwise, $N_H(H_\alpha) = 0$). Thus, $D(H_\alpha, H_\theta) \subseteq N_H(H_\alpha)$. It is easy to see that $D(H_\alpha, H_\theta)$ is a right $D(H)$-submodule of $N_H(H_\alpha)$.

We will made our discussion through two steps.

Step 1 $N_H(H_\alpha) = D(H_\alpha, H_\theta)$.

150

Step 2 $Q_H(H_\alpha) = N_H(H_\alpha)$.

The proof of these two steps are tedious but direct. So, we get the following:

Theorem 2.4 For a finite dimensional weak Hopf algebra H satisfying the same conditions in Theorem 2.3, its quantum double $D(H)$ over a field k is semisimple (resp. regular) if and only if H is a semisimple Hopf algebra (resp. regular).

References

C. I.G.Connell, On the group ring, Canad. J. Math. 15: 650-685 (1963).

Kap. I.Kaplansky, Bialgebras, Lecture Notes in Math. University of Chicago, 1975.

Kas. C.Kassel, Quantum Groups, Springer-Verlag, New York, 1995.

L1. F.Li, Weak Hopf algebras and some new solutions of quantum Yang-Baxter equation, J. Algebra 208: 72-100 (1998).

L2. F.Li, On quasi-bicrossed product of weak Hopf algeras, Acta Math. Sinica (English Series), 20(2): 305-318(2004).

L3. F.Li, Solutions of Yang-Baxter equation in endomorphism semigroups and quasi-(co)braided almost bialgebras, Comm. Algebra 28(5): 2253-2270 (2000).

L4. F.Li, The Structure of The Quantum Quasi-Double of A Finite Clifford Monoid and Its Application, to appear in Comm. Algebra.

L5. F.Li, S.Duplij, Weak Hopf algebras and singular solutions of quantum Yang-Baxter equation, Comm. Math. Phys. 225: 191-217 (2002).

L6. F.Li, Yao-zhong Zhang, Quantum double for a class of noncocommutative weak Hopf algebras, to appear in J. of Math. Phys. .

Mo1. S. Montgomery, Indecomposable coalgebras, simple comodules, and pointed Hopf algebras, Proceding of American Mathematical Society, 123(8):2343-2351 (1995).

Mo2. S. Montgomery, Hopf algebras and their actions on rings, CBMS Regional Conference Series in Mathematics, 82. American Mathematical Society, Providence, RI, 1993.

Pe. M.Petrich, Inverse Semigroups, John Wiley & Sons, New York, 1984.

Pi. R.S.Pierce, Associative Algebras, Springer-Verlag, New York, 1969.

Su. Michio Suzuki, Group theory, Springer-Verlag Berlin Heidelberg, New York, 1982.

Sw. M.E.Sweedler, Hopf Algebras, Benjamin, Elmsford, New York, 1980.

We. J.Weissglass, Semigroup rings and semilattice sums of rings, Proc. Amer. Math. Soc. 39(3): 471-478 (1973).

Wi. S.J.Witherspoon, The representation ring of the quantum double of a finite group, J. Algebra 179: 305-329 (1996).

NOTES ON *FP*-PROJECTIVE MODULES AND *FP*-INJECTIVE MODULES

LIXIN MAO

Department of Mathematics, Nanjing Institute of Technology
Nanjing 210013, P.R. China
Department of Mathematics, Nanjing University
Nanjing 210093, P.R. China
E-mail: maolx2@hotmail.com

NANQING DING

Department of Mathematics, Nanjing University
Nanjing 210093, P.R. China
E-mail: nqding@nju.edu.cn

In this paper, we study the FP-projective dimension under changes of rings, especially under (almost) excellent extensions of rings. Some descriptions of FP-injective envelopes are also given.

1. Introduction

Throughout this paper, all rings are associative with identity and all modules are unitary. We write M_R ($_RM$) to indicate a right (left) R-module, and freely use the terminology and notations of [1, 4, 9].

A right R-module M is called *FP-injective* [11] if $\operatorname{Ext}_R^1(N, M) = 0$ for all finitely presented right R-modules N.

The concepts of FP-projective dimensions of modules and rings were introduced and studied in [5]. For a right R-module M, the FP-projective dimension $fpd_R(M)$ of M is defined to be the smallest integer $n \geq 0$ such that $\operatorname{Ext}_R^{n+1}(M, N) = 0$ for any FP-injective right R-module N. If no such n exists, set $fpd_R(M) = \infty$. M is called *FP-projective* if $fpd_R(M) = 0$. We note that the concept of FP-projective modules coincides with that of *finitely covered* modules introduced by J. Trlifaj (see [12, Definition 3.3 and Theorem 3.4]). It is clear that $fpd_R(M)$ measures how far away a right R-module M is from being FP-projective. The right FP-projective dimension $rfpD(R)$ of a ring R is defined as $\sup\{fpd_R(M) : M$ is a finitely generated right R-module} and measures how far away a ring R is from being right noetherian (see [5, Proposition 2.6]).

Let $\mathcal{C}$ be a class of right R-modules and M a right R-module. A homomorphism $\phi : M \to F$ with $F \in \mathcal{C}$ is called a $\mathcal{C}$-*preenvelope* of M [4] if for any homomorphism $f\colon M \to F'$ with $F' \in \mathcal{C}$, there is a homomorphism $g : F \to F'$ such that $g\phi = f$. Moreover, if the only such g are automorphisms of F when $F' = F$ and $f = \phi$, the $\mathcal{C}$-preenvelope ϕ is called a $\mathcal{C}$-*envelope* of M. A $\mathcal{C}$-envelope $\phi : M \to F$ is said to have the *unique mapping property* [3] if for any homomorphism $f\colon M \to F'$ with $F' \in \mathcal{C}$, there is a unique homomorphism $g : F \to F'$ such that $g\phi = f$. Following [4, Definition 7.1.6], a monomorphism $\alpha : M \to C$ with $C \in \mathcal{C}$ is said to be a *special $\mathcal{C}$-preenvelope* of M if $\mathrm{coker}(\alpha) \in {}^{\perp}\mathcal{C}$, where ${}^{\perp}\mathcal{C} = \{F : \mathrm{Ext}^1_R(F, C) = 0 \text{ for all } C \in \mathcal{C}\}$. Dually we have the definitions of a *(special) $\mathcal{C}$-precover* and a $\mathcal{C}$-*cover* (with unique mapping property). Special $\mathcal{C}$-preenvelopes (resp., special $\mathcal{C}$-precovers) are obviously $\mathcal{C}$-preenvelopes (resp., $\mathcal{C}$-precovers).

Denote by $\mathcal{FP}_R$ (resp., $\mathcal{FI}_R$) the class of FP-projective (resp., FP-injective) right R-modules. In what follows, special $\mathcal{FP}_R$-(pre)covers (resp., $\mathcal{FI}_R$-(pre)envelopes) will be called special FP-projective (pre)covers (resp., FP-injective (pre)envelopes).

We note that $(\mathcal{FP}_R, \mathcal{FI}_R)$ is a cotorsion theory (for the category of right R-modules) which is cogenerated by the representative set of all finitely presented right R-modules (cf. [4, Definition 7.1.2]). Thus, by [4, Theorem 7.4.1 and Definition 7.1.5], every right R-module M has a special FP-injective preenvelope, i.e., there is an exact sequence $0 \to M \to F \to L \to 0$, where $F \in \mathcal{FI}_R$ and $L \in \mathcal{FP}_R$; and every right R-module has a special FP-projective precover, i.e., there is an exact sequence $0 \to K \to F \to M \to 0$, where $F \in \mathcal{FP}_R$ and $K \in \mathcal{FI}_R$. We observe that, if $\alpha : M \to F$ is an FP-injective envelope of M, then $\mathrm{coker}(\alpha)$ is FP-projective, and if $\beta : F \to M$ is an FP-projective cover of M, then $\mathrm{ker}(\beta)$ is FP-injective by Wakamatsu's Lemmas [4, Propositions 7.2.3 and 7.2.4].

A ring S is said to be an *almost excellent extension* of a ring R [14, 15] if the following conditions are satisfied:

(1) S is a finite normalizing extension of a ring R [10], that is, R and S have the same identity and there are elements $s_1, \cdots, s_n \in S$ such that $S = Rs_1 + \cdots + Rs_n$ and $Rs_i = s_i R$ for all $i = 1, \cdots, n$.

(2) ${}_R S$ is flat and S_R is projective.

(3) S is right R-projective, that is, if M_S is a submodule of N_S and M_R

is a direct summand of N_R, then M_S is a direct summand of N_S.

Further, S is an *excellent extension* of R if S is an almost excellent extension of R and S is free with basis $s_1, \cdots, s_n$ as both a right and a left R-module with $s_1 = 1_R$. The concept of excellent extension was introduced by Passman [7] and named by Bonami [2]. The notion of almost excellent extensions was introduced and studied in [14, 15] as a non-trivial generalization of excellent extensions.

In this paper, we first study the FP-projective dimension under changes of rings. Let R and S be right coherent rings (i.e., rings such that every finitely generated right ideal is finitely presented) and $\varphi : R \to S$ be a surjective ring homomorphism with S projective as a right R-module and flat as a left R-module. It is proven that $fpd_S(M) = fpd_R(M)$ for any right S-module M_S, and hence $rfpD(S) \le rfpD(R)$.

Let S be a finite normalizing extension (in particular, an (almost) excellent extension) of a ring R. It is well known that R is right noetherian if and only if S is right noetherian [8, Proposition 5]. It seems natural to generalize descent of right noetherianess to right FP-projective dimensions in the case when S is an (almost) excellent extension of a ring R. We show that if R and S are right coherent rings and S is an almost excellent extension of R, then $fpd_R(M) = fpd_S(M)$ for any right S-module M_S, and $rfpD(S) \le rfpD(R)$, the equality holds if $rfpD(R) < \infty$. We also show that, for a right coherent ring R, $rfpD(R) \le 2$ and every (resp. FP-injective) right R-module has an FP-projective envelope if and only if every (resp. FP-injective) right R-module has an FP-projective envelope with the unique mapping property.

Although the class of FP-injective R-modules is not enveloping (a class $\mathcal{C}$ is enveloping if every R-module has a $\mathcal{C}$-envelope) (see [12, Theorem 4.9]), an individual R-module may have FP-injective envelopes. Some descriptions of an FP-injective envelope of an R-module are given. For example, it is shown that, if M_R has an FP-injective envelope and is a submodule of an FP-injective right R-module L, then the inclusion $i : M \to L$ is an FP-injective envelope of M if and only if L/M is FP-projective and any endomorphism γ of L such that $\gamma i = i$ is a monomorphism if and only if L/M is FP-projective and there are no nonzero submodules N of L such that $M \cap N = 0$ and $L/(M \oplus N)$ is FP-projective. It is also shown that if R is a right coherent ring and M_R has an FP-projective cover, then M_R has a special FP-injective preenvelope $\alpha : M \to N$ such that N has an FP-projective cover. Finally we consider FP-projective precovers under almost excellent extensions of rings. Let S be an almost excellent extension of a ring R, it is proven that if $\theta : N_S \to M_S$ is an S-epimorphism, then $\theta : N_R \to M_R$ is a special FP-projective precover of M_R if and only if $\theta : N_S \to M_S$ is a special FP-projective precover of M_S.

2. Results

We start with

Lemma 2.1. *Let $\varphi : R \to S$ be a surjective ring homomorphism with S_R projective and M_S a right S-module (and hence a right R-module).*

(1) If M_S is finitely presented, then M_R is finitely presented.

(2) If M_S is FP-projective, then M_R is FP-projective.

Proof. (1). Since M_S is finitely presented, there is an exact sequence $0 \to K \to P \to M \to 0$ of right S-modules with K finitely generated and P finitely generated projective. Since $\varphi : R \to S$ is surjective, it is easy to see that K is a finitely generated right R-module and P is a finitely generated projective right R-module by [9, Theorem 9.32] (for S_R is projective). Therefore M is a finitely presented right R-module.

(2). If M_S is FP-projective, then M_S is a direct summand in a right S-module N such that N is a union of a continuous chain, $(N_\alpha : \alpha < \lambda)$, for a cardinal λ, $N_0 = 0$, and $N_{\alpha+1}/N_\alpha$ is a finitely presented right S-module for all $\alpha < \lambda$ (see [12, Definition 3.3]). By (1), $N_{\alpha+1}/N_\alpha$ is a finitely presented right R-module for all $\alpha < \lambda$. So M_R is FP-projective. $\blacksquare$

Lemma 2.2. *Suppose that $\varphi : R \to S$ is a ring homomorphism with S flat as a left R-module. If M_S is FP-injective, then M_R is FP-injective.*

Proof. If N is a finitely presented right R-module, then there is an exact sequence $0 \to K \to P \to N \to 0$ of right R-modules with K finitely generated and P finitely generated projective. Since $_RS$ is flat, we have the following right S-module exact sequence

$$0 \to K \otimes_R S_S \to P \otimes_R S_S \to N \otimes_R S_S \to 0.$$

Note that $K \otimes_R S_S$ is a finitely generated right S-module, $P \otimes_R S_S$ is a finitely generated projective right S-module, and so $N \otimes_R S_S$ is a finitely presented right S-module.

Since M_S is FP-injective, we have $\operatorname{Ext}^1_S(N \otimes_R S_S, M) = 0$ by definition. Therefore $\operatorname{Ext}^1_R(N, M) = 0$ by [9, Theorem 11.65], and so M_R is FP-injective, as desired. $\blacksquare$

Proposition 2.1. *Let R and S be right coherent rings. If $\varphi : R \to S$ is a surjective ring homomorphism with S flat as a left R-module and projective as a right R-module, then*

(1) $fpd_S(M) = fpd_R(M)$ for any right S-module M_S.
(2) $rfpD(S) \le rfpD(R)$.

Proof. (1). We first prove $fpd_S(M) \le fpd_R(M)$. Assume $fpd_R(M) = n < \infty$. Let F_S be an FP-injective right S-module, then F_R is an FP-injective right R-module by Lemma 2.2. By [9, Theorem 11.66], we have

$$\mathrm{Ext}_S^{n+1}(M_S, \mathrm{Hom}_R(S, F_R)) \cong \mathrm{Ext}_R^{n+1}(M_R, F_R) = 0.$$

Note that $F_S \cong \mathrm{Hom}_R(S, F_R)$ (for φ is surjective), so $\mathrm{Ext}_R^{n+1}(M_S, F_S) = 0$. Therefore $fpd_S(M) \le n$, and hence $fpd_S(M) \le fpd_R(M)$.

Conversely, assume $fpd_S(M) = n < \infty$. By [5, Proposition 3.1], there exists a right S-module exact sequence $0 \to P_n \to P_{n-1} \to \cdots \to P_1 \to P_0 \to M \to 0$, where each P_i is an FP-projective right S-module. By Lemma 2.1 (2), each P_i is FP-projective as a right R-module. Thus $fpd_R(M) \le n$ by [5, Proposition 3.1] again and so $fpd_R(M) \le fpd_S(M)$.

(2) follows from (1). ∎

Lemma 2.3. *Let S be an almost excellent extension of a ring R and M_S a right S-module. Then*

(1) M_S is finitely presented if and only if M_R is finitely presented.
(2) M_S is FP-injective if and only if M_R is FP-injective if and only if $\mathrm{Hom}_R(S, M)$ is an FP-injective right S-module.
(3) M_S is FP-projective if and only if M_R is FP-projective.

Proof. (1). " $\Rightarrow$ ". Since M_S is finitely presented, there is an exact sequence $0 \to K \to P \to M \to 0$ of right S-modules with K finitely generated and P finitely generated projective. Let $K_S = a_1 S + a_2 S + \cdots + a_m S$. Note that $S = s_1 R + \cdots + s_n R$, we have $\{a_i s_j : 1 \le i \le m, 1 \le j \le n\}$ is a generating set of K_R. Thus K_R is finitely generated, and so is P_R. On the other hand, P_R is projective since P_S and S_R are projective. Therefore M_R is finitely presented.

" $\Leftarrow$ ". If M_R is finitely presented, then there is an exact sequence $0 \to K \to P \to M \to 0$ of right R-modules with K finitely generated and P finitely generated projective. Since $_R S$ is flat, we have the following right S-module exact sequence

$$0 \to K \otimes_R S_S \to P \otimes_R S_S \to M \otimes_R S_S \to 0.$$

Note that $K \otimes_R S_S$ is finitely generated, $P \otimes_R S_S$ is finitely generated projective, and so $M \otimes_R S_S$ is finitely presented. Since M_S is isomorphic to a direct summand of $M \otimes_R S_S$ by [15, Lemma 1.1 (1)], M_S is finitely presented.

(2). Suppose that M_S is FP-injective. Let L be a finitely presented right R-module. Since $_R S$ is flat, we have the following isomorphism

$$\mathrm{Ext}_R^1(L, M) \cong \mathrm{Ext}_S^1(L \otimes_R S, M)$$

by [9, Theorem 11.65]. By the proof of (1), $L \otimes_R S$ is a finitely presented right S-module, and so $\mathrm{Ext}_S^1(L \otimes_R S, M) = 0$. Thus $\mathrm{Ext}_R^1(L, M) = 0$, and hence M_R is FP-injective.

Now suppose that M_R is FP-injective. Let N_S be a finitely presented right S-module, then N_R is a finitely presented right R-module by (1), and so $\mathrm{Ext}_R^1(N, M) = 0$. Since $\mathrm{Ext}_S^1(N \otimes_R S, M) \cong \mathrm{Ext}_R^1(N, M)$, we have $\mathrm{Ext}_S^1(N \otimes_R S, M) = 0$. Therefore $\mathrm{Ext}_S^1(N, M) = 0$ by [15, Lemma 1.1 (1)] , and so M_S is FP-injective. On the other hand, by [9, Exercise 9.21, p.258], we have the isomorphism

$$\mathrm{Ext}_R^1(N \otimes_S S, M) \cong \mathrm{Ext}_S^1(N, \mathrm{Hom}_R(S, M)).$$

Note that M_S is isomorphic to a direct summand of $\mathrm{Hom}_R(S, M)$ by [15, Lemma 1.1(2)]. So M_S is FP-injective if and only if M_R is FP-injective if and only if $\mathrm{Hom}_R(S, M)$ is an FP-injective right S-module.

(3). " $\Rightarrow$ " follows from the proof of Lemma 2.1 (2).

"$\Leftarrow$". Suppose that M_R is FP-projective. For any FP-injective right S-module N_S, we have $\mathrm{Ext}_R^1(M_R, N_R) = 0$ since N_R is FP-injective by (2), and so it follows that $\mathrm{Ext}_S^1(M_R \otimes_R S, N_S) = 0$ by the isomorphism $\mathrm{Ext}_S^1(M_R \otimes_R S, N_S) \cong \mathrm{Ext}_R^1(M_R, N_R)$. Thus $\mathrm{Ext}_S^1(M_S, N_S) = 0$ by [15, Lemma 1.1 (1)], and hence M_S is FP-projective. $\blacksquare$

Theorem 2.1. *Let R and S be right coherent rings and S an almost excellent extension of R. Then $fpd_R(M) = fpd_S(M) = fpd_S(M \otimes_R S)$ for any right S-module M_S.*

Proof. We first claim that $fpd_R(M) \leq fpd_S(M)$. Without loss of generality, we may assume that $fpd_S(M) = n < \infty$. Then, by [5, Proposition 3.1], there exists an exact sequence

$$0 \to P_n \to P_{n-1} \to \cdots \to P_1 \to P_0 \to M \to 0,$$

where each P_i is an FP-projective right S-module. Note that each P_i is also an FP-projective right R-module by Lemma 2.3 (3), and hence $fpd_R(M) \leq n$ by [5, Proposition 3.1] again.

Now we prove that $fpd_S(M \otimes_R S) \leq fpd_R(M)$. If $fpd_R(M) = n < \infty$, then there exists an exact sequence

$$0 \to P_n \to P_{n-1} \to \cdots \to P_1 \to P_0 \to M \to 0$$

of right R-modules, where each P_i is an FP-projective right R-module. Since $_RS$ is flat, we have the following exact sequence

$$0 \to P_n \otimes_R S \to P_{n-1} \otimes_R S \to \cdots \to P_1 \otimes_R S \to P_0 \otimes_R S \to M \otimes_R S \to 0$$

of right S-modules. Note that each $P_i \otimes_R S$ is an FP-projective right S-module by [5, Lemma 3.18], and so $fpd_S(M \otimes_R S) \leq n$.

On the other hand, we have $fpd_S(M) \leq fpd_S(M \otimes_R S)$ since M_S is isomorphic to a direct summand of $M \otimes_R S_S$. $\blacksquare$

Corollary 2.1. *Let R and S be right coherent rings.*

(1) If S is an almost excellent extension of R, then $rfpD(S) \leq rfpD(R)$.

(2) If S is an excellent extension of a ring R, then $rfpD(S) = rfpD(R)$.

Proof. (1) follows from Theorem 2.1.

(2). Since S is an excellent extension of R, R is an R-bimodule direct summand of S. Let $_RS_R = R \oplus T$, and M_R be any right R-module. Note that $M \otimes_R S \cong M_R \oplus (M \otimes_R T)$. Therefore by Theorem 2.1, we have

$$fpd_R(M) \leq fpd_R(M \otimes_R S) = fpd_S(M \otimes_R S) \leq rfpD(S)$$

and hence $rfpD(R) \leq rfpD(S)$. So we have the desired equality by (1). $\blacksquare$

Theorem 2.2. *Let S be an almost excellent extension of a ring R. If R and S are right coherent and $rfpD(R) < \infty$, then $rfpD(S) = rfpD(R)$.*

Proof. It is enough to show that $rfpD(R) \leq rfpD(S)$ by Corollary 2.1. Let $rfpD(R) = n < \infty$, there exists a right R-module M such that $fpd_R(M) = n$. Define a right R-homomorphism $\alpha : M \to M \otimes_R S$ via $\alpha(m) = m \otimes 1$ for any $m \in M$. Note that the exact sequence $0 \to \ker(\alpha) \to M$ gives rise to the exactness of the sequence $0 \to \ker(\alpha) \otimes_R S \to M \otimes_R S$ since $_RS$ is flat. So $\ker(\alpha) \otimes_R S = 0$, and hence $\ker(\alpha) = 0$ by [10, Proposition 2.1]. Thus α is monic, and so we have a right R-module exact sequence $0 \to M \to M \otimes_R S \to L \to 0$. Note that

$$n = fpd_R(M) \leq sup\{fpd_R(M \otimes_R S), fpd_R(L) - 1\} \leq rfpD(R) = n$$

by [5, Proposition 3.2 (2)]. Since $fpd_R(L) - 1 \leq n - 1$, $fpd_R(M \otimes_R S) = n$. On the other hand, by Theorem 2.1, we get $fpd_R(M \otimes_R S) = fpd_S(M \otimes_R S) \leq rfpD(S)$. Therefore $rfpD(R) \leq rfpD(S)$, as desired. ∎

Remark 2.1. We note that if S is an almost excellent extension of a ring R, then R is right coherent if and only if S is right coherent by [15, Theorem 1.9]. So the condition "R and S are right coherent" in the previous discussion can be replaced by "either R or S is right coherent".

It is known that every right R-module has an epic FP-projective envelope if and only if $rfpD(R) \leq 1$ and any direct product of FP-projective right R-modules is FP-projective (see [6, Theorem 6.3]). Now we have

Proposition 2.2. *If every right R-module has an epic FP-projective (pre)envelope, then $\mathcal{FP}_R$ is closed under inverse limits.*

Proof. Let $\{C_j, \varphi_j^l\}$ be any inverse system with C_j FP-projective. By hypothesis, $\varprojlim C_j$ has an epic FP-projective preenvelope $\alpha : \varprojlim C_j \to E$. Let $\alpha_j : \varprojlim C_j \to C_j$ with $\alpha_i = \varphi_i^j \alpha_j$ whenever $i \leq j$. Then there exists $f_i : E \to C_i$ such that $\alpha_i = f_i \alpha$ for any $i \leq j$. It follows that $f_i \alpha = \varphi_i^j f_j \alpha$, and so $f_i = \varphi_i^j f_j$ (for α is epic). Therefore, by the definition of inverse limits, there exists $\beta : E \to \varprojlim C_j$ such that the following diagram is commutative

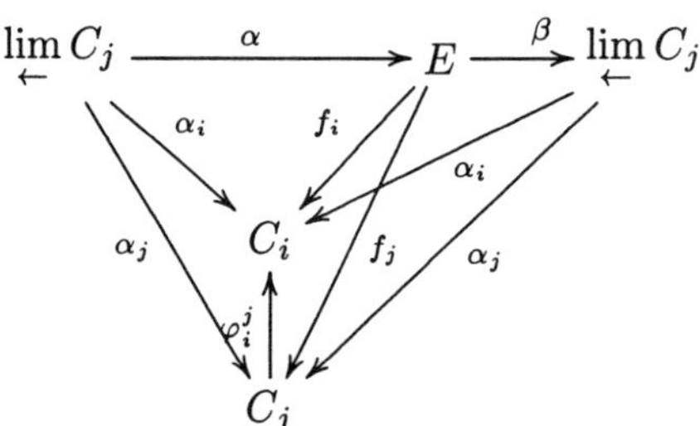

Thus $f_i = \alpha_i \beta$, and so $\alpha_i(\beta\alpha) = (\alpha_i\beta)\alpha = f_i\alpha = \alpha_i$ for any $i \leq j$. Therefore $\beta\alpha = 1_{\varprojlim C_j}$ by the definition of inverse limits, and hence α is an isomorphism. So $\varprojlim C_j$ is FP-projective. ∎

Next we consider when every right R-module has an FP-projective envelope with the unique mapping property.

Theorem 2.3. *The following are equivalent for a right coherent ring R:*

(1) Every (resp., FP-injective) right R-module has an FP-projective envelope with the unique mapping property;

(2) $rfpD(R) \leq 2$ and every (resp., FP-injective) right R-module has an FP-projective envelope.

Proof. $(2) \Rightarrow (1)$. Let M be any (resp., FP-injective) right R-module. Then M has an FP-projective envelope $f : M \to F$ by (2). It is enough to show that, for any FP-projective right R-module G and any homomorphism $g : F \to G$ such that $gf = 0$, we have $g = 0$. In fact, there exists $\beta : M \to \ker(g)$ such that $i\beta = f$ since $\mathrm{im}(f) \subseteq \ker(g)$, where $i : \ker(g) \to F$ is the inclusion. Note that $\ker(g)$ is FP-projective by [5, Proposition 3.1] since $fpd_R(G/\mathrm{im}(g)) \leq 2$. Thus there exists $\alpha : F \to \ker(g)$ such that $\beta = \alpha f$, and so we get the following exact commutative diagram

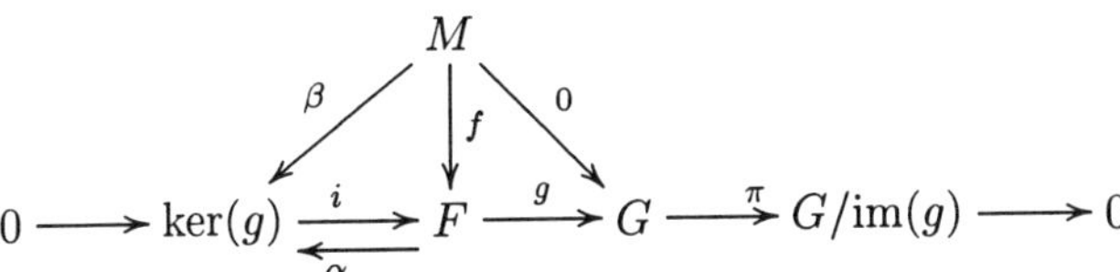

Note that $(i\alpha)f = f$, and hence $i\alpha$ is an isomorphism since f is an envelope. Therefore i is epic, and so $g = 0$.

$(1) \Rightarrow (2)$. Let M be any right R-module. Then we have the following exact sequences $0 \longrightarrow C \xrightarrow{\ i\ } F_0 \xrightarrow{\ \alpha\ } M \longrightarrow 0$ and $0 \longrightarrow F_2 \xrightarrow{\ \psi\ } F_1 \xrightarrow{\ \beta\ } C \longrightarrow 0$, where $\alpha : F_0 \to M$ and $\beta : F_1 \to C$ are special FP-projective precovers respectively, then C and F_2 are FP-injective. Thus we get an exact sequence

$$0 \longrightarrow F_2 \xrightarrow{\ \psi\ } F_1 \xrightarrow{\ \varphi=i\beta\ } F_0 \xrightarrow{\ \alpha\ } M \longrightarrow 0.$$

Let $\theta : F_2 \to H$ be an FP-projective envelope with the unique mapping property. Then there exists $\delta : H \to F_1$ such that $\psi = \delta\theta$. Thus $\varphi\delta\theta = \varphi\psi = 0$, and hence $\varphi\delta = 0$, which implies that $\mathrm{im}(\delta) \subseteq \ker(\varphi) = \mathrm{im}(\psi)$. So there exists $\gamma : H \to F_2$ such that $\psi\gamma = \delta$, and hence we get the following exact commutative diagram

Note that $\psi\gamma\theta = \psi$, and so $\gamma\theta = 1_{F_2}$ since ψ is monic. Thus F_2 is isomorphic to a direct summand of H, and hence F_2 is FP-projective. Therefore $fpd_R(M) \leq 2$ by [5, Proposition 3.1], and so $rfpD(R) \leq 2$. ∎

Following [11], the FP-injective dimension of a right R-module M, denoted by FP-$id(M)$, is defined to be the smallest integer $n \geq 0$ such that $\mathrm{Ext}_R^{n+1}(F, M) = 0$ for all finitely presented right R-modules F (if no such n exists, set FP-$id(M) = \infty$), and $r.FP$-$\dim(R)$ is defined as $\sup\{FP$-$id(M) : M$ is a right R-module$\}$.

It is well known that for a right coherent ring R, every (FP-projective) right R-module has a monic FP-injective cover if and only if R is right semi-hereditary (see [6, Corollary 4.2]). The next result may be regarded as a dual of Theorem 2.3.

Proposition 2.3. *The following are equivalent for a right coherent ring R:*

(1) Every (resp., FP-projective) right R-module has an FP-injective cover with the unique mapping property;

(2) $r.FP$-$\dim(R) \leq 2$, and every (resp., FP-projective) right R-module has an FP-injective cover.

For an individual module M, it is well known that an injective module N containing M as a submodule is an injective envelope of M if and only if N is an essential extension of M. As is known to all, every module has an injective envelope. However, FP-injective envelopes may not exist in general (see [12]). If M has an FP-injective envelope, we get the following descriptions of an FP-injective envelope of M.

Theorem 2.4. *Suppose that a right R-module M has an FP-injective envelope. Let M be a submodule of an FP-injective right R-module L. Then the following are equivalent:*

(1) $i : M \to L$ is an FP-injective envelope (here i is regarded as the inclusion);

(2) L/M is FP-projective, and there are no direct summands L_1 of L with $L_1 \neq L$ and $M \subseteq L_1$;

(3) L/M is FP-projective, and for any epimorphism $\alpha : L/M \to N$ such that $\alpha\pi$ is split, $N = 0$, where $\pi : L \to L/M$ is the canonical map;

(4) L/M is FP-projective, and any endomorphism γ of L such that $\gamma i = i$ is a monomorphism;

(5) L/M is FP-projective, and there are no nonzero submodules N of L such that $M \cap N = 0$ and $L/(M \oplus N)$ is FP-projective.

Proof. (1) $\Leftrightarrow$ (2) follows from [13, Corollary 1.2.3] and Wakamatsu's Lemma [4, Proposition 7.2.4]. (1) $\Rightarrow$ (4) is clear.

(2) $\Rightarrow$ (3). Since $\alpha\pi$ is split, there is a monomorphism $\beta : N \to L$ such that $L = \ker(\alpha\pi) \oplus \beta(N)$. Note that $M \subseteq \ker(\alpha\pi)$, and so $L = \ker(\alpha\pi)$ by (2). Thus $\beta(N) = 0$, and hence $N = 0$.

(3) $\Rightarrow$ (2). If $L = L_1 \oplus N$ with $M \subseteq L_1$. Let $p : L \to N$ be the canonical projection. Then $M \subseteq \ker(p)$, and so there is $\alpha : L/M \to N$ such that $\alpha\pi = p$. Therefore $N = 0$ by (3), and hence $L = L_1$, as required.

(4) $\Rightarrow$ (1). Since L/M is FP-projective, i is a special FP-injective preenvelope. Let $\sigma_M : M \to \mathcal{FI}(M)$ be an FP-injective envelope of M. There exist $\mu : L \to \mathcal{FI}(M)$ and $\nu : \mathcal{FI}(M) \to L$ such that $\mu i = \sigma_M$ and $\nu\sigma_M = i$. Hence $\mu\nu\sigma_M = \sigma_M$ and $i = \nu\mu i$. Thus $\mu\nu$ is an isomorphism, and so μ is epic. In addition, by (4), $\nu\mu$ is monic, and hence μ is monic. Therefore μ is an isomorphism, and so i is an FP-injective envelope of M.

(5) $\Rightarrow$ (1). Let $\sigma_M : M \to \mathcal{FI}(M)$ be an FP-injective envelope of M. Since L/M is FP-projective, i is a special FP-injective preenvelope. Thus we have the following commutative diagram with an exact row.

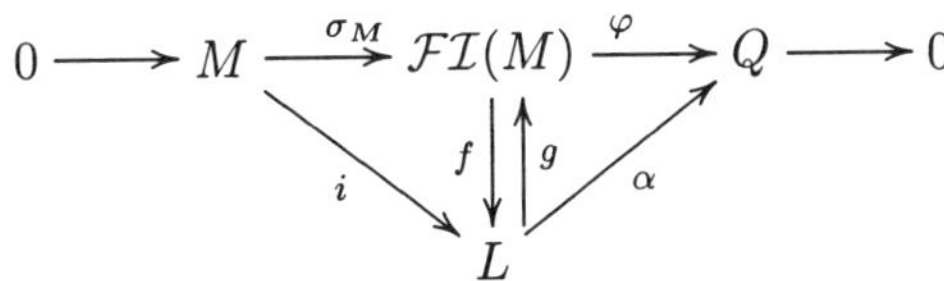

i.e., $f\sigma_M = i, gi = \sigma_M$. So $gf\sigma_M = \sigma_M$. Note that σ_M is an FP-injective envelope, and hence gf is an isomorphism. Without loss of generality, we may assume $gf = 1$. Write $\alpha = \varphi g : L \to Q$. It is clear that α is epic and $M \cap \ker(g) = 0$. Next we show that $M \oplus \ker(g) = \ker(\alpha)$. Indeed, $M \oplus \ker(g) \subseteq \ker(\alpha)$ is obvious. Let $x \in \ker(\alpha)$. Then $\alpha(x) = \varphi g(x) = 0$. It follows that $g(x) = \sigma_M(m)$ for some $m \in M$, and hence $fg(x) = f\sigma_M(m) = m, g(x) = gfg(x) = g(m)$. Thus $x \in M \oplus \ker(g)$, and so $\ker(\alpha) \subseteq M \oplus \ker(g)$, as desired.

Consequently, $L/(M \oplus \ker(g)) = L/\ker(\alpha) \cong Q$ is FP-projective by Wakamatsu's Lemma. Thus $\ker(g) = 0$ by hypothesis, and hence g is an isomorphism. So $i : M \to L$ is an FP-injective envelope.

(1) $\Rightarrow$ (5). It is obvious that L/M is FP-projective. Suppose there is a nonzero submodules $N \subseteq L$ such that $M \cap N = 0$ and $L/(M \oplus N)$ is FP-projective. Let $\pi : L \to L/N$ be the canonical map. Since $L/(N \oplus M)$

is FP-projective and L is FP-injective, there is $\beta : L/N \to L$ such that the following row exact diagram

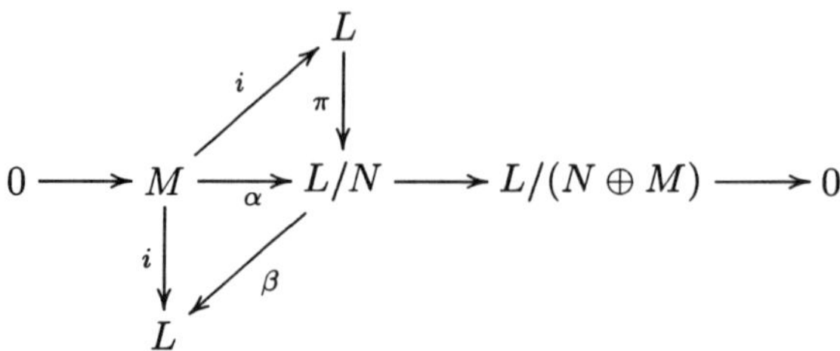

is commutative. Hence $\beta\pi i = i$. Note that i is an envelope, and so $\beta\pi$ is an isomorphism, whence π is an isomorphism. But this is impossible since $\pi(N) = 0$. ∎

We note that the equivalence of (1) and (5) in Theorem 2.4 is motivated by [13, Theorem 3.4.5] which gives a description of a cotorsion envelope of M.

Recall that a *minimal injective extension* of an R-module M is a monomorphism $i : M \to E$ with E injective such that for every R-monomorphism $f : M \to Q$ with Q injective there is a monomorphism $g : E \to Q$ such that $f = gi$. It is well known that $i : M \to E$ is an injective envelope of M if and only if i is a minimal injective extension of M (see [1, Corollary 18.11]). Similarly, we have the concept of the *minimal FP-injective extension*, and obtain the following

Corollary 2.2. *Let N be a submodule of an FP-injective right R-module M such that M/N is FP-projective.*

(1) If N is an essential submodule of M, then the inclusion $i : N \to M$ is an FP-injective envelope of N.

(2) If $i : N \to M$ is a minimal FP-injective extension of N, then $i : N \to M$ is an FP-injective envelope of N.

Proof. (1) follows from Theorem 2.4.

(2). Note that the injective envelope of N is an essential FP-injective extension of N, so N is an essential submodule of M by [1, Exercise 5.14 (1), p.77]. Thus (2) holds by (1). ∎

Assume that R is a Prüfer domain and the quotient field Q of R has projective dimension greater than or equal to 2. Let M be a free R-module. Clearly, M is FP-projective, but M has no FP-injective envelopes by [12, Theorem 4.9]. This shows that a right R-module which has an FP-

projective cover may have no FP-injective envelopes. However, we have the following

Theorem 2.5. *Let R be a right coherent ring. If a right R-module M has an FP-projective cover. Then M has a special FP-injective preenvelope $\alpha : M \to N$ such that N has an FP-projective cover.*

Proof. Let $\theta : Q \to M$ be an FP-projective cover of M. Then there is an exact sequence $0 \longrightarrow K \longrightarrow Q \overset{\theta}{\longrightarrow} M \longrightarrow 0$, where K is FP-injective by Wakamatsu's Lemma. Note that Q has a special FP-injective preenvelope, so there is an exact sequence $0 \longrightarrow Q \overset{f}{\longrightarrow} D \overset{g}{\longrightarrow} L \longrightarrow 0$, where D is FP-injective and L is FP-projective. Thus we have the following pushout diagram

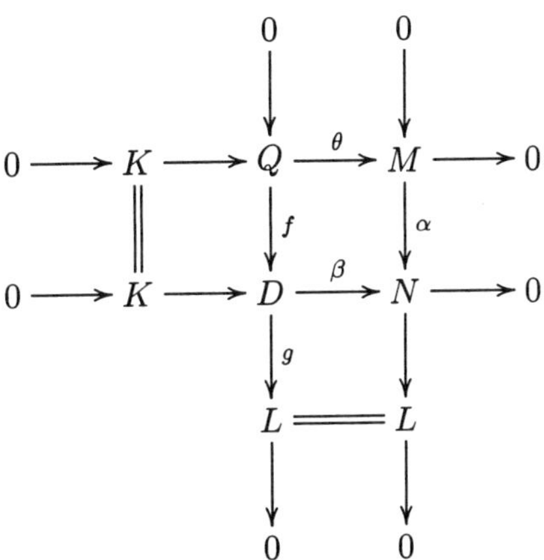

which is also a pullback diagram. Note that N is FP-injective by [11, Lemma 3.1] since R is right coherent. So α is a special FP-injective preenvelope of M. In addition, D is FP-projective since Q and L are. Therefore β is a special FP-projective precover of N.

Now let γ be an endomorphism of D with $\beta\gamma = \beta$. Then $\beta(\gamma f) = \beta f = \alpha\theta$. By the property of pullback, there exists $h : Q \to Q$ such that $\theta h = \theta$ and $fh = \gamma f$. Thus h is an isomorphism since θ is an FP-projective cover. Let $\gamma(d) = 0$ for some $d \in D$, then $\beta(d) = \beta\gamma(d) = 0$, and so $d = f(q)$ for some $q \in Q$. Thus $fh(q) = \gamma f(q) = 0$, and hence $q = 0$. Therefore $d = 0$, and so γ is monic. On the other hand, for any $t \in D$, $\beta\gamma(t) = \beta(t)$, and so $\gamma(t) - t = f(s)$ for some $s \in Q$. Then $t = \gamma(t) + f(s) = \gamma(t + fh^{-1}(s))$. Thus γ is epic, and hence an isomorphism. So β is an FP-projective cover of N. ∎

Finally, we consider FP-projective precovers (FP-injective preenvelopes) under almost excellent extensions of rings.

Theorem 2.6. *Let S be an almost excellent extension of a ring R and $\theta : N_S \to M_S$ an S-epimorphism, then the following are equivalent:*

(1) $\theta : N_R \to M_R$ is a special FP-projective precover of M_R;

(2) $\theta : N_S \to M_S$ is a special FP-projective precover of M_S.

> *Moreover, if S is an excellent extension of R, then the above conditions are also equivalent to*

(3) $\theta_* : \mathrm{Hom}_R(S, N) \to \mathrm{Hom}_R(S, M)$ is a special FP-projective precover of $\mathrm{Hom}_R(S, M)$;

(4) $\theta \otimes I_S : N \otimes_R S \to M \otimes_R S$ is a special FP-projective precover of $M \otimes_R S$.

Proof. (2) $\Rightarrow$ (1). Suppose that $\theta : N_S \to M_S$ is a special FP-projective precover of M_S. Then there is an exact sequence $0 \longrightarrow K \longrightarrow N \overset{\theta}{\longrightarrow} M \longrightarrow 0$ of right S-modules with $K \in \mathcal{FI}_S$ and $N \in \mathcal{FP}_S$. By Lemma 2.3, $N \in \mathcal{FP}_R$ and $K \in \mathcal{FI}_R$. Thus $\theta : N_R \to M_R$ is a special FP-projective precover of M_R.

(1) $\Rightarrow$ (2). Assume that $\theta : N_R \to M_R$ is a special FP-projective precover of M_R, i.e., there is an exact sequence $0 \longrightarrow K \longrightarrow N \overset{\theta}{\longrightarrow} M \longrightarrow 0$ of right R-modules with $K \in \mathcal{FI}_R$ and $N \in \mathcal{FP}_R$. Since S_R is projective, we have the exactness of the right S-module sequence

$$0 \longrightarrow \mathrm{Hom}_R(S, K) \longrightarrow \mathrm{Hom}_R(S, N) \overset{\theta_*}{\longrightarrow} \mathrm{Hom}_R(S, M) \longrightarrow 0.$$

Note that M_S (resp., N_S) is isomorphic to a direct summand of $\mathrm{Hom}_R(S, M)$ (resp., $\mathrm{Hom}_R(S, N)$) by [15, Lemma 1.1 (2)], and so we have the following exact commutative diagram

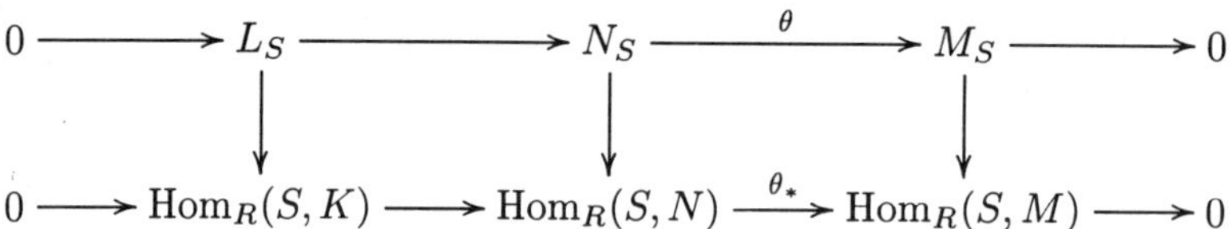

where $L_S = \ker(\theta)$. Note that $\mathrm{Hom}_R(S, K) \in \mathcal{FI}_S$ by Lemma 2.3 (2) since $K \in \mathcal{FI}_R$. It is easy to verify that L_S is isomorphic to a direct summand of $\mathrm{Hom}_R(S, K)$, and so L_S is FP-injective. In addition, N_S is FP-projective by Lemma 2.3 (3). Thus (2) holds.

$(1) \Rightarrow (3)$. By the proof of $(1) \Rightarrow (2)$, we have that $\mathrm{Hom}_R(S, K) \in \mathcal{FI}_S$. Note that $\mathrm{Hom}_R(S, N) \in \mathcal{FP}_R$ since N_R is FP-projective, S_R and $_RS$ are both finitely generated free. Thus $\mathrm{Hom}_R(S, N) \in \mathcal{FP}_S$ by Lemma 2.3 (3), and hence (3) follows.

$(3) \Rightarrow (2)$. Suppose $\theta_* : \mathrm{Hom}_R(S, N) \to \mathrm{Hom}_R(S, M)$ is a special FP-projective precover of $\mathrm{Hom}_R(S, M)$, then there exists a right S-module exact sequence

$$0 \longrightarrow Q_S \longrightarrow \mathrm{Hom}_R(S, N) \xrightarrow{\ \theta_*\ } \mathrm{Hom}_R(S, M) \longrightarrow 0$$

with $Q_S \in \mathcal{FI}_S$ and $\mathrm{Hom}_R(S, N) \in \mathcal{FP}_S$. The rest is similar to that of $(1) \Rightarrow (2)$.

The proof of $(1) \Leftrightarrow (4)$ is similar to that of $(1) \Leftrightarrow (3)$. ∎

Corollary 2.3. *Let S be an almost excellent extension of a ring R, and $\theta : N_S \to M_S$ an S-epimorphism. Then θ is an FP-projective cover of M_S if θ is an FP-projective cover of M_R.*

Proof. By Theorem 2.6, it is enough to prove the second condition of an FP-projective cover. Suppose $\alpha\theta = \theta$, where α is an S-module endomorphism of N_S. Then the equality is still true when α and θ are viewed as R-homomorphisms. So α is an R-isomorphism since N_R is an FP-projective cover of M_R. Therefore $\alpha_* : \mathrm{Hom}_R(S, N) \to \mathrm{Hom}_R(S, N)$ is an S-isomorphism. Note that N_S is isomorphic to a direct summand of $\mathrm{Hom}_R(S, N)$, it follows that α is an S-isomorphism, as required. ∎

We conclude the paper with the following proposition which is the dual of Theorem 2.6 and Corollary 2.3.

Proposition 2.4. *Let S be an almost excellent extension of R, and $\theta : M_S \to N_S$ an S-monomorphism, then*

    (1) $\theta : M_R \to N_R$ is a special FP-injective preenvelope of M_R if and only if $\theta : M_S \to N_S$ is a special FP-injective preenvelope of M_S.

    (2) $\theta : M_S \to N_S$ is an FP-injective envelope of M_S if $\theta : M_R \to N_R$ is an FP-injective envelope of M_R.

Acknowledgments

This research was partially supported by Specialized Research Fund for the Doctoral Program of Higher Education of China (No. 20020284009, 20030284033), EYTP and NNSF of China (No. 10331030) and the Nanjing Institute of Technology of China.

References

1. F.W. Anderson and K.R. Fuller, Rings and Categories of Modules; Springer-Verlag: New York, 1974.
2. L.Bonami, On the Structure of Skew Group Rings; Algebra Berichte 48, Verlag Reinhard Fisher: Munchen, 1984.
3. N.Q. Ding, On envelopes with the unique mapping property. *Comm. Algebra* **24**(4) (1996), 1459-1470.
4. E.E. Enochs and O.M.G. Jenda, Relative Homological Algebra; Walter de Gruyter: Berlin-New York, 2000.
5. L.X. Mao and N.Q. Ding, FP-projective dimensions. *Comm. Algebra* (to appear).
6. L.X. Mao and N.Q. Ding, Relative FP-projective modules. *Comm. Algebra* (to appear).
7. D.S. Passman, The Algebraic Structure of Group Rings; Wiley-Interscience, New York, 1977.
8. R. Resco, Radicals of finite normalizing extensions. *Comm. Algebra* **9** (1981), 713-725.
9. J.J. Rotman, An Introduction to Homological Algebra; Academic Press: New York, 1979.
10. A. Shamsuddin, Finite normalizing extensions. *J. Algebra* **151** (1992), 218-220.
11. B. Stenström, Coherent rings and *FP*-injective modules. *J. London Math. Soc.* **2** (1970), 323-329.
12. J. Trlifaj, Covers, Envelopes, and Cotorsion Theories; Lecture notes for the workshop, "Homological Methods in Module Theory". Cortona, September 10-16, 2000.
13. J. Xu, Flat Covers of Modules, Lecture Notes in Math. 1634; Springer Verlag: Berlin-Heidelberg-New York, 1996.
14. W.M. Xue, On a generalization of excellent extensions. *Acta Math. Vietnam* **19** (1994), 31-38.
15. W.M. Xue, On almost excellent extensions. *Algebra Colloq.* **3** (1996), 125-134.

A SURVEY OF MORPHIC MODULES AND RINGS

W.K. Nicholson
Department of Mathematics
University of Calgary
Calgary T2N 1N4, Canada
wknichol@ucalgary.ca

November 10, 2004

Abstract

An endomorphism α of a module $_RM$ is called morphic if $M/M\alpha \cong ker(\alpha)$, that is if the dual of the isomorphism theorem holds for α. The module $_RM$ is called a morphic module if every endomorphism is morphic, and we call a ring R left morphic if $_RR$ is morphic. This paper is a survey of what is presently known about these rings.

In [7] Erlich showed that an endomorphism α of a module $_RM$ is unit regular if and only if it is regular and $M/M\alpha \cong ker(\alpha)$. We call α morphic if $M/M\alpha \cong ker(\alpha)$, that is if the dual of the isomorphism theorem holds for α. The module $_RM$ is called a morphic module if every endomorphism is morphic, and a ring R is called left morphic if $_RR$ is a morphic module. This paper is primarily a survey of the work in [13], [14] and [15]. Most proofs are omitted, although some new proofs and results are included. Our focus is on the module case, with applications to rings.

Throughout this paper every ring R is associative with unity and all modules are unitary. We write morphisms of left modules on the right. If M is an R-module we write $J(M)$, $soc(M)$ and $Z(M)$ for the Jacobson radical, the socle, and the singular submodule of M, respectively. The uniform (Goldie) dimension of a module is denoted by $dim(M)$. We often abbreviate $J(R) = J$. We write $N \subseteq^{ess} M$ if N is an essential submodule of M, and $N \subseteq^{\oplus} M$ if N is a direct summand of M. We denote left and right annihilators of a subset $X \subseteq R$ by $\mathbf{1}(X)$ and $\mathbf{r}(X)$ respectively, and we write $\mathbb{Z}$ for the ring of integers and $\mathbb{Z}_n$ for the ring of integers modulo n. If R is a ring and $_RM_R$ is a bimodule the trivial extension of R by M is denoted $R \propto M = R \oplus M$ (with multiplication $(a, m)(b, n) = (ab, an + mb)$).

1. MORPHIC MODULES

We begin with a fundamental characterization of morphic endomorphisms [15].

Lemma 1.1. The following conditions are equivalent for $\alpha \in end(_RM)$:
   (1) α is morphic, that is $M/M\alpha \cong ker(\alpha)$.
   (2) There exists $\beta \in end(M)$ such that $M\beta = ker(\alpha)$ and $ker(\beta) = M\alpha$.

Corollary 1.2. A morphic endomorphism is monic if and only if it is epic. In particular, every left morphic ring is directly ønite ($ab = 1$ implies $ba = 1$).

Thus, for example, an inønite direct sum of copies of a nonzero module cannot be morphic.

Corollary 1.3. An element $a \in R$ is left morphic (as the endomorphism $r \mapsto ra$) if and only if $Ra = \mathbf{1}(b)$ and $\mathbf{1}(a) = Rb$ for some $b \in R$.

Hence no polynomial ring $R[x]$ is left morphic, and left morphic domains are division rings.

Corollary 1.4.   A direct product $\Pi_i R_i$ of rings is left morphic if and only if R_i is left morphic.

A ring R is called right P-injective if, for each principal right ideal aR of R, each R-morphism $aR \to R$ extends to R; equivalently if $\mathbf{rl}(a) = Ra$.

Corollary 1.5.   [13], [16] Let R be a left morphic ring. Then:
    (1) R is right P-injective.
    (2) $Z(R_R) = J(R)$.
    (3) $soc(R_R) \subseteq soc(_R R)$.
    (4) If aR is simple, $a \in R$ then Ra is simple.
    (5) R is right morphic if and only if it is left P-injective.

If $R = end(V)$ where V is a vector space of countably inønite dimension, then R is right and left P-injective (being regular), but it is neither left nor right morphic by Corollary 1.2.

If R is left morphic then $Z(R_R) = J$ by Corollary 1.5; here are some observations on the left singular ideal $Z(_R R)$. Recall that a ring R is reduced if it has no nonzero nilpotent elements.

Proposition 1.6.   [13] Let R be a left morphic ring.
    (1) $Z(_R R) \subseteq J$.
    (2) If R is reduced then $Z(_R R) = 0$ and R is a left duo ring.
    (3) The following are equivalent:
        (a) $_R R$ is uniform.
        (b) $Z(_R R)$ is the set of nonunits.
        (c) R is local and $Z(_R R) = J$.

Note that the ring R in Example 3.1 below is left morphic and satisøes $Z(_R R) = J$ but R is not right morphic. This ring also has the property that $Z(_R R) = J = Z(R_R)$ but R_R is not uniform.

Question 1. If R is a semiprime, left morphic ring, is $J(R) = 0$?

Returning to modules, if $\alpha, \beta \in end(M)$, write $\alpha \frown \beta$ when $M\beta = ker(\alpha)$ and $ker(\beta) = M\alpha$. Hence every idempotent ε and automorphism τ in $end(M)$ is morphic because $\varepsilon \frown 1 - \varepsilon$ and $\tau \frown 0$. Recall that an element a in a ring R is called regular if $aua = a$ for some $u \in R$, and a is called unit regular if u can be chosen to be a unit.

Proposition 1.7.   If $\alpha \in end(M)$ is morphic and τ is an automorphism of M, then both $\alpha\tau$ and $\tau\alpha$ are morphic. In particular every unit regular morphism is morphic.

Proof. Using the notation above, if $\alpha \frown \beta$, then $\alpha\tau \frown \tau^{-1}\beta$ and $\tau\alpha \frown \beta\tau^{-1}$. If $\alpha\sigma\alpha = \alpha$ then $\alpha = \varepsilon\sigma^{-1}$ where $\varepsilon = \alpha\sigma$ satisøes $\varepsilon^2 = \varepsilon$. $\square$

Much of our motivation stems from Erlich's characterization of unit regular endomorphisms. This is included in the next lemma, along with Azumaya's theorem in the regular case.

Lemma 1.8.   Let α be an endomorphism of $_R M$.
    (1) Azumaya [1]. α is regular if and only if $M\alpha$ and $ker(\alpha)$ are both direct summands of M.
    (2) Erlich [7]. α is unit regular if and only if it is both regular and morphic.

Thus every semisimple artinian ring is left and right morphic (it is unit regular). Note that the ring $\mathbb{Z}_4$ is left and right morphic but it is not unit regular.

Question 2. If R is left and right morphic and $J = 0$, is R (unit) regular?

It was proved by Camillo and Yu [3] that every unit regular ring is clean (that is each element is the sum of an idempotent a unit), so a natural question (see [13, Page 393]) is whether every left and right morphic ring is clean. The answer is ɪno ɟ: Chen and Zhou [5] show that the trivial extension $\mathbb{Z} \propto (\mathbb{Q}/\mathbb{Z})$ is a commutative morphic ring that is not clean. It would be interesting to see an example with zero Jacobson radical.

The next result characterizes morphic modules in terms of submodules and factors.

Theorem 1.9. [15] A module M is morphic if and only if whenever $M/K \cong N$ where K and N are submodules of M, then $M/N \cong K$.

Corollary 1.2 shows that $\mathbb{Z}$ is not morphic as a $\mathbb{Z}$-module. However, since ønite cyclic groups are isomorphic if and only if they have the same order, Theorem 1.9 shows that $\mathbb{Z}_n$ is morphic as a $\mathbb{Z}$-module for each $n \geq 2$. In fact we have

Theorem 1.10. [15] A ønitely generated abelian group is morphic if and only if it is ønite and, for each prime p, each p-primary component has the form $(\mathbb{Z}_{p^k})^n$ for some $n \geq 0$ and $k \geq 0$.

We remark in passing that, for integers $n \geq 1$ and $m \geq 1$, $\mathbb{Z}_m \oplus \mathbb{Z}_n$ is morphic if and only if $m = da$ and $n = db$ where $gcd(d,a) = 1$, $gcd(d,b) = 1$, and $gcd(a,b) = 1$. Note that every proper image of $\mathbb{Z}$ is morphic, but $\mathbb{Z}$ itself is not morphic. Let $len(M)$ denote the composition length of M.

Lemma 1.11. [15] A module $_RM$ of ønite length is morphic if either (1) or (2) holds:
  (1) (a) Every submodule of M is isomorphic to an image of M; and
       (b) If $len(K) = len(K')$ where $K, K' \subseteq M$, then $M/K \cong M/K'$.
  (2) (c) Every image of M is isomorphic to a submodule of M; and
       (d) If $len(M/K) = len(M/K')$ where $K, K' \subseteq M$, then $K \cong K'$.

A module is called uniserial if its submodule lattice is a chain. Note that both (b) and (d) in Lemma 1.11 hold in a uniserial module of ønite length. The module $_\mathbb{Z}M = \mathbb{Z}_2 \oplus \mathbb{Z}_3$ is morphic and has ønite length, but (b) and (d) both fail for M.

Example 1.12. The $\mathbb{Z}$-module $M = \mathbb{Z}_2 \oplus \mathbb{Z}_4$ is a non-morphic module of length 8 in which both (a) and (c) hold (and so both (b) and (d) fail).

Proof. By the fundamental theorem of ønite abelian groups, the only images of M are M, $\mathbb{Z}_2 \oplus \mathbb{Z}_2$, $\mathbb{Z}_2$, $\mathbb{Z}_4$ and 0, each is isomorphic to a submodule, and these are the only submodules. However M is not morphic. In fact, if $K = \mathbb{Z}_2 \oplus 2\mathbb{Z}_4$ then and $N = \mathbb{Z}_2 \oplus 0$, then $M/K \cong \mathbb{Z}_2 \cong N$ but $M/N \cong \mathbb{Z}_4 \not\cong K$. $\square$

If $_RK$ and $_RN$ are morphic and $hom_R(K,N) = 0 = hom_R(N,K)$, it is easy to verify that $K \oplus N$ is morphic. It follows that every semisimple module $_RM$ of ønite length is morphic. In fact a semisimple module is morphic if and only if every homogeneous component has ønite length. We return this result in Proposition 2.5 below.

Example 1.13. [15] If the composition length of $_RM$ is at most 2, then M is morphic.

Theorem 1.14. [15] Every direct summand of a morphic module is again morphic.

On the other hand, Example 1.12 shows that the class of morphic modules is not closed under taking direct sums. This points to a diŒcult problem:

Question 3. When is the direct sum $_RK \oplus {_RN}$ morphic?

Question 4. When is $_RK \oplus {_RK}$ morphic?

The following necessary condition that $K \oplus N$ is morphic will be used several times.

Lemma 1.15.   [15] Let $M = K \oplus N$ be a morphic module. If $\lambda : K \to N$ is R-linear then
$$K \oplus (N/K\lambda) \cong ker(\lambda) \oplus N.$$
Hence:

(1) If λ is monic then $N \cong K \oplus (N/K\lambda)$.

(2) If λ is epic then $K \cong ker(\lambda) \oplus N$.

In particular, if K is isomorphic to either a submodule or an image of N, then K is isomorphic to a direct summand of N.

Note that Lemma 1.15 gives immediately that $\mathbb{Z}_2 \oplus \mathbb{Z}_4$ is not morphic because $\mathbb{Z}_2$ is isomorphic to a submodule of $\mathbb{Z}_4$, but it is not a summand.

It is useful to reformulate Lemma 1.15 as follows: If $M = K \oplus N$ is morphic, $X \subseteq K$ and $Y \subseteq N$, and then

$$\text{If} \quad K/X \cong Y \quad \text{then} \quad K \oplus (N/Y) \cong X \oplus N.$$

In fact, if $\sigma : K/X \to Y$ is an isomorphism and we deøne $\lambda : K \to N$ by $k\lambda = (k + x)\sigma$, then $K\lambda = Y$ and $ker(\lambda) = X$, so Lemma 1.15 applies.

While every ønite length semisimple module is morphic, the uniserial case is more interesting.

Example 1.16.   [15] If $R = \left\{ \begin{bmatrix} a & b & c \\ 0 & a & d \\ 0 & 0 & a \end{bmatrix} \mid a, b, c \in F \right\}, M = \begin{bmatrix} F \\ F \\ F \end{bmatrix}, P = \begin{bmatrix} F \\ F \\ 0 \end{bmatrix}, \text{and } Q = \begin{bmatrix} F \\ 0 \\ 0 \end{bmatrix},$
where F is a øeld, then M is a non-morphic module M with submodule lattice $0 \subset Q \subset P \subset M$.

Thus $\mathbb{Z}_8$ and the module M in Example 1.16 are uniserial modules with isomorphic submodule lattices, but $\mathbb{Z}_8$ is morphic while M is not. The Prfer group $\mathbb{Z}_{p^\infty}$ is uniserial, injective and artinian but it is not morphic by Theorem 1.9. Thus the injective hull of the (simple) morphic module $\mathbb{Z}_p$ is not morphic. A ring is called left duo if every left ideal is two-sided.

Proposition 1.17.   [15] Let $_R M$ be a uniserial module of ønite length.

(1) If every submodule of M is an image of M then M is morphic.

(2) In particular, M is morphic if $M = Rm$ where $1(m)$ is an ideal of R.

(3) Hence every uniserial left module of ønite length over a left duo ring is morphic.

The converse of (2) in Proposition 1.17 is false: Take $M = R/L$ and $m = 1 + L$, where L is a maximal left ideal of R that is not an ideal. The converse of (1) is also false:

Example 1.18.   [15] If D is a division ring and $R = \begin{bmatrix} D & D \\ 0 & D \end{bmatrix}$, let $M = \begin{bmatrix} 0 & D \\ 0 & D \end{bmatrix}$. Then $_R M$ is uniserial of length 2, but not every submodule is an image.

It would be interesting to see an example of a non-morphic module of ønite length in which every submodule is isomorphic to an image and every image is isomorphic to a submodule. However, if the module is morphic (not necessarily of ønite length), these two conditions are equivalent.

Theorem 1.19.   [15] The following are equivalent for a morphic module $_R M$:

(1) Every submodule of M is isomorphic to an image of M.

(2) Every image of M is isomorphic to a submodule of M.

In this case, the following hold:

(a) If N and N' are submodules of M then $M/N \cong M/N'$ if and only if $N \cong N'$.

(b) M is ønitely generated if and only if M is noetherian.

A left morphic ring R satisøes conditions (1) and (2) in Theorem 1.19 if and only if every left ideal is principal. Accordingly, we call a module P-morphic if it is morphic and satisøes conditions (1) and (2) in Theorem 1.19. A ring R is left P-morphic if $_R R$ is a P-morphic module; these rings are left noetherian by (b) in Theorem 1.19.

A semisimple module is P-morphic if and only if it is morphic. The morphic $\mathbb{Z}$-module $\mathbb{Z}_n$ is P-morphic since $\mathbb{Z}_n$ has a subgroup of every order dividing n. The module M in Example 1.18 is morphic and noetherian but not P-morphic.

A ring R is called left Kasch if every simple left module embeds in $_R R$.

Corollary 1.20.   [15] The following are equivalent for a ring R :
   (1) R is left P-morphic.
   (2) R is left morphic and every principal left R-module embeds in $_RR$.
   (3) R is left morphic and each left ideal has the form $L = \mathbf{1}(a)$ for some $a \in R$.
In this case R is left Kasch.

Thus semisimple artinian rings are left and right P-morphic, as are the rings $\mathbb{Z}_{p^n}$, p a prime. A product $R = \Pi_{i-1}^n R_i$ of rings is left P-morphic if and only if each R_i is left P-morphic. Example 3.9 is a commutative, morphic, left Kasch ring that is not P-morphic.
   The class of P-morphic modules is not closed under taking direct sums ($\mathbb{Z}_2 \oplus \mathbb{Z}_4$). However:

Theorem 1.21.   [15] Every direct summand of a P-morphic module is again P-morphic.

2. ENDOMORPHISM RINGS

It is natural to enquire into the relationship between when $_RM$ is a morphic module and when $E = end(_RM)$ is a left morphic ring. The result gives information about both objects.
   A module $_RM$ will be called image-projective if, whenever $M\gamma \subseteq M\alpha$ where $\alpha, \gamma \in E = end(M)$, then $\gamma \in E\alpha$, that is if the map δ exists in the diagram

when α and γ are given. Hence every quasi-projective module is image-projective. In a dioeerent direction, [17, Proposition 5.18] shows that $_RM$ is image-projective if $E = end(M)$ is right P-injective, and that the converse holds if M cogenerates $M/M\beta$ for every $\beta \in E$.
   We say that M generates its kernels if M generates $ker(\beta)$ for each $\beta \in E$, that is $ker(\beta) = \Sigma\{M\lambda \mid \lambda \in E,\ \lambda\beta = 0\}$.

Lemma 2.1.   [15] Let $_RM$ be a module and write $E = end(_RM)$.
   (1) If E is left morphic then M is image-projective.
   (2) If M is morphic and image-projective, then E is left morphic.
   (3) If M is morphic then it generates its kernels.
   (4) If E is left morphic and M generates its kernels, then M is morphic.

Combining these we get a characterization of the image-projective, morphic modules.

Theorem 2.2.   [15] The following are equivalent for a module $_RM$:
   (1) $_RM$ is morphic and image-projective.
   (2) $end(_RM)$ is left morphic and $_RM$ generates its kernels.

Corollary 2.3.   Let $_RM$ be a module and assume that $E = end(M)$ is regular. Then M is morphic and image-projective if and only if E is unit regular.

Theorem 2.4.   [15] Let R be a ring.
   (1) If $n \geq 1$, $M_n(R)$ is left morphic if and only if $_RR^n$ is morphic.
   (2) If R is left morphic and $e^2 = e \in R$ then eRe is left morphic.

Question 5. If R is left P-morphic and $e^2 = e \in R$, is eRe is left P-morphic? What if $ReR = R$?

Note that, in Question 5, Re is a P-morphic module by Theorem 1.21, and that in [13, Lemma 14] it is proved that $a \in eRe$ is left morphic in eRe if and only if $a + (1 - e)$ is left morphic in R.
   Recall that $_RM$ is morphic if $end(M)$ is unit regular (Proposition 1.7); we now describe several situations when the converse holds.

Proposition 2.5.   [15] The following are equivalent for a semisimple module M :
  (1) M is morphic.
  (2) $end(M)$ is unit regular.
  (3) Each homogeneous component of M is artinian.
In this case $end(M)$ is a direct product of matrix rings over division rings.

Note that, as M is semisimple, we can replace ɩmorphicɟ by ɩP-morphicɟ in (1) of Proposition 2.5.

Proposition 2.6.   [15] A ring R is semisimple artinian if and only if every ønitely generated (respectively every 2-generated) left module is morphic.

Zelmanowitz [18] calls a module ${}_R M$ regular if for any $m \in M$ there exists $\lambda \in hom_R(M, R)$ such that $(m\lambda)m = m$. In this case, if we write $e = m\lambda$, then $e^2 = e$, $\lambda : Rm \to Re$ is an isomorphism (so Rm is projective), and $M = Rm \oplus W$ where $w = \{w \in M \mid (w\lambda)m = 0\}$. Zelmanowitz proves [18, Theorem 1.6] that every ønitely generated submodule of a regular module M is a projective direct summand of M. Our interest lies in a larger class of modules wherein $Rm \subseteq^{\oplus} M$ for each $m \in M$ (equivalently [18, Corollary 1.3] if every ønitely generated submodule is a summand).

Corollary 2.7.   Assume that $Rm \subseteq^{\oplus} M$ for all $m \in {}_R M$ (for example if M is regular).
  (1) M is morphic and image-projective if and only if $end(M)$ is left morphic.
  (2) If M is ønitely generated then M is morphic if and only if $end(M)$ is unit regular.
  (3) In particular, every ønite-dimensional regular module is morphic.

Corollary 2.8.   Let M be a ønitely generated module over a commutative ring. Then M is regular and morphic if and only if M is projective and $end(M)$ is unit regular.

One situation when a module M generates its kernels is when $ker(\alpha) \subseteq^{\oplus} M$ for every $\alpha \in end(M)$. We say that M is kernel-direct in this case, and call M image-direct if $im(\alpha) \subseteq^{\oplus} M$ for each $\alpha \in end(M)$. Modules with a regular endomorphism ring (and hence all semisimple modules) enjoy both properties. Note that, by Lemma 1.1, a morphic module is kernel direct if and only if it is image direct.

Lemma 2.9.   [15] Every kernel-direct module is image-projective.

Since kernel-direct modules generate their kernels, Theorem 2.2 gives

Corollary 2.10.   If M is kernel-direct then M is morphic if and only if $end(M)$ is left morphic.

Theorem 2.11.   [15] The following are equivalent for a module M :
  (1) $end(M)$ is unit regular.
  (2) M is morphic and kernel-direct.
  (3) M is morphic and image-direct.

If R is a ring then ${}_R R$ is image direct if and only if R is regular, so Theorem 2.11 shows again that the unit regular rings are just the regular, left morphic rings. On the other hand, ${}_R R$ is kernel-direct if and only if $1(a) \subseteq^{\oplus} {}_R R$ for all $a \in R$, that is if and only if every principal left ideal Ra is projective. These are called left PP rings, and Theorem 2.11 gives

Corollary 2.12.   A ring R is unit regular if and only if it is a left morphic, left PP ring.

Corollary 2.13.   The following are equivalent for a ønite dimensional module M :
  (1) M is morphic and kernel-direct.
  (2) M is morphic and image-direct.
  (3) $end(M)$ is semisimple artinian.

Lemma 2.14. [17, Proposition 5.18] Let $_RM$ be a module with $E = end(M)$.
  (1) If E is right P-injective then M is right image-projective.
  (2) The converse holds if M cogenerates $M/M\beta$ for each $\beta \in E$.

Theorem 2.15. The following are equivalent for a ring R :
  (1) Every left module is image-projective.
  (2) Every 2-generated left module is image-projective.
  (3) R is semisimple artinian.

Proof. (3)$\Rightarrow$(1)$\Rightarrow$(2) are clear. Given (2), let $L \subseteq^{max}$ $_RR$ and let $\theta : R \to R/L \to 0$ be the coset map. Then θ splits by Lemma 2.16 below, so $L \subseteq^{\oplus}$ $_RR$, and (3) follows. $\qquad\square$

Lemma 2.16. Let $P \xrightarrow{\theta} M \to 0$ be epic. If $P \oplus M$ is image-projective then θ splits.

Proof. Let $\sigma_P, \pi_P, \sigma_M$ and π_M be canonical for $P \oplus M$.

Then $\pi_P\theta : P \oplus M \to M$ is epic so there exists
$\lambda : P \oplus M \to P \oplus M$ such that $\lambda\pi_P\theta = \pi_M$.
Deøne $\phi = \sigma_M\lambda\pi_P : M \to P$. Then

$\phi\theta = \sigma_M\lambda\pi_P\theta = \sigma_M\pi_M = 1_M$, so $P = ker(\theta) \oplus M\phi$. $\qquad\square$

We conclude this section with a look at when $end(_RM)$ is right morphic. We call a module $_RM$ image-injective if R-linear maps $M\beta \to M$ extend to M for each $\beta \in end(_RM)$, and we say that M cogenerates its cokernels if it cogenerates $M/M\beta$ for each $\beta \in end(_RM)$. Note that $_RR$ is image-injective if and only if R is left P-injective, and $_RR$ cogenerates its cokernels if and only if R is right P-injective. With this, we can obtain ıdualj versions of Lemma 2.1 and Theorem 2.2.

Lemma 2.17. [15] Let $_RM$ be a module and write $E = end(_RM)$.
  (1) If E is right morphic then M is image-injective.
  (2) If M is morphic and image-injective, then E is right morphic.
  (3) If M is morphic then it cogenerates its cokernels.
  (4) If E is right morphic and M cogenerates its cokernels, then M is morphic.

Theorem 2.18. [15] The following are equivalent for a module $_RM$.
  (1) M is morphic and image-injective.
  (2) $end(M)$ is right morphic and M cogenerates its cokernels.

If R is left and right P-injective and we take $M = {}_RR$ then this shows (again) that R is left morphic if and only if R is right morphic. Note ønally that the ıdualj of Lemma 2.9 (every kernel-direct module is image-projective) is true: Every image-direct module is clearly image-injective.

3. LEFT SPECIAL RINGS

We begin with an example of Bjrk [2] (see [17, Example 2.5]).

Example 3.1. Let F be a øeld with an isomorphism $x \mapsto \bar{x}$ from F to a subøeld $\bar{F} \neq F$. Let R denote the left F-space on basis $\{1, c\}$ where $c^2 = 0$ and $cx = \bar{x}c$ for all $x \in F$. Then R is a left artinian, local, left P-morphic ring that is not right morphic. Moreover, if $dim(_{\bar{F}}F) < \infty$ then R is right artinian (for example, if $F = \mathbb{Z}_p(x)$ and $\bar{w} = w^p$, p a prime).

The ring in Example 3.1 turns out to be a prototype for all local, left morphic rings with nilpotent Jacobson radical. We need a technical lemma about local rings.

Lemma 3.2. [14] Let R denote a local ring in which $J = Rc$ for $c \in R$. Then:
 (1) $J^m = Rc^m$ for every $m \geq 0$.
 (2) If $Rc^{m+1} \subset Rc^m$ then $Rc^m - Rc^{m+1} = Uc^m$ for every $m \geq 0$.
 (3) If L is a left ideal and $L \not\subseteq \cap_{n \geq 0} J^n$, then $L = J^m$ for some $m \geq 0$.

Theorem 3.3. [13] The following conditions are equivalent for a ring R :
 (1) R is left morphic, local and J is nilpotent.
 (2) R is local and $J = Rc$ for some $c \in R$ with $c^n = 0$, $n \geq 1$.
 (3) There exists $c \in R$ and $n \geq 1$ such that $c^{n-1} \neq 0$ and $R \supset Rc \supset Rc^2 \supset \cdots \supset Rc^n = 0$
     are the only left ideals of R.
 (4) R is left uniserial of ønite composition length.
 (5) There exists $c \in R$ such that $c^n = 0$, $n \geq 1$, and $R = \{uc^k \mid k \geq 0,\ u \in U\}$.
If c is as in (3) then:
 (a) $1(c^k) = Rc^{n-k}$ and $Rc^k - Rc^{k+1} = Uc^k$ for $0 \leq k < n$.
 (b) $soc(_R R) = Rc^{n-1}$ is simple and essential in $_R R$.
 (c) $Rc^k = J^k$ for $0 \leq k \leq n$.

We refer to the rings in Theorem 3.3 as left special rings. These rings are all left P-morphic. Note that the left special rings with $J = 0$ are just the division rings, and the ring in Example 3.1 is left special of left composition length 2. If p is a prime, the ring $\mathbb{Z}_{p^n}$ is left and right special for every $n \geq 1$. Note that every left special ring R is a left duo ring. However if F is a øeld then $M_2(F)$ is a left and right morphic ring (it is unit regular), but is neither left nor right duo.

Corollary 3.4. Let R be left special with $J = Rc$ as in part (2) of Theorem 3.3. If R is also right special, then $J = cR$ (and so the left-right analogues of the properties in Theorem 3.3 hold).

Example 3.5. [13] The ring R in Example 3.1 is left special but not right special.

Every left P-morphic ring is left Kasch by Corollary 1.20; however Example 3.1 is left and right Kasch but not right morphic.

Proposition 3.6. [13] The following are equivalent for a left morphic ring R :
 (1) R is left Kasch.
 (2) Every maximal left ideal of R is an annihilator.
 (3) Every maximal left ideal of R is principal.

Question 6. If a ring R is left morphic and left Kasch, is R right Kasch?

To characterize the left special rings among the left P-morphic rings, we need:

Lemma 3.7. [14] Let R be a local left morphic ring with a simple left ideal, in which J is not nilpotent. If $Ra \subseteq R$ is simple choose $c \in R$ such that $Rc = 1(a)$ and $1(c) = Ra$. Then $1(c^t) \subset 1(c^{t+1})$ for every $t \geq 0$.

With this we can characterize the local, left P-morphic rings.

Theorem 3.8. [14] The following are equivalent for a ring R :
 (1) R is local and left P-morphic.
 (2) R is local, left morphic, with a simple left ideal and ACC on left annihilators.
 (3) R is left special.

Example 3.9. Clark [6] gives an example of a commutative local ring R with ideal lattice

$$0 \subset Rv_1 \subset Rv_2 \subset \cdots \subset V \subset \cdots \subset Rc^2 \subset Rc \subset R.$$

This example is a morphic ring with exactly one non-principal ideal.

The details are complex and the reader is referred to [14, Theorem 18] where it is proved that a ring with such a left ideal lattice is left morphic if and only if $\mathbf{r}(J) = \mathbf{l}(J)$, equivalently if and only if $soc(_R R) = soc(R_R)$. Moreover, in [14, Theorem 23] it is shown that if R is a local, left morphic ring in which $S_r \neq 0$ and J is not nilpotent, and if R contains a unique non-principal left ideal, then there exists $c \in R$ such that the left ideal lattice is $0 \subset \mathbf{l}(c) \subset \mathbf{l}(c^2) \subset \cdots \subset V \subset \cdots \subset Rc^2 \subset Rc \subset R$.

We saw in Theorem 2.15 that a ring R is semisimple artinian if and only if every (every 2-generated) left module is morphic.

Theorem 3.10. [15] If R is left special then every principal left module is morphic.

Question 7. For which rings is every principal left module morphic?

We conclude this section with some examples due to Chen and Zhou [5].

Example 3.11. Let $R \propto M$ be the trivial extension of the ring R by the bimodule $_R M_R$.
(1) If R is a PID with ring Q of fractions, then the trivial extension $R \propto (Q/R)$ is morphic. Moreover, if $\mathbb{Z} \propto M$ is morphic then $M \cong \mathbb{Q}/\mathbb{Z}$.
(2) If $n = dm > 0$ in $\mathbb{Z}$ where $d > 1$ then $\mathbb{Z}_n \propto \mathbb{Z}_d$ is morphic if and only if d and m are relatively prime and d is square-free.

4. MATRIX RINGS

If R is a ring we know (Theorem 2.4) that:
(1) If $n \geq 1$, $M_n(R)$ is left morphic if and only if $_R R^n$ is morphic.
(2) If R is left morphic and $e^2 = e \in R$ then eRe is left morphic.

If R^2 is left morphic then R is left morphic by Theorem 1.14, but the converse is not true.

Example 4.1. [13] If R is the ring in Example 3.1 then R is left special but $M_2(R)$ is not left morphic. Hence neither being left morphic nor being left P-morphic are Morita invariants.

Question 8. When is $M_n(R)$ left morphic (left P-morphic)?

The next result identiøes an important situation where $M_n(R)$ is left and right morphic.

Theorem 4.2. [13] Let R be a left and right special ring. Then $M_n(R)$ is left and right morphic for each $n \geq 1$.

Question 9. If R is left and right morphic, is the same true of $M_2(R)$?

This is true if R is unit regular [11, Corollary 3], but see Example 4.1. With Theorem 2.4, Question 9 asks whether ıleft and right morphicȷ is a Morita invariant?
The next result extends Theorem 2.4 to the case of left P-morphic rings.

Theorem 4.3. [15] Let R be a ring. Then $M_n(R)$ is left P-morphic if and only if $_R R^n$ is P-morphic.

As we have seen, the property of being left morphic (or being left P-morphic) does not pass to matrix rings. In fact, Example 3.1 exhibits a left and right artinian, left P-morphic ring R such that $M_2(R)$ is not left morphic. Accordingly, the following classes of rings are of interest. A ring R is called strongly left morphic (respectively strongly left P-morphic) if every matrix ring $M_n(R)$ is left morphic (respectively left P-morphic). The left and right special rings are all strongly left and right P-morphic by Theorem 4.2. Note that Example 3.1 is a left special ring R for which $M_2(R)$ is not left morphic. Chen and Zhou [5, Theorem 7] show that, if R is semisimple, the trivial extension $R \propto R$ is strongly left and right morphic. Every unit regular ring is strongly left and right morphic (unit regularity is a Morita invariant by [11, Corollary 3]).

Question 10. If a ring R is strongly left and right morphic and $J(R) = 0$, is R unit regular?

Theorem 4.4. [15] The following are equivalent for a ring R :
  (1) R is strongly left morphic (respectively strongly left P-morphic).
  (2) $_R R^n$ is morphic (respectively P-morphic) for each $n \geq 1$.
  (3) Every ønitely generated projective left R-module is morphic (respectively P-morphic).

Theorem 4.5. [15] If R is strongly left morphic the same is true of eRe for any idempotent $e \in R$.

We do not know if Theorem 4.5 holds for strongly left P-morphic rings because we do not know if the left P-morphic property passes from R to eRe, $e^2 = e$, even if $ReR = R$.

Theorem 4.6. [15] Being strongly left morphic is a Morita invariant.

Proposition 4.7. Direct products of strongly left morphic rings, and ønite direct products of strongly left P-morphic rings, are again of the same type.

A ring R is said to be stably ønite if $M_n(R)$ is directly ønite for every $n \geq 1$. Hence Corollary 1.2 gives:

Proposition 4.8. Every strongly left morphic ring is stably ønite.

Question 11. If $M_2(R)$ is left morphic, is R strongly left morphic?

A ring R is called right FP-injective if every R-morphism from a ønitely generated submodule of a free right R-module F to R extends to F. Every strongly left morphic ring R is right FP-injective by [17, Theorem 5.41] because every left morphic ring is right P-injective by Corollary 1.5.

Example 4.9. [15] or [17, Example 2.6] There exists a commutative, local, FP-injective ring R with $J^3 = 0$ and J^2 simple and essential in R, but which is not morphic. In fact, $R = F[x_1, x_2, \cdots]$ where F is a øeld and the x_i are commuting indeterminants satisfying the relations $x_i^3 = 0$ for all i, $x_i x_j = 0$ for all $i \neq j$, and $x_i^2 = x_j^2$ for all i and j.

5. STRUCTURE THEOREMS

We begin with a result that gives insight into when a matrix ring is left morphic. Recall that a Morita context is a four-tuple (R, V, W, S) where R and S are rings, and $V = {}_R V_S$ and $W = {}_S W_R$ are bimodules with multiplications $V \times W \to R$ and $W \times V \to S$ such that $C = \begin{bmatrix} R & V \\ W & S \end{bmatrix}$ is an associative ring matrix operations (the context ring).

Proposition 5.1. [13] Let $C = \begin{bmatrix} R & V \\ W & S \end{bmatrix}$ be a context ring and assume that C is left morphic. If either $VW \subseteq J(R)$ or $WV \subseteq J(S)$, then $V = 0$ and $W = 0$.

An idempotent e in a ring R is called local if eRe is a local ring, and e is called full (in R) if $ReR = R$.

Corollary 5.2. [13] Let e and f be idempotents in a left morphic ring R.
  (1) If e and f are orthogonal and $eRf \subseteq J$ then $eRf = 0 = fRe$.
  (2) e is central if and only if $eR(1 - e) = 0$.
  (3) If $e^2 = e \in R$ is local, then $1 - e$ is either full or central.

Theorem 5.3. [13] Let e and f be idempotents in the left morphic ring R. Then:
  (1) $eRf = 0$ if and only if $fRe = 0$.
  (2) If e and f are orthogonal and local, then $eRf \neq 0$ if and only if $eR \cong fR$.

If mild øniteness conditions are applied to a left morphic ring, we obtain some structure results. To begin, Theorem 5.3 leads to the following theorem in the semiperfect case.

Theorem 5.4.   [13] A ring R is semiperfect and left morphic if and only if

$$R \cong M_{n_1}(R_1) \times M_{n_2}(R_2) \times \cdots \times M_{n_k}(R_k)$$

where each $M_{n_i}(R_i)$ is left morphic and $R_i \cong e_i R e_i$ for some local idempotent $e_i \in R$.

We hasten to note that $M_n(R)$ need not be left morphic even if R is left special as Example 4.1 shows. What we want in Theorem 5.4 is a condition such that R is semiperfect and left morphic if and only if $R \cong M_{n_1}(R_1) \times M_{n_2}(R_2) \times \cdots \times M_{n_k}(R_k)$ where each R_i is local, left morphic and satisøes the condition.

Question 12. If R is local and left morphic, when is $M_2(R)$ left morphic?

We do get a better theorem for semiprimary, left and right morphic rings.

Corollary 5.5.   [13] The following are equivalent for a ring R :
   (1) R is a semiprimary ring that is left and right morphic.
   (2) $R \cong M_{n_1}(R_1) \times M_{n_2}(R_2) \times \cdots \times M_{n_k}(R_k)$ where each R_i is left and right special.

For convenience, the rings in Corollary 5.5 are called semispecial. Recall that a ring R is called right selønjective if every R-linear map $\gamma : T \to R_R$, T a right ideal of R, extends to $R_R \to R_R$, equivalently if $\gamma = c\cdot$ is left multiplication by some $c \in R$. A left and right selønjective ring R is called quasi-Frobenius if it is left and right artinian.

Proposition 5.6.   [15] Every semispecial ring R is quasi-Frobenius.

The converse to Proposition 5.6 is false.

Example 5.7.   [13] If C_2 denotes the group of order 2, the group ring $R = \mathbb{Z}_4 C_2$ is a commutative, local quasi-Frobenius ring which is not morphic.

We return to these semispecial rings later.

Theorem 5.8.   [15] A ring R is strongly left morphic and semiperfect if and only if R is a ønite product of matrix rings over local, strongly left morphic rings.

The next result is part of the proof of [14, Theorem 13].

Lemma 5.9.   If $R = M_n(S)$ is left P-morphic and S is local then S is left special.

Recall that a ring R is called an exchange ring if $_R R$ (equivalently R_R) has the ønite exchange property. This is a large class of rings, containing every semiregular ring R (that is, R/J is regular and idempotents can be lifted modulo J). However, we have

Theorem 5.10.   [14] The following conditions are equivalent for a left P-morphic ring R :
   (1) R is an exchange ring.
   (2) R is a semiperfect ring.
   (3) $R \cong \Pi_{i=1}^{k} M_{n_i}(S_i)$ where each S_i is left special.
   (4) R is left artinian.

The semispecial rings in Corollary 5.5 are all left and right artinian (this is true of left and right special rings), and we present several characterizations of these rings below. This entails an examination of the eœect on a left morphic ring of various øniteness conditions. We begin with the ascending chain condition on right annihilators.

Theorem 5.11. [13] Let R be a left morphic ring with ACC on right annihilators. Then:
(1) eRe is left special for every local idempotent $e \in R$.
(2) R is left artinian.
(3) R is right and left Kasch.
(4) $soc(R_R) = soc(_RR)$.
(5) $Z(_RR) = J = Z(R_R)$.

Note that every left special ring is left duo and satisøes the ACC on right annihilators (it is left artinian). Hence Theorem 3.3 gives:

Corollary 5.12. A left duo, left morphic ring has ACC on right annihilators if and only if it is a ønite direct product of special left morphic rings.

The converse to Theorem 5.11 is not true. In fact if R is the ring in Example 3.1 then $M_2(R)$ enjoys properties $(1)^{\smile}(5)$ in Theorem 5.11 but it is not left morphic by Example 4.1.

The ring R in Example 3.1 is left artinian and left P-morphic but $M_2(R)$ is not left morphic by Example 4.1. Hence the rings identiøed in Theorem 5.10 do not form a Morita invariant class. However, being left and right P-morphic is a Morita invariant property, and we now determine the structure of these rings. The following result will be needed and is of interest in itself.

Theorem 5.13. [14], see also [9] Let R be a left and right special ring. If $0 \neq {}_RM \subseteq R^n$ then M is a direct sum of at most n principal submodules.

Note that the ring in Example 3.1 is left special but not left selønjective (not even left P-injective). Moreover, by Corollary 1.5 a left morphic ring R is left selønjective if and only if it is left P-injective, if and only if it is right morphic.

We can now give the main structure theorem for left and right P-morphic rings.

Theorem 5.14. [14] A ring R is left and right P-morphic if and only if it is semispecial.

The proof of Theorem 5.16 below requires the following lemma.

Lemma 5.15. [13] The following are equivalent for a semiperfect, left morphic ring R :
(1) J is nilpotent.
(2) J is nil and $soc(R_R) \subseteq^{ess} R_R$.
(3) R has ACC on principal left ideals and $soc(R_R) \subseteq^{ess} R_R$.

We can now prove a structure theorem for left perfect, left and right morphic rings.

Theorem 5.16. [13] The following are equivalent for a ring R :
(1) R is left artinian and left and right morphic.
(2) R is semiprimary and left and right morphic.
(3) R is left perfect and left and right morphic.
(4) R is a semiperfect, left and right morphic ring in which J is nil and $soc(R_R) \subseteq^{ess} R_R$.
(5) R is a semiperfect, left and right morphic ring with ACC on principal left ideals in which $soc(R_R) \subseteq^{ess} R_R$.
(6) R is semispecial.

Corollary 5.17. Being semispecial is a Morita invariant. In addition, if R is semispecial the same is true of eRe for any idempotent $e \in R$.

6. INTERNAL CANCELLATION

A module $_RM$ is said to have internal cancellation (IC) if, whenever $M = N \oplus K = N_1 \oplus K_1$ and $N \cong N_1$, it follows that $K \cong K_1$. Each indecomposable module M has IC, and we have

Proposition 6.1. [15] Every direct summand of an IC module has IC.

We say that a ring R has left internal cancellation (left IC) if $_RR$ has IC. This holds if and only if $Re \cong Rf$, $e^2 = e$, $f^2 = f$, implies that $R(1 - e) \cong R(1 - f)$. In this case, we have $f = u^{-1}eu$ for some unit $u \in R$.

If $\pi^2 = \pi$ and $\tau^2 = \tau$ in $E = end(_RM)$, it is routine to verify that $M\pi \cong M\tau$ as R-modules if and only if $E\pi \cong E\tau$ as left E-ideals. It follows that $_RM$ has IC if and only if $E = end(_RM)$ has left IC. Hence Proposition 6.1 gives

Corollary 6.2. If R has left IC then eRe has left IC for every idempotent $e \in R$.

Goodearl [10] shows that for a module M with $end(M)$ regular, internal cancellation is equivalent to $end(M)$ being unit regular. In fact

Theorem 6.3. [15] A module $_RM$ has IC if and only if every regular element in $end(_RM)$ is morphic.

Corollary 6.4. Every morphic module has IC.

The converse to Corollary 6.4 is false: Every local ring has left (and right) IC, but need not be left morphic. In fact the localization $\mathbb{Z}_{(p)}$ of the integers at the prime p is a counterexample that is a local integral domain. Indeed, Example 4.9 shows that the counterexample can actually be chosen to be commutative and P-injective. For an artinian example, the $\mathbb{Z}$-module $\mathbb{Z}_2 \oplus \mathbb{Z}_4$ can be veriøed to have IC but is not morphic by Example 1.12.

Corollary 6.5. Given $_RM$, $end(M)$ is unit regular if and only if M has IC and $end(M)$ is regular.

The next result extends Proposition 2.5.

Corollary 6.6. A semisimple module M is morphic if and only if it has IC.

Proof. If M has IC, let N and K be submodules with $M/K \cong N$. Since M is semisimple let $M = K \oplus K' = N \oplus N'$. Then $N \cong M/K \cong K'$ so, because M has IC, $K \cong N' \cong M/N$. Hence M is morphic. The converse is by Corollary 6.5 because M is semisimple. $\square$

A ring R is said to have stable range 1 if $aR + bR = R$ implies that $a + bt$ is a unit in R for some t. Evans [8] showed that if $end(M)$ has stable range 1 then M is cancellable in the sense that $M \oplus A \cong M \oplus B$ implies $A \cong B$. Camillo and Yu [4, Theorem 3] show that an exchange ring R has stable range 1 if and only if every regular element of R is unit regular (extending the same result of Kaplansky in the regular case).

Corollary 6.7. Every injective, morphic module is cancellable.

Proof. Mohamed and Mller [12, Theorem 1.29] show that an injective module is cancellable if and only if it is directly ønite. $\square$

Corollary 6.8. If M is morphic with the ønite exchange property then M is cancellable.

Proof. Mohamed and Mller [12, Proposition 1.23] show that if M has the ønite exchange property, then M is cancellable if and only if M has IC. Now use Proposition 6.6. $\square$

Acknowledgement: This research was supported by NSERC Grant A8075.

180

References

[1] G. Azumaya, On generalized semi-primary rings and Krull-Remak-Schmidt's theorem, Japan J. Math. 19 (1960), 525-547.

[2] J.-E. Bjrk, Rings satisfying certain chain conditions, J. Reine Angew. Math. 245 (1970), 63-73.

[3] V. Camillo and H.-P. Yu, Exchange rings, units and idempotents, Comm. in Algebra 22 (1994), 4737-4749.

[4] V. Camillo and H.-P. Yu, Stable range 1 for rings with many idempotents, Trans. A.M.S. 347 (1995), 3141-3147.

[5] J. Chen and Y. Zhou, Morphic rings as trivial extensions, to appear in Glasgow M. J.

[6] J. Clark, On a question of Faith in commutative endomorphism rings, Proc. A.M.S. 98 (1986), 196-198.

[7] G. Erlich, Units and one-sided units in regular rings, Trans. A.M.S. 216 (1976), 81-90.

[8] E.G. Evans, Krull-Schmidt and cancellation over local rings, Paciøc J. Math. 46 (1973), 115-121.

[9] A. Facchini, ıModule Theoryȷ, Progress in Mathematics, Volume 167. Birkhuser, Basel, 1998.

[10] K.R. Goodearl, ıVon Neumann Regular Ringsȷ, Second Edition. Krieger, Malabar, Florida, 1991.

[11] D. Handelman, Perspectivity and cancellation in regular rings, J. Algebra 48 (1977), 1-16.

[12] S.H. Mohamed and B.J. Mller, ıContinuous and Discrete Modulesȷ, London Mathematical Society Lecture Notes 147. Cambridge, 1990.

[13] W.K. Nicholson and E. Snchez Campos, Rings with the dual of the isomorphism theorem, J. Algebra 271 (2004), 391-406.

[14] W.K. Nicholson and E. Snchez Campos, Principal rings with the dual of the isomorphism theorem, Glasgow M. J. 46 (2004), 181-191.

[15] W.K. Nicholson and E. Snchez Campos, Morphic modules, to appear.

[16] W.K. Nicholson and M.F. Yousif, Principally injective rings, J. Algebra 174 (1995), 77-93.

[17] W.K. Nicholson and M.F. Yousif, ıQuasi-Frobenius Ringsȷ, Cambridge Tracts No. 158. Cambridge University Press, London, New York, 2003.

[18] J. Zelmanowitz, Regular modules, Trans. A.M.S. 163 (1972), 341-355.

CLEAN RINGS: A SURVEY

W. KEITH NICHOLSON*

Department of Mathematics and Statistics
University of Calgary
Calgary T2N 1N4, Canada
E-mail: wknichol@ucalgary.ca

YIQIANG ZHOU[†]

Department of Mathematics and Statistics
Memorial University of Newfoundland
St.John's A1C 5S7, Canada
E-mail: zhou@math.mun.ca

A ring is called clean if each element is the sum of a unit and an idempotent. All semiperfect and unit regular rings are clean, and all clean rings are exchange rings. This survey contains a current account of the various results known about clean rings.

Rings will be associative with identity unless specified otherwise. Certainly the units and idempotents of a ring are key elements determining the structure of the ring. A ring R is unit regular if, for any $a \in R$, $a = aua$ for a unit u in R, equivalently $a = eu$ for some idempotent e and unit u [if $a = aua$ then $a = eu^{-1}$ where $e = au$; if $a = eu$ then $a = au^{-1}a$]. The "sum" analog of the above condition is the notion of a clean ring. An element of a ring is called clean if it is the sum of an idempotent and a unit. A ring R is called clean if every element of R is clean. This notion was introduced by Nicholson [24] in 1977 in a study of exchange rings. Since then various results on this notion have been obtained. In this survey paper, we intend to bring out a up to date account of the study of this class of rings. We write $J(R)$ and $U(R)$ for the Jacobson radical and the group of units of R respectively. The left and right annihilators of an element $a \in R$ are

---

*Work partially supported by NSERC grant A8075

[†]Work partially supported by NSERC grant OGP0194196

182

1. Connections with other notions

Clean rings are closely connected to some important notions in ring theory.
The first is that of an exchange ring.

A module $_RM$ has the (full) exchange property if for every module $_RA$
and any two decompositions $A = M' \oplus N = \oplus_{i \in I} A_i$ with $M' \cong M$, there
exist submodules $A_i' \subseteq A_i$ such that $A = M' \oplus (\oplus_{i \in I} A_i')$. The module $_RM$
has the finite exchange property if the above condition is satisfied whenever
the index set I is finite. Warfield [41] called a ring R an exchange ring if
$_RR$ has the finite exchange property and showed that this definition is left-
right symmetric using a duality argument. A short, elementary proof of
the left-right symmetry of exchange rings is given by Nicholson [25]. The
first element-wise characterization of exchange rings was given by Monk
[23] which says that R is an exchange ring if and only if $\forall a \in R$, $\exists b, c \in R$
such that $bab = b$ and $c(1 - a)(1 - ba) = 1 - ba$. Independently, Goodearl
[18] and Nicholson [24] obtained the very useful characterization that R is
an exchange ring if and only if $\forall a \in R$, $\exists e^2 = e \in R$ such that $e \in aR$ and
$1 - e \in (1 - a)R$. Nicholson [24] also shows that R is an exchange ring if
and only if idempotents can be lifted modulo every left (equivalently, right)
ideal of R if and only if $R/J(R)$ is an exchange ring and idempotents can
be lifted modulo $J(R)$.

Theorem 1.1. *[24] Every clean ring R is an exchange ring; the converse
holds if all idempotents of R are central.*

As observed by Camillo and Yu [11], the ring in the next example constructed by Bergman (see [20, Example 1]) is an exchange ring which is not
clean.

Example 1.1. Let k be a field, and $A = k[[x]]$ the power series ring. Let
K be the field of fractions of A. Define $R = \{r \in end(A_k) : \exists q \in K$ and
$\exists n > 0$ with $r(a) = qa$ for all $a \in (x^n)\}$. Then R is an exchange ring but
not a clean ring.

The second part of Theorem 1.1 has been extended to a larger class of
rings by Yu [42] where it is proved that any exchange ring whose maximal
left (or right) ideals are two-sided ideals is a clean ring. Later, Chen [13]
proved that any exchange ring with artinian primitive factors (for example
an exchange ring satisfying a polynomial identity) is clean.

Theorem 1.2. *[11] A ring R is semiperfect if and only if R is a clean ring containing no infinite set of orthogonal idempotents.*

Theorem 1.3. *[10, 11] A ring R is unit regular if and only if every element a of R can be written as $a = e + u$ such that $aR \cap eR = 0$, where e is an idempotent and u is a unit in R.*

Question 1.1. *[10]* Which von Neumann regular rings are clean?

A ring is said to have the n-sum property if every element of the ring is the sum of n units.

Theorem 1.4. *[11] If R is a clean ring with $2 \in U(R)$, then every element of R is the sum of a unit and a square root of 1. In particular, R has the 2-sum property.*

2. Clean endomorphism rings

It is observed in [24, page 272] that the $n \times n$ matrix ring $M_n(R)$ is clean for any algebraically closed field R. Later, Camillo and Yu [11] proved that if $R/J(R)$ is a unit regular ring such that idempotents of $R/J(R)$ lift to idempotents of R, then $M_n(R)$ is clean. Lastly, Han and Nicholson [19] proved that $M_n(R)$ is clean for any clean ring R.

On the other hand, Ó Searcóid [35] showed that for any vector space V over a field F, the linear transformation ring $end_F V$ is clean; a result due to Nicholson and Varadarajan [27] states that the linear transformation ring $end_D V$ of a vector space V of countable infinite dimension over a division ring D is clean. The next theorem of Nicholson-Varadarajan-Zhou [28] extends the two results and answers affirmatively the question raised in [27] which asks whether the linear transformation ring of a vector space of arbitrary infinite dimension over a division ring is again clean.

Theorem 2.1. *[28] For any projective left module over a left perfect ring R, $end_R P$ is a clean ring.*

If $C(R)$ denotes the center of a ring R and $g(x)$ is a polynomial in $C(R)[x]$, we say that R is $g(x)$-clean if every element r of R has the form $r = s + u$ where $g(s) = 0$ and u is a unit. The $(x^2 - x)$-clean rings are precisely the clean rings. If V is a vector space of countable infinite dimension over a division ring D, Camillo and Simón [12] proved that $end_D V$ is $g(x)$-clean provided that $g(x)$ has two distinct roots in $C(D)$. Recently, this result has been extended as the following.

184

Theorem 2.2. *[31] Let R be a ring, let $_RM$ be a semisimple module over R, and write $C = C(R)$. If $g(x) \in (x - a)(x - b)C[x]$ where $a, b \in C$ are such that b and $b - a$ are both units in R, then $\mathrm{end}_R M$ is $g(x)$-clean.*

The following corollary extends a theorem of Camillo and Simón [12] who obtained the countable infinite dimensional case.

Corollary 2.1. *[31] Let $_DV$ be a vector space over a division ring D. If $g(x)$ is a polynomial in $C(D)[x]$ with at least two roots in $C(D)$, then $\mathrm{end}_D V$ is $g(x)$-clean.*

Corollary 2.2. *[31] If $_RM$ is a semisimple module over a ring R, then $\mathrm{end}(_RM)$ is clean.*

In February 2004, attending a talk on clean rings by the first author, Dr. Guil Asensio asked if every left self-injective ring is clean. The answer to this question is "Yes" by the next theorem. Consider the following conditions for a module M:

(C1) Every submodule of M is essential in a direct summand of M.

(C2) Every submodule that is isomorphic to a direct summand of M is itself a direct summand.

(C3) If N and K are direct summands of M with $N \cap K = 0$, then $N \oplus K$ is a direct summand of M.

Dually, there are following conditions:

(D1) For every submodule X of M, there exists a decomposition $M = A \oplus B$ such that $A \subseteq X$ and $X \cap B$ is small in M.

(D2) If $A \subseteq M$ such that M/A is isomorphic to a direct summand of M, then A is a direct summand of M.

(D3) If N and K are direct summands of M with $N + K = M$ then $N \cap K$ is a direct summand of M.

A module is called continuous if it satisfies both (C1) and (C2), and a module is called quasi-continuous if it satisfies (C1) and (C3). A module is called discrete if it satisfies (D1) and (D2), and a module is called quasi-discrete if it satisfies (D1) and (D3).

A module M is called pure-injective if for any module A and any pure submodule B of A, every homomorphism $f : B \to M$ extends to a homomorphism $g : A \to M$. A module M is called cotorsion if $\mathrm{Ext}_R^1(F, M) = 0$ for every flat R-module F. The ring R is called left cotorsion if $_RR$ is cotorsion.

The next result extends Theorem 2.1 and Corollary 2.2.

Theorem 2.3. *[33] If a module $_RM$ is continuous or discrete or pure-injective or flat cotorsion, then $\mathrm{end}(_RM)$ is clean.*

The next result extends the result of Camillo and Yu that every semiperfect ring is clean.

Corollary 2.3. *[33] If $R/J(R)$ is a left self-injective ring and idempotents lift modulo $J(R)$, then R is clean.*

Corollary 2.4. *[33] Every left cotorsion ring is clean.*

Theorem 2.4. *[33] Let $M = \oplus_{i\in I}M_i$ where each M_i is indecomposable. Then $\mathrm{end}(_RM)$ is clean if and only if each $\mathrm{end}(_RM_i)$ is local and the decompsition $M = \oplus_{i\in I}M_i$ complements direct summands.*

Theorem 2.5. *[33] Let R be a ring and $F = R^{(\mathbb{N})}$. The following are equivalent:*

(1) For every projective module $_RP$, $\mathrm{end}(_RP)$ is clean and $\mathrm{end}(_RP)/J(\mathrm{end}(_RP))$ is regular.

(2) $\mathrm{end}(_RF)$ is clean and $\mathrm{end}(_RF)/J(\mathrm{end}(_RF))$ is regular.

(3) $\mathrm{end}(_RF)$ is clean and R is semilocal.

(4) R is right perfect.

Corollary 2.5. *[33] Let R be a semilocal ring. The following are equivalent:*

(1) For every projective module $_RP$, $\mathrm{end}(_RP)$ is clean.

(2) $\mathrm{end}(_R(R^{(\mathbb{N})}))$ is clean.

(3) R is left perfect.

In [19], the authors proved that if R is a clean ring then so is the matrix ring $M_n(R)$, and they further asked whether the endomorphism ring $\mathrm{end}(_RF)$ of a countably generated free module $_RF$ over a clean ring R is again clean (see [19, Question 2]). Corollary 2.5 shows that the answer to this question is negative. Indeed, if R is a semiperfect ring which is not left perfect then R is clean by Theorem 1.2, but $\mathrm{end}(_R(R^{(\mathbb{N})}))$ is not clean by Corollary 2.5.

The following questions remain open:

Question 2.1. *[19] If R is a clean ring and $e^2 = e \in R$ with $ReR = R$, is the ring eRe clean?*

Question 2.2. *[27]* Is the ring of countably infinite, row and column finite matrices over a division ring clean? This is a question of Ara. This ring is exchange by O'Meara [34].

Question 2.3. Can the assumption that R is a semilocal ring be deleted from Corollary 2.5?

3. The center of a clean ring

It is interesting to know if the center of a ring shares the same property with the ring. The center of a regular ring is again regular [17]; but the center of an exchange ring need not be exchange [21]. So one raises the following:

Question 3.1. Is the center of a clean ring necessarily clean?

Since semiperfect rings are precisely those clean rings containing no infinite set of orthogonal idempotents, an affirmative answer to Question 3.1 will imply an affirmative answer to the next question:

Question 3.2. Is the center of a semiperfect ring necessarily semiperfect?

4. Strongly clean rings

We call an element a in a ring R strongly clean if $a = e + u$ where $e^2 = e$ and $u \in U(R)$ and $eu = ue$. The ring R is called a strongly clean ring if every element is strongly clean. Units are clearly strongly clean, as are idempotents $e = e^2$ (since $e = (2e - 1) + (1 - e)$) and elements a in the Jacobson radical ($a = (a - 1) + 1$). As an easy consequence all local rings are strongly clean. Strongly clean rings were introduced and studied by Nicholson [26]. The interest in this notion stems from its connection with strongly π-regular rings and hence its relationship to Fitting's lemma.

An element $a \in R$ is called right π-regular if it satisfies the following equivalent conditions:

(1) $a^n \in a^{n+1}R$ for some integer $n \geq 1$.
(2) $a^n R = a^{n+1}R$ for some integer $n \geq 1$.
(3) The chain $aR \supseteq a^2 R \supseteq \cdots$ terminates.

The left π-regular elements are defined analogously. These conditions were studied separately for nearly 25 years before the following remarkable result was proved.

Lemma 4.1. *[15] If every element of a ring R is right π-regular then every element is left π-regular.*

An element $a \in R$ is called strongly π-regular if it is both left and right π-regular, and R is called a strongly π-regular ring if every element is strongly π-regular. Clearly every algebraic algebra is strongly π-regular. Moreover, every left or right perfect ring R is strongly π-regular because R is left (right) perfect if and only if it has the DCC on principal right (left) ideals.

Theorem 4.1. *[9] Every strongly π-regular ring is strongly clean.*

In particular, every left (or right) perfect ring is strongly clean. The converse of Theorem 4.1 is false. If $R = \{\frac{m}{n} \in \mathbb{Q} : n \text{ is odd}\}$, then R is local, hence strongly clean, but R is not strongly π-regular because $J(R)$ is not nil. The equivalence of $(1) \Leftrightarrow (2)$ of the next theorem is due to Armendariz, Fisher and Snider [6].

Theorem 4.2. *The following are equivalent for $\alpha \in E = end(_R M)$:*

(1) α is strongly π-regular in E.
(2) $\exists n \geq 1$ such that $M = M\alpha^n \oplus ker(\alpha^n)$.
(3) $M = P \oplus Q$ where P and Q are α-invariant, $\alpha|_P$ is a unit in $end(P)$ and $\alpha|_Q$ is nilpotent in $end(Q)$.
(4) There exists $\pi^2 = \pi \in E$ such that $\pi\alpha = \alpha\pi$, $\alpha\pi$ is a unit in $\pi E\pi$ and $\alpha(1 - \pi)$ is nilpotent in $(1 - \pi)E(1 - \pi)$.

It is interesting to compare Theorem 4.3(3) with Theorem 4.2(3).

Theorem 4.3. *[26] Let $E = end(_R M)$. Then following are equivalent for $\alpha \in E$:*

(1) α is strongly clean in E.
(2) $\exists \pi^2 = \pi \in E$ such that $\alpha\pi = \pi\alpha$, $\alpha\pi$ is a unit in $\pi E\pi$ and $(1 - \alpha)(1 - \pi)$ is a unit in $(1 - \pi)E(1 - \pi)$.
(3) $M = P \oplus Q$ where P and Q are α-invariant, and both $\alpha|_P$ and $(1 - \alpha)|_Q$ are isomorphisms.
(4) $M = P \oplus Q$ where P and Q are α-invariant, $ker(\alpha) \subseteq Q \subseteq M(1 - \alpha)$ and $ker(1 - \alpha) \subseteq P \subseteq M\alpha$.
(5) $M = P_1 \oplus \cdots \oplus P_n$ for some $n \geq 1$ where P_i is α-invariant and $\alpha|_{P_i}$ is strongly clean in $end(P_i)$ for each i.

A module $_RM$ is said [6] to satisfy Fitting's lemma if, for all $\alpha \in end(_RM)$, there exists an integer $n \geq 1$ such that $M = M\alpha^n \oplus ker(\alpha^n)$. In this case α satisfies Theorem 4.3(3) with $P = M\alpha^n$ and $Q = ker(\alpha^n)$, so it is natural to say that $_RM$ satisfies a general Fitting's lemma if Theorem 4.3(3) holds. Thus, a module satisfies a general Fitting's lemma if and only if its endomorphism ring is strongly clean.

Responding to two questions in [26], it was proved in [40] that $M_2(\mathbb{Z}_{(2)})$ is not strongly clean where $\mathbb{Z}_{(2)}$ is the localization of the ring $\mathbb{Z}$ of integers at the prime 2. This is also proved in [38] where it is shown that if R is strongly clean so also is eRe for any idempotent e in R. Thus, 'strongly clean' is not a Morita invariant and a semiperfect ring need not be strongly clean. Hence a clean ring need not be strongly clean (see Theorem 1.2). The following example is contained in [40].

Example 4.1. If R is a commutative local ring with $R/J(R) \cong \mathbb{Z}_2$, then $T_n(R)$ is strongly clean for every $n \geq 1$.

A ring R is said to have stable range 1 if, whenever $aR + bR = R$ where $a, b \in R$, $a + by$ is a unit for some $y \in R$. A ring R is called directly finite if $ab = 1$ in R always implies $ba = 1$. Every unit regular ring is clean by Theorem 1.3, and every strongly π-regular ring has stable range 1 ([2]) and is directly finite. But the following questions, all raised in [26], remain open.

Question 4.1. Does every strongly clean ring have stable range 1?

Question 4.2. Is every strongly clean ring directly finite?

Question 4.3. Is every unit regular ring strongly clean?

5. Uniquely clean rings

An element a in a ring R is called uniquely clean if $a = e + u$ where $e^2 = e$ and $u \in U(R)$, and the representation is unique. A ring R is called a uniquely clean ring if every element is uniquely clean. Uniquely clean rings were first considered by Anderson and Camillo [1] in the commutative case where the following facts are observed: Any commutative clean ring R with $R/M \cong \mathbb{Z}_2$ for each maximal ideal M of R is uniquely clean, so a commutative local ring is uniquely clean if and only if $R/J(R) \cong \mathbb{Z}_2$; a commutative ring R is uniquely clean if and only if so is $R[[x]]$ if and only if so is $R/\sqrt{0}$ where $\sqrt{0}$ is the nil radical of R; a zero-dimensional commutative ring R is uniquely clean if and only if $R/\sqrt{0}$ is a boolean ring if and only if $R/M \cong \mathbb{Z}_2$ for each maximal ideal M of R. A study of noncommutative uniquely clean rings is carried out in [29] where the following are proved. Recall that a ring R is called I-finite if R contains no infinite orthogonal sets of idempotents.

Proposition 5.1. *[29] The following statements hold:*

(1) Central idempotents and central nilpotents are uniquely clean in a ring; so every boolean ring is uniquely clean.

(2) Every idempotent in a uniquely clean ring is central; so if R is uniquely clean then R is directly finite and eRe is again uniquely clean for each $e^2 = e \in R$.

(3) R is local and uniquely clean if and only if $R/J(R) \cong \mathbb{Z}_2$; consequently, R is a uniquely clean, I-finite ring if and only if $R \cong R_1 \times \cdots \times R_n$ for some $n \geq 1$ where $R_i/J(R_i) \cong \mathbb{Z}_2$ for each i.

Thus, no matrix ring $M_n(R)$, and no triangular matrix ring $T_n(R)$, is uniquely clean if $n \geq 2$. For an ideal $I \lhd R$ we say that idempotents lift uniquely modulo I if, whenever $a^2 - a \in I$ there exists a unique idempotent $e \in R$ such that $a - e \in I$.

Theorem 5.1. *[29] The following are equivalent for a ring R:*

(1) R is uniquely clean.

(2) $R/J(R)$ is boolean and idempotents lift uniquely modulo $J(R)$.

(3) $R/J(R)$ is boolean, idempotents lift modulo $J(R)$, and idempotents in R are central.

(4) For every $a \in R$ there exists a unique idempotent $e \in R$ such that $e - a \in J(R)$.

Thus, R is a regular, uniquely clean ring if and only if R is boolean.

In [1] the authors ask whether a commutative uniquely clean ring R must have $R/M \cong \mathbb{Z}_2$ for each maximal ideal M of R, or equivalently whether the homomorphic image of a commutative uniquely clean ring is again uniquely clean. The answer is affirmative by the following theorem.

Theorem 5.2. *[29] Every factor ring of a uniquely clean ring is again uniquely clean.*

The next examples are given in [29].

Example 5.1. If R is a ring and $\alpha : R \to R$ is a ring endomorphism, then $R[[x, \alpha]]$ is uniquely clean if and only if R is uniquely clean and $e = \alpha(e)$ for all $e^2 = e \in R$.

Example 5.2. Let R be uniquely clean and let $S = \{(a_{ij}) \in T_n(R) : a_{11} = \cdots = a_{nn}\}$. Then S is uniquely clean and is noncommutative if $n \geq 3$.

6. Group rings

If G is a group, we denote the group ring over R by RG. If RG is clean (or uniquely clean) then R must be clean (or uniquely clean), being an image of RG. But it is difficult to determine conditions on R and G which imply that RG is clean (or uniquely clean). The next example answers, in the negative, a question of J.K.Park whether the group ring RG is clean in case R is clean and G is a finite group such that $|G|$ is a unit in R. We write C_n for the cyclic group of order n.

Example 6.1. [19] If $R = \{\frac{m}{n} \in \mathbb{Q} : 7 \text{ does not divide } n\}$, then RC_3 is not clean.

The two positive results below are contained in [19].

Proposition 6.1. *If R is a semiperfect ring, then RC_2 is clean.*

Proposition 6.2. *If R is a boolean ring and G is a locally finite group, then RG is clean.*

It is well known that if RG is regular then G is locally finite (see [14, Theorem 3]). Thus, one raises the following question.

Question 6.1. *[19]* If R is a commutative von Neumann regular ring and G is a locally finite group, is RG clean?

For uniquely clean group rings, the following results are obtained in [32]. A group G is called a 2-group if, for every element $g \in G$, the order of g is equal to 2^k for some $k \geq 0$.

Proposition 6.3. *Let G be a locally finite group.*

(1) If D is a division ring, then DG is uniquely clean if and only if $D \cong \mathbb{Z}_2$ and G is a 2-group.

(2) If R is a boolean ring, then RG is uniquely clean if and only if G is a 2-group.

Theorem 6.1. *If R is a semiperfect, uniquely clean ring, and if G is a locally finite 2-group, then RG is uniquely clean.*

Question 6.2. If R is a uniquely clean ring and G is a finite 2-group, is RG uniquely clean?

Theorem 6.2. *If R is a ring and G is an abelian group, then RG is uniquely clean if and only if R is uniquely clean and G is a 2-group.*

Example 6.2. If R is boolean and $n \geq 3$ is odd, then RC_n is clean but not uniquely clean.

Example 6.3. RD_∞ is not uniquely clean for any ring R, where D_∞ is the infinite dihedral group.

7. The extension questions of clean and strongly clean rings

Results in this section and in next section are contained in [30]. A result of Han and Nicholson [19] says that, for any ideal I of R with $I \subseteq J(R)$, R is clean if and only if $R/J(R)$ is clean and idempotents lift modulo I. This is a motivation of the following question: For which ideals I of R, R/I being clean implies R being clean? (note that a homomorphic image of a clean ring is obviously clean.) The consideration of this question leads one to extend clean rings to rings without identity.

From now on, by a general ring we mean an associative ring with or without identity. For a general ring A and $a, b \in A$, let $a * b = a + b + ab$ and let
$$Q(A) = \{q \in A : \exists p \in A \text{ such that } p * q = 0 = q * p\}.$$
It is well known that $(Q(A), *)$ is a group. If A has 1 then $(Q(A), *) \cong (U(A), \cdot)$ as groups via $q \mapsto 1 + q$. The Jacobson radical of a general ring A is denoted by $J(A)$. The next lemma is easy to prove.

Lemma 7.1. *A ring R is clean if and only if, $\forall a \in R$, $a = e + q$ where $e^2 = e$ and $q \in Q(R)$.*

Hence we call a general ring A clean if, for any $a \in A$, $a = e + q$ where $e^2 = e$ and $q \in Q(A)$.

Lemma 7.2. *[3] A general ring A is called an exchange ring if the following equivalent conditions hold:*

(1) $\forall x \in A$, $\exists r, s \in A$ and $e^2 = e \in A$ such that $e = xr = s + x - xs$.

192

(2) $\forall x \in A$, $\exists r, s \in A$ and $e^2 = e \in A$ such that $e = rx = s + x - sx$.

Theorem 7.1. *Let A be a general ring.*

(1) If A is clean then the following hold:

    (a) $M_n(A)$ is clean for every $n \geq 1$.

    (b) A is exchange.

    (c) Idempotents lift modulo every left or right ideal of A.

    (d) Every one-sided ideal not contained in $J(A)$ contains a nonzero idempotent.

(2) If A is exchange with idempotents central then A is clean.

(3) If $I \lhd A$ with $I \subseteq J(A)$, then A is clean if and only if A/I is clean and idempotents lift modulo I.

Theorem 7.2. *Let A be a general ring and let $I \lhd A$.*

(1) If A is clean then I and A/I are both clean and idempotents lift modulo I.

(2) The converse is true if $I \subseteq J(A)$ or if all primitive factors of A are artinian.

Question 7.1. If $I \lhd A$, both I and A/I are clean, and idempotents lift modulo I, is A clean?

If I and A are as in Question 7.1, then A is exchange by Ara [3, Theorem 2.2].

We do not know if right ideals of a clean general ring are again clean. We do have:

Corollary 7.1. *If R is a ring and $e^2 = e \in R$, then eR is a clean general ring if and only if eRe is a clean ring.*

It is an open question whether eRe is clean if R is a clean ring and $e^2 = e \in R$. Using Corollary 7.1, this becomes:

Question 7.2. If $e^2 = e \in R$ where R is a clean ring, is eR a clean general ring?

In contrast to the ideals, subrings of clean rings need not be clean. In fact, the ring $R[[x]]$ is clean if and only if R is clean [19, Proposition 5]; and the polynomial ring $R[x]$ is never clean if $R \neq 0$ [29, Proposition 13].

The notion of a uniquely clean ring can be similarly extended to a general ring: A general ring A is called uniquely clean if every element of A

can be uniquely written as the sum of an idempotent and an element from $Q(A)$.

Thus, for a general ring A, A is boolean if and only if it is uniquely clean and $Q(A) = 0$; and A is radical if and only if it is uniquely clean and has no nonzero idempotents.

Proposition 7.1. *Let A be a uniquely clean general ring. Then the following hold:*

(1) Every idempotent of A is central.
(2) eAe is a uniquely clean ring whenever $e^2 = e \in A$.
(3) $2a \in J(A)$ for any $a \in A$.
(4) $Q(A) = J(A)$.

We say that idempotents lift uniquely modulo an ideal I of a general ring A if, whenever $x^2 - x \in I$, $x \in A$, there exists a unique idempotent $e \in A$ such that $x - e \in I$.

Theorem 7.3. *The following are equivalent for a general ring A :*

(1) A is uniquely clean.
(2) For each $x \in A$, there exists a unique $e^2 = e \in A$ such that $x - e \in J(A)$.
(3) $A/J(A)$ is boolean and idempotents lift uniquely modulo $J(A)$.
(4) $A/J(A)$ is boolean, idempotents lift modulo $J(A)$, and idempotents in A are central.

An extension theorem on uniquely clean rings can be proved.

Theorem 7.4. *Let A be a general ring and $I \triangleleft A$. Then A is uniquely clean if and only if the following conditions hold:*

(1) I and A/I are uniquely clean.
(2) Every idempotent of A/I can be lifted to a central idempotent of A.
(3) $J(A/I) = (I + J(A))/I$.

A very useful special case of Theorem 7.4 is the construction of ideal extensions. Let R be a ring and let $_R V_R$ be a bimodule which is itself a general ring in which $(vw)r = v(wr)$, $(vr)w = v(rw)$ and $(rv)w = r(vw)$ hold for all $v, w \in V$ and $r \in R$. Then the ideal-extension $\mathbb{E}(R; V)$ of R by V is defined to be the abelian group $\mathbb{E}(R; V) = R \oplus V$ with multiplication $(r, v)(s, w) = (rs, rw + vs + vw)$; and $\mathbb{E}(R; V)$ is clearly a ring. More

194

examples of uniquely clean rings can now be constructed using the next corollary.

Corollary 7.2. *Let $S = \mathbb{E}(R; V)$. Then S is uniquely clean if and only if the following conditions hold:*

(1) R and V are uniquely clean.
(2) If $a^2 = a \in R$, then $ab = ba$ and $ax = xa$ for all $b \in R$ and all $x \in V$.
(3) $\forall a \in J(R)$, there exists $x \in V$ such that $(a, x) \in J(S)$.

Example 7.1. Let R be a uniquely clean ring and let $\{I_j\}_j$ be a family of ideals of R. Then $\mathbb{E}(R; \oplus_j I_j)$ is a uniquely clean ring.

8. Semiboolean rings

This section is devoted to an important notion, identified in the following lemma, that lies between being clean and being uniquely clean.

Lemma 8.1. *The following are equivalent for a general ring A :*

(1) Each $x \in A$ has the form $x = e + a$ where $e^2 = e$ and $a \in J(A)$.
(2) A is clean and $Q(A) = J(A)$.
(3) $A/J(A)$ is boolean and idempotents lift modulo $J(A)$.

With an eye on condition (3), we call a general ring A semiboolean if it satisfies the conditions in Lemma 8.1. Thus boolean general rings and radical rings are semiboolean. The ring $T_2(\mathbb{Z}_2)$ is semiboolean by Example 8.1 below, but it is not uniquely clean because idempotents are not central. Since $\mathbb{Z}_9$ is clean but not semiboolean, the implications

$$\text{uniquely clean} \quad \Rightarrow \quad \text{semiboolean} \quad \Rightarrow \quad \text{clean}$$

are both non-reversible (even for artinian rings).

Example 8.1. Let A and A_j denote general rings.

(1) Every uniquely clean general ring is semiboolean.
(2) Every image of a semiboolean ring is again semiboolean.
(3) A direct product $\Pi_j A_j$ or a direct sum $\oplus_j A_j$ of general rings is semiboolean if and only if each A_i is semiboolean.
(4) If $n \geq 1$ then $T_n(A)$ is semiboolean if and only if A is semiboolean.
(5) If $n \geq 2$ then $M_n(A)$ is semiboolean if and only if A is radical.
(6) A is uniquely clean if and only if A is semiboolean and all idempotents of A are central.

Proposition 8.1. *The following hold for a semiboolean general ring A :*

(1) Every ideal $I \lhd A$ is semiboolean.

(2) eAe is semiboolean for every $e^2 = e \in A$.

(3) For $a, b \in A$, $a * b = 0$ implies $b * a = 0$.

It is interesting to compare the following theorem with Theorem 7.4.

Theorem 8.1. *Let A be a general ring and $I \lhd A$. Then A is semiboolean if and only if the following conditions hold:*

(1) I and A/I are semiboolean.

(2) Every idempotent of A/I can be lifted to an idempotent of A.

(3) $J(A/I) = (I + J(A))/I$.

Corollary 8.1. *Let $S = \mathbb{E}(R; V)$. Then S is semiboolean if and only if the following conditions hold:*

(1) R and V are semiboolean.

(2) For all $a \in J(R)$, there exists $x \in V$ such that $(a, x) \in J(S)$.

Proposition 8.2. *Let R be a ring.*

(1) R is local and semiboolean if and only if $R/J(R) \cong \mathbb{Z}_2$.

(2) The following are equivalent:

(a) R is semiperfect and semiboolean.

(b) There exist orthogonal idempotents $\{e_1, \cdots, e_n\}$ in R such that $1 = e_1 + \cdots + e_n$ and, for any i and any $a \in R$, either $ae_i \in J(R)$ or $ae_i - e_i \in J(R)$.

(c) There exists a chain $J(R) = A_0 \subset A_1 \subset \cdots \subset A_n = R$ of ideals of R such that there exists $e_i = e_i^2 \in A_i \backslash A_{i-1}$ and $A_i/A_{i-1} \cong \mathbb{Z}_2$ for $i = 1, \cdots, n$.

9. Connections with C^*-Algebras and clean rings of continuous functions

The theory of rings of continuous functions and that of Operator Algebras display the connections of Ring Theory with Topology and Functional Analysis. It was proved by Ara, Goodearl, O'Meara and Pardo [5] that the C^*-algebras of real rank zero (see Brown and Pedersen [8]) are precisely the C^*-algebras which are exchange rings. This important result opened the way for a transfer of technology between Ring Theory and Operator Algebras, which has been exploited already in both directions (see [4], [36] and [37]). It would be interesting to know which C^*-algebras (of real rank zero) are clean rings, so that new examples of clean rings may occur and, on the other hand, methods in clean rings may help to study this class of C^*-algebras.

Let P be a topological space and S be a ring endowed with the discrete topology. Then a continuous f of P to S is said to have compact carrier if $f = 0$ on the complement of a compact subset of P. The well known Stone Theorem [39] states that any boolean ring is isomorphic to the ring of continuous functions with compact carriers from a suitable totally disconnected locally compact space to the field $\mathbb{Z}_2$ (see [22]).

Question 9.1. Is every commutative clean (or uniquely clean) ring representable as the ring of certain continuous functions on a certain topological space?

We conclude by presenting a result of Azarpanah [7] on when the ring of continuous functions on a completely regular Hausdorff space is clean. A topological space X is called strongly zero-dimensional if X is a nonempty completely regular Hausdorff space and every finite functionally open cover $\{U_i\}_{i=1}^k$ of the space X has a finite open refinement $\{V_i\}_{i=1}^k$ such that $V_i \cap V_j = \emptyset$ whenever $i \neq j$. This is equivalent to the condition that for every pair A, B of completely separated subsets of the space X, there exists a clopen (closed and open) set U in X such that $A \subseteq U \subseteq X \backslash B$ (see [16]).

Let X be a completely regular Hausdorff space and let $C(X)$ be the ring of all continuous real valued functions on X and $C^*(X)$ be the subring of $C(X)$ consisting of all bounded functions in $C(X)$.

Theorem 9.1. *[7] The following statements are equivalent:*

(1) $C(X)$ is a clean ring.
(2) $C^*(X)$ is a clean ring.
(3) The set of clean elements in $C(X)$ is a subring of $C(X)$.
(4) X is strongly zero-dimensional.
(5) Every zero-divisor in $C(X)$ is clean.
(6) $C(X)$ has a clean prime ideal.

For the Stone Čech compactification βX of X, $C(\beta X) \cong C^*(X)$. It follows from this and Theorem 9.1 that X is strongly zero-dimensional if and only if βX of X is strongly zero-dimensional (see [7]).

References

1. D.D.Anderson and V.P.Camillo, Commutative rings whose elements are a sum of a unit and idempotent, *Comm. Alg.* **30**(2002), 3327-3336.
2. P.Ara, Strongly π-regular rings have stable range one, *Proc. AMS.* **124**(1996), 3293-3298.
3. P.Ara, Extensions of Exchange rings, *J. Algebra* **197**(1997), 409-423.
4. P.Ara, K.R.Goodearl, K.C.O'Meara and R.Raphael, K_1 of separative exchange rings and C^*-algebras with real rank zero, *Pacific J.Math.* **195**(2000), 261-275.
5. P.Ara, K.R.Goodearl, C.O'Meara and E.Pardo, Separative cancellation for projective modules over exchange rings, *Israel J.Math.* **105**(1998), 105-137.
6. E.P.Armendariz, J.W.Fisher and R.L.Snider, On injective and surjective endomorphisms of finitely generated modules, *Comm. Alg.* **6**(1978), 659-672.
7. F.Azarpanah, When is $C(X)$ a clean ring?, *Acta Math. Hungar.* **94**(2002), 53-58.
8. L.G.Brown and G.K.Pedersen, C^*-algebras of real rank zero, *J. Functional Analysis* **99**(1991), 131-149.
9. W.D.Burgess and P.Menal, On strongly π-regular rings and homomorphisms into them, *Comm. Alg.* **16**(1988), 1701-1725.
10. V.P.Camillo and D.Khurana, A characterization of unit regular rings, *Comm. Alg.* **29**(2001), 2293-2295.
11. V.P.Camillo and H.-P.Yu, Exchange rings, units and idempotents, *Comm. Alg.* **22**(1994), 4737-4749.
12. V.P.Camillo and J.J.Simón, The Nicholson-Varadarajan theorem on clean linear transformations, *Glasgow Math. J.* **44** (2002), 365-369.
13. H.Chen, Exchange rings with artinian primitive factors, *Algebras and Representation Theory* **2**(1999), 201-207.
14. I.G.Connell, On the group ring, *Can. Math.J.* **15**(1963), 656-685.
15. M.F.Dischinger, Sur les anneaux fortement π-réguliers, *C.R.Aca.Sc.Paris* **283**(1976), 571-573.
16. R.Engelking, General Topology, PWN Polish Scientific Publishers, 1977.
17. K.R.Goodearl, Von Neumann Regular Rings, Pitman, 1979 (Second Edition, Kreiger, 1991).
18. K.R.Goodearl and R.B.Warfield, Jr., Algebras over zero-dimensional rings, *Math. Ann.* **223**(1976), 157-168.
19. J.Han and W.K.Nicholson, Extensions of clean rings, *Comm. Alg.* **20**(2001), 2589-2596.
20. D.Handelman, Perspectivity and cancellation in regular rings, *J. Algebra* **48**(1977), 1-16.
21. C.Y.Hong, N.K.Kim and Y.Lee, Exchange rings and their extensions, *J.Pure Appl. Alg.* **179**(2003), 117-126.
22. N.Jacobson, Structure of Ring Theory, AMS Colloquium Publications, Vol. XXXVII, 1956
23. G.S.Monk, A characterization of exchange rings, *Proc. AMS.* **35**(1972), 344-353.

24. W.K.Nicholson, Lifting idempotents and exchange rings, *Trans. AMS.* **229**(1977), 269-278.

25. W.K.Nicholson, On exchange rings, *Comm. Alg.* **25**(6)(1997),1917-1918.

26. W.K.Nicholson, Strongly clean rings and Fitting's lemma, *Comm. Alg.* **27**(1999), 3583-3592.

27. W.K.Nicholson and K.Varadarajan, Countable linear transformations are clean, *Proc.AMS.* **126**(1998), 61-64.

28. W.K.Nicholson, K.Varadarajan and Y. Zhou, Clean endomorphism rings, *Archiv der Mathematik (Basel)*, in press.

29. W.K.Nicholson and Y. Zhou, Rings in which elements are uniquely the sum of an idempotent and a unit, *Glasgow Math. J.* **46**(2004), 227-236.

30. W.K.Nicholson and Y. Zhou, Clean general rings, preprint, 2004.

31. W.K.Nicholson and Y. Zhou, Endomorphisms that are the sum of a unit and a root of a fixed polynomial, preprint, 2004.

32. W.K.Nicholson and Y. Zhou, On uniquely clean group rings, preprint, 2004.

33. W.K.Nicholson and Y. Zhou, When are endomorphism rings clean?, preprint, 2004.

34. K.C.O'Meara, The exchange property for row and column-finite matrix rings, *J. Algebra* **268**(2003), 744-749.

35. M.Ó Searcóid, Perturbation of linear operators by idempotents, *Irish Math. Soc.Bull.* **39**(1997), 10-13.

36. F.Perera, Lifting units modulo exchange ideals and C^*-algebras with real rank zero, *J.Reine Angew. Math.* **522**(2000), 51-62.

37. F.Perera, Ideal structure of multiplier algebras of simple C^*-algebras with real rank zero, *Canad.J.Math.* **53**(3)(2001), 592-630.

38. E.Sánchez Campos, On strongly clean rings, unpublished.

39. M.H.Stone, Applications of the theory of Boolean rings to general topology, *Trans. AMS.* **41**(1937), 375-381.

40. Z.Wang and J.Chen, On two open problems about strongly clean rings, *Bull. Austral. Math.Soc.*, to appear.

41. R.B.Warfield, Jr., Exchange rings and decompositions of modules, *Math. Ann.* **199**(1972), 31-36.

42. H.-P.Yu, On quasi-duo rings, *Glasgow Math.J.* **37**(1995), 21-31.

FLAT COVER AND COTORSION ENVELOPE COMMUTE

PHILIPP ROTHMALER

Department of Mathematics
The Ohio State University at Lima
4240 Campus Drive
Lima, OH 45804, USA
E-mail: rothmaler.1@osu.edu

The statement of the title is proved in two, more specific ways using pushouts and pullbacks.

1. Introduction

It has been known since [1] that flat covers, and therefore also cotorsion envelopes [4, Thm. 3.4.6], exist (see also [2, Thm. 7.4.4]). As envelopes are unique up to isomorphism, if $M \to C$ is a cotorsion envelope, one may denote the cotorsion module C by $\mathcal{C}E(M)$. Similarly, as covers are unique up to isomorphism, if $F \to M$ is a flat cover, one may denote the flat module F by $\mathcal{F}C(M)$. The main result can then be stated as an equation: $\mathcal{C}E \cdot \mathcal{F}C = \mathcal{F}C \cdot \mathcal{C}E$, which stands for $\mathcal{C}E(\mathcal{F}C(M)) \cong \mathcal{F}C(\mathcal{C}E(M))$ *for all modules M*. I prove this in two different ways, once via pushouts, once via pullbacks, this giving two, more specific results, see the theorem below.

In fact, I prove this fact for general cotorsion theories $(\mathcal{F}, \mathcal{C})$ enjoying some natural closure properties. Once and for all, I fix such a *cotorsion theory* $(\mathcal{F}, \mathcal{C})$, that is, $\mathcal{F}$ and $\mathcal{C}$ are classes of modules (over a fixed ring) such that $\mathcal{F}^{\perp} = \mathcal{C}$ and $\mathcal{F} = {}^{\perp}\mathcal{C}$. Here *orthogonality* is defined in terms of the vanishing of Ext. More precisely, $\mathcal{F}^{\perp}$ is the intersection of the kernels of the map $\mathrm{Ext}(F, -)$ where F runs through $\mathcal{F}$, while ${}^{\perp}\mathcal{C}$ is the intersection of the kernels of the map $\mathrm{Ext}(-, C)$ where C runs through $\mathcal{C}$. Clearly, the pairs (projectives, all modules) and (all modules, injectives) are cotorsion theories. The most prominent nontrivial cotorsion theory is obtained when $\mathcal{F}$ is the class **Flat** of all flat modules and $\mathcal{C}$ is $\mathbf{Cot} = \mathbf{Flat}^{\perp}$. The elements of **Cot** are called *cotorsion modules*. That (**Flat**, **Cot**) is indeed a cotorsion theory follows from the stronger result that any module left orthogonal to all pure-injective modules is already flat, see [4, [Lemma 3.4.1].

I assume familiarity with the concepts of cover, precover, envelope, and preenvelope as developed by Enochs. For the purposes of this paper, *(pre)cover* means $\mathcal{F}$-(pre)cover, and *(pre)envelope* means $\mathcal{C}$-(pre)envelope. A *special precover* of a module M is an epimorphism from a module from $\mathcal{F}$ onto M whose kernel is in $\mathcal{C}$. Dually, a *special preenvelope* of M is an embedding of M into a module from $\mathcal{C}$ whose cokernel is in $\mathcal{F}$. Every special precover is a precover and every special preenvelope is a preenvelope, cf. [2, remarks after Def. 7.1.5] or [4, Propositions 2.1.3 and 2.1.4]. Further, every epic cover is itself a special precover, provided $\mathcal{F}$ is closed under extension, see [2, Lemma 5.3.25 or Cor. 7.2.3] or [4, Lemma 2.1.1], while every monic envelope is itself a special preenvelope, whenever $\mathcal{C}$ is closed under extension, see [2, Prop. 7.2.4] or [4, Lemma 2.1.2]. These last two results are special cases of what are known as Wakamatsu's Lemmas. Note that the existence of special preenvelopes (resp. special precovers) implies that *every* preenvelope (resp. precover) is a monomorphism (resp. epimorphism). (It follows from the definitions that one precover of a given module is epic if and only if all of them are, and that one preenvelope of a given module is monic if and only if all of them are.)

This and all other notation and terminology can be found in [2] or [4]. All maps are written on the left of the argument.

I would like to thank Nanqing Ding, his colleagues and their students from Nanjing and Southwest Universities at Nanjing for the wonderful hospitality during the time when the version of the proof presented here was prepared.

2. The result

For the purpose of presentation, call a pushout (resp. pullback) diagram

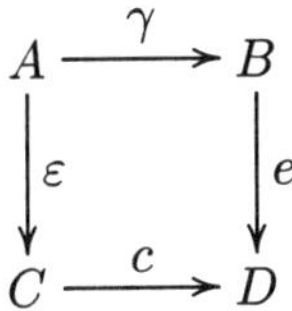

special if γ (resp. c) is a special precover and ε (resp. e) is a special preenvelope. Call it an $(\mathcal{F}, \mathcal{C})$-*pushout* (resp. $(\mathcal{F}, \mathcal{C})$-*pullback*) *diagram* if γ (resp. c) is an epic $\mathcal{F}$-cover and ε (resp. e) is a monic $\mathcal{C}$-envelope.

An $\mathcal{F}$-*pure embedding* is a monomorphism whose cokernel is in $\mathcal{F}$. This term is motivated by the fact that a **Flat**-pure embedding is pure in the usual sense (as a matter of fact, this can be taken as a definition of *flat*).

It is well known that $\mathcal{F} = \textbf{Flat}$ has all the properties required for $\mathcal{F}$ in the following lemma. [4, Prop. 3.1.2] shows that $\mathcal{C} = \textbf{Cot}$ satisfies all those required for $\mathcal{C}$.

Lemma.
(1) *If $\mathcal{F}$ is closed under extensions and $\mathcal{C}$ is closed under under homomorphic images with kernels in $\mathcal{C}$, then every special pushout diagram is a special pullback diagram.*
(2) *If $\mathcal{C}$ is closed under extensions and $\mathcal{F}$ is closed under $\mathcal{F}$-pure submodules, then every special pullback diagram is a special pushout diagram.*

Proof. It is well known that a pushout diagram of two maps one of which is surjective or injective is at the same time a pullback diagram. Dually, a pullback diagram of two maps one of which is surjective or injective is at the same time a pushout diagram. See [3, Ch. IV, §5, Example 3] for these kinds of result.

Complete the given pushout or pullback diagram to the following commutative diagram with exact rows and columns.

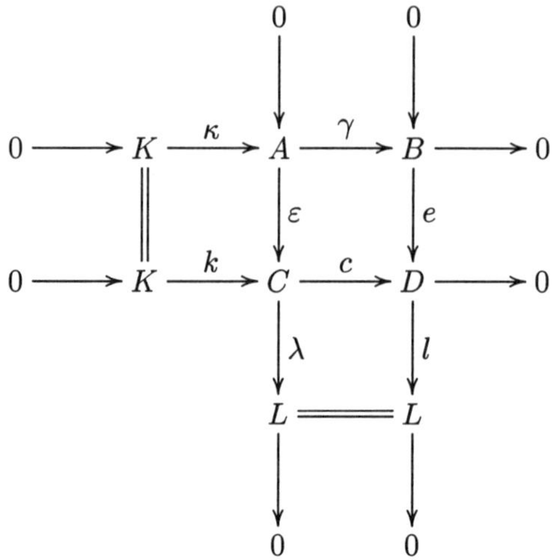

To prove (1), suppose ε is a special preenvelope and γ is a special precover. Then $C \in \mathcal{C}$, $L \in \mathcal{F}$, $A \in \mathcal{F}$, and $K \in \mathcal{C}$. Since $\mathcal{C}$ is closed under factor modules with kernel in $\mathcal{C}$, also $D \in \mathcal{C}$, and therefore e is a special preenvelope. Since $\mathcal{F}$ is closed under extension, $C \in \mathcal{F}$, and therefore c is a special precover.

For (2), suppose e is a special preenvelope and c is a special precover. Then $D \in \mathcal{C}$, $L \in \mathcal{F}$, $C \in \mathcal{F}$, and $K \in \mathcal{C}$. Since $\mathcal{C}$ is closed under extension, $C \in \mathcal{C}$, and therefore ε is a special preenvelope. Since $\mathcal{F}$ is closed under $\mathcal{F}$-pure submodules, $A \in \mathcal{F}$, and therefore γ is a special precover. $\qquad\square$

Our goal is the following similar commutation result for $(\mathcal{F},\mathcal{C})$-pushouts and $(\mathcal{F},\mathcal{C})$-pullbacks as introduced before the lemma.

Theorem.

(1) *If $\mathcal{F}$ is closed under extensions and $\mathcal{C}$ is closed under under homomorphic images with kernels in $\mathcal{C}$, then every $(\mathcal{F},\mathcal{C})$-pushout diagram is an $(\mathcal{F},\mathcal{C})$-pullback diagram.*

(2) *If $\mathcal{C}$ is closed under extensions and $\mathcal{F}$ is closed under $\mathcal{F}$-pure submodules, and $\mathcal{F}$-covers always exist, then every $(\mathcal{F},\mathcal{C})$-pullback diagram is*

an $(\mathcal{F},\mathcal{C})$-pushout diagram.

Proof. (1) Let

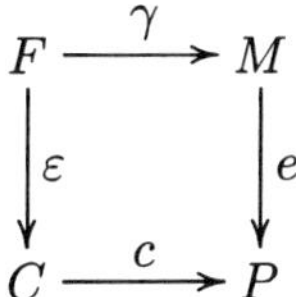

be an $(\mathcal{F},\mathcal{C})$-pushout, where $\gamma : F \to M$ is an $\mathcal{F}$-cover and $\varepsilon : F \to C$ is a $\mathcal{C}$-envelope. By the lemma (and Wakamatsu's Lemma), e is a $\mathcal{C}$-preenvelope and c an $\mathcal{F}$-precover. So, in order to show that e is a $\mathcal{C}$-envelope and c is an $\mathcal{F}$-cover, we have to verify the automorphism property, which will be done in the next two claims.

But first of all, extend the pushout to the standard commutative diagram with exact rows and columns (which is possible, because ε is injective and γ is surjective).

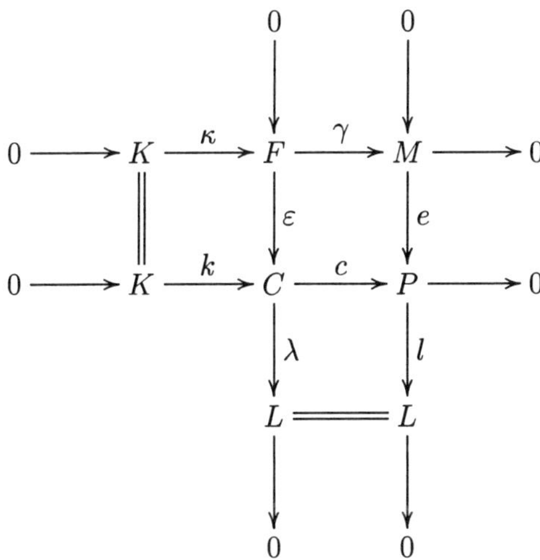

Claim c. If $g \in \operatorname{End} C$ and $c = cg$, then $g \in \operatorname{Aut} C$. Thus, c is an $\mathcal{F}$-cover.

204

Proof. Consider the diagram

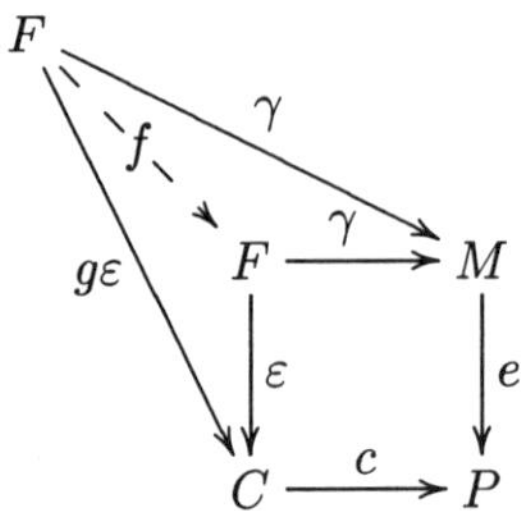

Since $c(g\varepsilon) = (cg)\varepsilon = c\varepsilon = e\gamma$, it commutes, and by the lemma, it is a pullback diagram and can therefore be completed by a map f as shown. As γ is a cover, $f \in \mathrm{Aut}\, F$. Hence $\varepsilon = g\varepsilon f^{-1}$ and therefore $\mathrm{im}\,\varepsilon \subseteq \mathrm{im}\, g$ and $\ker g\varepsilon f^{-1} = 0$.

By exactness of the big diagram, $\ker c = \mathrm{im}\, k = \mathrm{im}\,\varepsilon\kappa \subseteq \mathrm{im}\,\varepsilon = \mathrm{im}\,\varepsilon f^{-1}$, which, together with $\ker g \subseteq \ker cg = \ker c$, yields $\ker g \subseteq \mathrm{im}\,\varepsilon f^{-1}$. From this $\ker g = 0$ can be derived as follows. Write $x \in \ker g$ as $x = \varepsilon f^{-1}(y)$. Then $0 = g(x) = g\varepsilon f^{-1}(y)$. As $\ker g\varepsilon f^{-1} = 0$, we have $y = 0$ and therefore $x = 0$, as claimed.

To verify $\mathrm{im}\, g = C$, let $y \in C$. Then $c(y) = cg(y)$, hence $y - g(y) \in \ker c \subseteq \mathrm{im}\,\varepsilon \subseteq \mathrm{im}\, g$, and so $y \in \mathrm{im}\, g$, as desired. $\square$

Claim e. If $h \in \mathrm{End}\, P$ and $e = he$, then $h \in \mathrm{Aut}\, P$. Thus, e is a C-envelope.

Proof. As c is a precover, hc must factor through it, whence we get a map $g \in \mathrm{End}\, C$ with $cg = hc$. This gives us a diagram as in the proof of the previous claim with $cg\varepsilon = (hc)\varepsilon = h(c\varepsilon) = h(e\gamma) = (he)\gamma = e\gamma$, and so we may use the pullback property again to obtain a map $f \in \mathrm{Aut}\, F$ as before. Note that $g\varepsilon = \varepsilon f$ is monic.

As ε is a preenvelope, εf^{-1} must factor through it, this giving $\bar{g} \in \mathrm{End}\, C$ with $\bar{g}\varepsilon = \varepsilon f^{-1}$. Then $\varepsilon = \varepsilon(f^{-1}f) = \bar{g}(\varepsilon f) = (\bar{g}g)\varepsilon$ and $(g\bar{g})\varepsilon = (g\varepsilon)f^{-1} = \varepsilon(ff^{-1}) = \varepsilon$.

As ε is an envelope, this implies that both, $g\bar{g}$ and $\bar{g}g$ are in $\mathrm{Aut}\, C$, hence so are g and $\bar{g}$ themselves. Consequently, g is onto, and as so too is c, it follows that $hc = cg$, and therefore h is onto as well.

We are left with showing $\ker h = 0$. To this end, let $h(x) = 0$. As c is onto, we may write $x = c(y)$. Then $0 = h(x) = hc(y) = cg(y)$, so $g(y) \in \ker c = \operatorname{im} k = \operatorname{im} \varepsilon\kappa$. Write $g(y) = \varepsilon\kappa(z)$ accordingly. Then $y = g^{-1}\varepsilon\kappa(z) = \varepsilon f^{-1}\kappa(z)$, for $g\varepsilon = \varepsilon f$ implies $\varepsilon f^{-1} = g^{-1}\varepsilon$.

Further, $\gamma f = \gamma$ implies $\gamma = \gamma f^{-1}$, hence $\gamma f^{-1}\kappa(z) = \gamma\kappa(z)$. Since the latter is 0 by exactness, $f^{-1}\kappa(z) \in \ker \gamma = \operatorname{im} \kappa$. Write $f^{-1}\kappa(z) = \kappa(z')$ accordingly.

Altogether we now have $y = \varepsilon f^{-1}\kappa(z) = \varepsilon\kappa(z') = k(z')$, and hence $x = c(y) = ck(z')$, which is 0 by exactness, as desired. $\qquad\square$

This completes the proof of (1).

(2) Even though the proof is not entirely dual to the previous, I will keep the diagrams as close to the previous ones as possible. So let

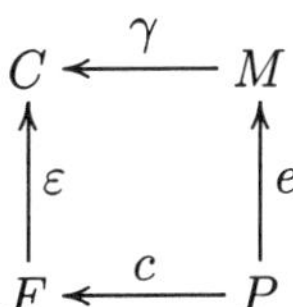

be an $(\mathcal{F}, \mathcal{C})$-pullback, where $\gamma : M \to C$ is a $\mathcal{C}$-envelope and $\varepsilon : F \to C$ is an $\mathcal{F}$-cover. By the lemma, c is a $\mathcal{C}$-preenvelope and e an $\mathcal{F}$-precover. So, in order to show that c is a $\mathcal{C}$-envelope and e is an $\mathcal{F}$-cover, we have to verify the automorphism property. For c this is done in the first claim below (and dual to its counterpart in (1)). For e only one half of our proof is dual to its counterpart above, see the second claim below. For the other half we have to invoke the existence of an $\mathcal{F}$-cover of M, as done at the end of the proof below.

But first extend the pullback of the theorem to the following commutative diagram with exact rows and columns (which is possible, because ε is surective and γ is injective).

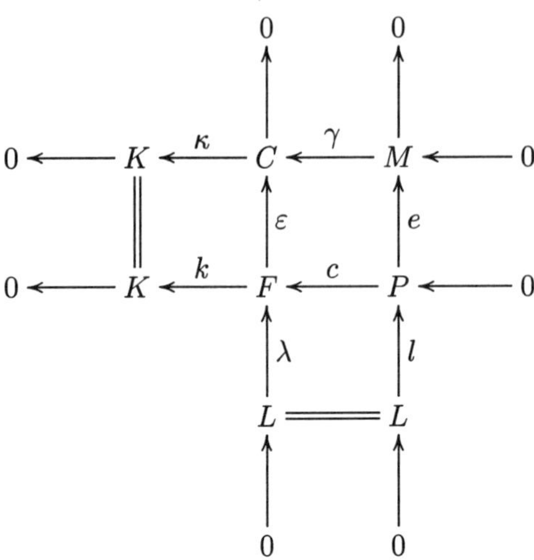

Claim c*. If $g \in \operatorname{End} F$ and $c = gc$, then $g \in \operatorname{Aut} F$. Thus, c is a $\mathcal{C}$-envelope.

Proof. As before, the diagram

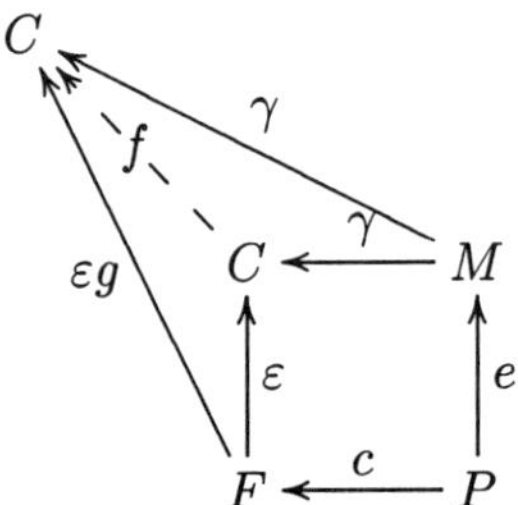

clearly commutes and is, by the lemma, a pushout diagram and can therefore be completed by a map f as shown. As γ is an envelope, $f \in \operatorname{Aut} C$. Hence $\varepsilon = f^{-1}\varepsilon g$ and therefore $\ker g \subseteq \ker \varepsilon$ and $C = \operatorname{im} \varepsilon = \operatorname{im} f^{-1}\varepsilon g = \operatorname{im} \varepsilon g$.

By exactness of the big diagram above, $\ker \varepsilon \subseteq \ker \kappa\varepsilon = \ker k = \operatorname{im} c$, which, together with $\ker g \subseteq \ker \varepsilon$, yields $\ker g \subseteq \operatorname{im} c$ and, together with $\operatorname{im} c = \operatorname{im} gc \subseteq \operatorname{im} g$, yields $\ker \varepsilon \subseteq \operatorname{im} g$.

To verify $\ker g = 0$, let $x \in \ker g$. Write $x = c(y)$, where $y \in F$. Then $0 = g(x) = gc(y) = c(y) = x$, as desired.

To verify $\operatorname{im} g = F$, let $y \in F$. As $C = \operatorname{im} \varepsilon g$, there is $z \in F$ with $\varepsilon(y) = \varepsilon g(z)$, hence $y - g(z) \in \ker \varepsilon \subseteq \operatorname{im} g$. Then $y \in g(z) + \operatorname{im} g \subseteq \operatorname{im} g$, which completes the proof of the claim. $\qquad\square$

Claim e*. If $h \in \operatorname{End} P$ and $e = eh$, then $\ker h = 0$.

Proof. As c is a C-(pre)envelope, ch must factor through it, whence we get a map $g \in \operatorname{End} F$ with $gc = ch$. This gives us a commutative diagram as in the proof of the previous claim, and so we may use the pushout property again to obtain a map $f \in \operatorname{Aut} C$ as before. Note that $\varepsilon g = f\varepsilon$ is epic, as so are f and ε.

Since ε is a (pre)cover, $f^{-1}\varepsilon$ must factor through it, this giving $\overline{g} \in \operatorname{End} F$ with $\varepsilon \overline{g} = f^{-1}\varepsilon$. Then $\varepsilon = (ff^{-1})\varepsilon = f(\varepsilon\overline{g}) = (f\varepsilon)\overline{g} = \varepsilon(g\overline{g})$ and $\varepsilon(\overline{g}g) = f^{-1}(\varepsilon g) = (f^{-1}f)\varepsilon = \varepsilon$.

As ε is a cover, this implies that both, $g\overline{g}$ and $\overline{g}g$ are in $\operatorname{Aut} F$, hence so are g and $\overline{g}$ themselves. Consequently, g is monic, and as so too is c, it follows that $ch = gc$ and therefore h is monic as well. $\qquad\square$

To finish off the proof of the theorem, choose an $\mathcal{F}$-cover $\alpha : F' \to M$. Then e factors through α, hence $\alpha\rho = e$ for some $\rho : P \to F'$. Further, α must also factor through e, hence $e\sigma = \alpha$ for some $\sigma : F' \to P$. Now, $\sigma\rho \in \operatorname{End} P$ and $e(\sigma\rho) = (e\sigma)\rho = \alpha\rho = e$, hence $\ker \sigma\rho = 0$ by Claim e*, and so ρ is monic. On the other hand, $\rho\sigma \in \operatorname{End} F'$ and $\alpha(\rho\sigma) = (\alpha\rho)\sigma = e\sigma = \alpha$, hence $\rho\sigma \in \operatorname{Aut} F'$, for α is a cover. Then ρ is also epic and thus an isomorphism. Consequently, e is an $\mathcal{F}$-cover too, which concludes the proof of (2) and thus the theorem. $\qquad\square$

Corollary. *Suppose $\mathcal{F}$ and $\mathcal{C}$ satisfy the hypotheses of (1) or satisfy the hypotheses of (2) of the theorem.*

If $\mathcal{F}$-covers are epic and $\mathcal{C}$-envelopes are monic, then $\mathrm{CE}\cdot\mathcal{F}\mathrm{C} = \mathcal{F}\mathrm{C}\cdot\mathrm{CE}$, that is, $\mathrm{CE}(\mathcal{F}\mathrm{C}(M)) \cong \mathcal{F}\mathrm{C}(\mathrm{CE}(M))$ for every module M.

As mentioned before, the proof of (2) of the theorem is not entirely dual to the one of (1), the difference being that in the last step of (2) the existence of an $\mathcal{F}$-cover of M is invoked (while in the proof of (1) neither the existence of a $\mathcal{C}$-envelope of M nor that of an $\mathcal{F}$-cover of P has to be assumed). I do not know if this existence condition is necessary.

References

1. L. Bican, R. El Bashir, E. E. Enochs, All Modules have Flat Covers, Bull. London Math. Soc. **33** (2001), 385–390.
2. E. E. Enochs, O. M. G. Jenda, Relative Homological Algebra, *de Gruyter Expositions in Mathematics* **30**, Walter de Gruyter, 2000.
3. B. Stenström, Rings of Quotients, Springer–Verlag, Berlin 1975
4. J. Xu, Flat Covers of Modules, *Lecture Notes in Mathematics* **1634**, Springer-Verlag, 1996.

A GENERALIZATION OF THE DEMEYER THEOREM FOR CENTRAL GALOIS ALGEBRAS

G. SZETO AND L. XUE

Department of Mathematics, Bradley University,
Peoria, Illinois 61625, USA
E-mail: szeto@bradley.edu; lxue@bradley.edu

Let A be an Azumaya algebra over a semi-local ring R, and M and N finitely generated projective left A-modules such that $\text{rank}_M = \text{rank}_N$. Then $M \cong N$. Thus it can be shown that a central Galois algebra over R is a projective group algebra, and a Galois algebra is a direct sum of projective group algebras.

1. Introduction

Let A be an Azumaya algebra over a semi-local ring R with no idempotents but 0 and 1, and M and N indecomposable finitely generated projective left A-modules. Then it was shown that $M \cong N$ ([3], Theorem 1). Thus the Noether-Skolem theorem can be generalized from central simple algebras to Azumaya algebras over a semi-local ring with no idempotents but 0 and 1, that is, any automorphism of A is inner ([1], page 122). Consequently, any central Galois algebra over a semi-local ring with no idempotents but 0 and 1 is a projective group algebra ([1], Theorem 6). The purpose of the present paper is to generalize the above result to an Azumaya algebra A over a semi-local ring R (not necessarily with no idempotents but 0 and 1). Let M and N be finitely generated projective left A-modules. If the rank functions of M and N over R are equal, then $M \cong N$, where $\text{rank}_M(p) =$ the rank of the free R_p-module M_p over the local ring R_p at the prime ideal p of R. Then we shall show that the Noether-Skolem theorem holds for A, and a central Galois algebra over R with Galois group G is a projective group algebra of G over R, RG_f, with a factor set $f : G \times G \longrightarrow \{\text{units of } R\}$ as defined by F. R. DeMeyer in [1]. Thus a Galois algebra (not necessarily central) over R can be shown to be a direct sum of projective group algebras.

2. Basic Definitions and Notations

Throughout this paper, B will represent a ring with 1, G a finite automorphism group of B, C the center of B, and B^G the set of elements in B fixed under each element in G.

Let A be a subring of a ring B with the same identity 1. We call B a separable extension of A if there exist $\{a_i, b_i$ in B, $i = 1, 2, ..., m$ for some integer $m\}$ such that $\sum a_i b_i = 1$, and $\sum ba_i \otimes b_i = \sum a_i \otimes b_i b$ for all b in B where $\otimes$ is over A. An Azumaya algebra is a separable extension of its center. A ring B is called a Galois extension of B^G with Galois group G if there exist elements $\{a_i, b_i$ in B, $i = 1, 2, ..., m\}$ for some integer m such that $\sum_{i=1}^{m} a_i g(b_i) = \delta_{1,g}$ for each $g \in G$, a Galois algebra over R if B is a Galois extension of R which is contained in C, and a central Galois extension if B is a Galois extension over its center C.

Let P be a projective module over a commutative ring R. Then for a prime ideal p of R, $P_p(= P \otimes_R R_p)$ is a free module over $R_p(=$ the local ring of R at p), and the rank of P_p over R_p is the number of copies of R_p in P_p. We denote the rank function associated with P from the prime spectrum of R to nonnegative integers by rank_P, that is, $\mathrm{rank}_P(p) =$ the number of copies of R_p in P_p.

3. Galois Extensions

Let R be a commutative ring with 1, M a finitely generated projective R-module. We recall that the rank function associated with M from the prime spectrum of R to nonnegative integers is denoted by rank_M. Let A be an Azumaya algebra over a semi-local ring R. We shall characterize a finitely generated projective left A-module M in terms of rank_M. This derives the Noether-Skolem theorem for A. Consequently, it can be shown that any central Galois algebra over R is a projective group algebra, and a Galois algebra over R is a direct sum of projective group algebras where a projective group algebra is defined by F. R. DeMeyer in [1]. We begin with a classification of finitely generated and projective modules over an Azumaya algebra by the rank function.

Lemma 3.1. *Let M and N be finitely generated projective modules over a semi-local ring R. If $rank_M = rank_N = k$ for some integer k, then $M \cong N \cong F_k$ which is a free R-module of rank k.*

Proof. Since R is semi-local, there are minimal idempotents $\{e_i \mid i = 1, 2, ..., m$ for some integer $m\}$ summing to 1. Hence Re_i is a semi-local ring with no idempotents but 0 and e_i such that $\text{rank}_{Me_i} = \text{rank}_{Ne_i} = k$ for each i. Let J be the Jacobson radical of Re_i. Then $Me_i/JMe_i \cong Ne_i/JNe_i$. Thus $Me_i \cong Ne_i \cong F_k e_i$ by using the Nakayama Lemma. This implies that $M \cong N \cong F_k$.

Theorem 3.1. *Let A be an Azumaya algebra over a semi-local ring R, and M and N finitely generated projective left A-modules. If $\text{rank}_M = \text{rank}_N = k$ for some integer k, then $M \cong N$ as left A-modules.*

Proof. Let $\{e_i \,|\, i = 1, 2, ..., m$ for some integer $m\}$ be the set of minimal idempotents in R summing to 1. We claim that $Me_i \cong Ne_i$ for each i. In fact, Let J be the Jacobson radical of Re_i. Noting that $\text{rank}_{Me_i} = \text{rank}_{Ne_i} = k$ (for $\text{Spec}(R) = \cup_{i=1}^m \text{Spec}(Re_i)$), we have that $Me_i \cong Ne_i \cong F_k e_i$ by Lemma 3.1. Thus $Me_i/JMe_i \cong Ne_i/JNe_i$ as left Ae_i/JAe_i-modules (for Ae_i/JAe_i is a direct sum of central simple algebras). Let $\pi : Me_i \longrightarrow Ne_i/JNe_i (\cong Me_i/JMe_i)$ be the surjection homomorphism. Since Ne_i is a finitely generated projective left Ae_i-module such that $Ne_i \longrightarrow Ne_i/JNe_i$ is surjective, there exists a homomorphism $\alpha : Ne_i \longrightarrow Me_i$ such that $Me_i = \alpha(Ne_i) + JMe_i$. But then $Me_i = \alpha(Ne_i)$ by the Nakayama Lemma. This implies that α is a surjection. Let $K = \ker(\alpha)$. Then $0 \longrightarrow K \longrightarrow Ne_i \longrightarrow Me_i \longrightarrow 0$ is a split exact sequence. Since Me_i is a finitely generated projective left Ae_i-module, $Ne_i \cong Me_i \oplus K$. But $\text{rank}_{Me_i} = \text{rank}_{Ne_i}$, so $K_p = 0$ for each $p \in \text{Spec}(Re_i)$. Thus $K = 0$. Therefore $Ne_i \cong Me_i$; and so $N \cong M$.

As a consequence of Theorem 3.1, we have a classification of finitely generated projective left A-modules.

Corollary 3.1. *Let A be an Azumaya algebra over a semi-local ring R, and M and N finitely generated projective left A-modules. If $\text{rank}_M = \text{rank}_N$, then $M \cong N$.*

Proof. Let Q be a finitely generated projective left A-module. Then Q is a finitely generated projective left R-module (for A is an Azumaya algebra over R). Noting that Re_i is a semi-local ring with no idempotents but 0 and e_i, $\text{rank}_{Me_i} = \text{rank}_{Ne_i} = k_i$ for some integer k_i for each i ([3], Theorem 1). Moreover Ae_i is an Azumaya algebra over the semi-local ring Re_i, we have that $Me_i \cong Ne_i$ for each i by Theorem 3.1. Thus $N \cong M$.

Now we show that the Noether-Skolem theorem for Azumaya algebras over a semi-local ring.

Theorem 3.2. *Let A be an Azumaya algebra over a semi-local ring R. If α is an automorphism of A, then α is an inner automorphism.*

Proof. Let A^o be the opposite algebra of A and $A^e = A \otimes_R A^o$. Then A is a left A^e-module by $(x \otimes y)(a) = xay$ for each $x \otimes y \in A^e$ and $a \in A$, which is denoted by A_1. Also, A is a left A^e-module by $(x \otimes y)(a) = \alpha(x)ay$ for each $x \otimes y \in A^e$ and $a \in A$, which is denoted by A_2. Noting that A^e is an Azumaya R-algebra (for A is an Azumaya R-algebra) and that both A_1 and A_2 are finitely generated projective left A^e-modules ([4], Proposition 1.1, page 40) such that $\text{rank}_{A_1} = \text{rank}_{A_2}$, we have that $\pi : A_1 \cong A_2$ as left A^e-modules by Corollary 3.1. Thus for each $a \in A$, $\pi(a) = \pi((a \otimes 1) \cdot 1) = \pi((1 \otimes a) \cdot 1)$, that is, $(a \otimes 1) \cdot \pi(1) = (1 \otimes a) \cdot \pi(1)$. This implies that $\alpha(a) \cdot \pi(1) = \pi(1) \cdot a$. Moreover, since $\pi : A_1 \cong A_2$, there exists an element $b \in A_1$ such that $\pi(b) = \pi(1) \cdot b = 1 = \alpha(b) \cdot \pi(1)$. Thus $\pi(1)$ is a unit in A such that $\alpha(a) = (\pi(1))a(\pi(1))^{-1}$ for each $a \in A$. This implies that α is an inner automorphism of A.

As an application of Theorem 3.2, the structure of a central Galois algebra over a semi-local ring can be derived. As defined by F. R. DeMeyer ([3]), RG_f is called a projective group algebra of a finite group G over a commutative ring R with a factor set $f : G \times G \longrightarrow \{\text{units of } R\}$ if RG_f is a free R-module with a basis $\{x_i \mid g_i \in G, i = 1, 2, ..., m$ for some integer $m\}$ such that $rx_i = x_i r$ for each $r \in R$ and $x_i x_j = x_k \cdot f(g_i, g_j)$ where $g_i g_j = g_k$ for $g_i, g_j \in G$.

Corollary 3.2. *If A is a central Galois algebra over a semi-local ring R with Galois group G, then A is isomorphic with a projective group algebra RG_f with a factor set $f : G \times G \longrightarrow$ the units of R.*

Proof. By Theorem 3.2, G is an inner Galois group of A, so $A \cong RG_f$ ([1], Theorem 6).

By Theorem 3.2, we have the following classes of Galois algebras (not necessarily central) which are also projective group algebras. Thus Theorem 6 in [1] is generalized to Galois algebras over a semi-local ring.

Theorem 3.3. *If B is a Galois algebra with Galois group G over a semi-local ring R with no idempotents but 0 and 1, then B is a projective group algebra.*

Proof. Let C be the center of B and $H = \{g \in G \mid g(c) = c$ for each $c \in C\}$. Then B is a central Galois algebra with Galois group H ([2],

Theorem 1). Moreover, since R is semi-local, C is a semi-local ring. Hence H is inner by Theorem 3.2; and so $B = CH_f$ which is a projective group algebra ([1], Theorem 6).

Theorem 3.4. *Let B be a Galois algebra over a semi-local ring R with Galois group G, C the center of B, $H = \{g \in G \mid g(c) = c$ for each $c \in C\}$, and $J_g = \{a \in B \mid ax = g(x)a$ for every $x \in B\}$. If $J_g = \{0\}$ for each $g \notin H$, then B is a projective group algebra.*

Proof. Since $J_g = \{0\}$ for each $g \notin H$, by Proposition 3 in [5], B is a central Galois algebra with Galois group H. Noting that C is a semi-local ring and that H is inner by Theorem 3.2, we have that B is a projective group algebra ([1], Theorem 6).

In general, for any Galois algebra over a semi-local ring, we shall show that B is a direct sum of projective group algebras. The following lemma for a Galois extension with finitely many central idempotents plays an important role.

Lemma 3.2. *Let B be a Galois extension of B^G with Galois group G. If B contains only finitely many central idempotents, then for any minimal central idempotent e, $(Be)^{G(e)} = B^G e$ where $G(e) = \{g \in G \mid g(e) = e\}$.*

Proof. Since e is minimal, $e \cdot g(e) = e$ or 0 for any $g \in G$. Thus $(Be)^{G(e)} = B^G e$ ([6], Lemma 9).

Theorem 3.5. *Let B be a Galois algebra over a semi-local ring R with Galois group G. If $G(e_i) \neq \{1\}$ for each minimal central idempotent, then B is a direct sum of projective group algebras.*

Proof. Let C be the center of B. Since B is a Galois algebra over a semi-local ring R, C is also a semi-local ring. Hence B has only finitely many central idempotents. Let e be a minimal central idempotent. Then Be is a Galois extension of $(Be)^{G(e)}$ with Galois group $G(e)$ where $G(e) = \{g \in G \mid g(e) = e\}$ ([7], Lemma 3.7). By Lemma 3.2, $(Be)^{G(e)} = B^G e = Re$, so Be is a Galois algebra over Re with Galois group $G(e)$. Noting that Re is a semi-local ring with no idempotents but 0 and e, we conclude that Be is a projective group algebra by Theorem 3.3. But B contains only finitely many central idempotents, so $B = \oplus \sum_{i=1}^{m} Be_i$ where $\{e_i \mid i = 1, 2, ..., m$ for some integer $m\}$ are all minimal central idempotents of B. Therefore B is a direct sum of projective group algebras.

We note that the condition in Theorem 3.5, $G(e_i) \neq \{1\}$, is important to have a nontrivial Galois algebra Be_i over Re_i. In case $G(e_i) = \{1\}$ for some i, we shall employ the structure theorem as given in [7] for B to avoid this situation.

Theorem 3.6. *If B is a Galois algebra over a semi-local ring R with Galois group G, then $B = A \oplus B'$ where A is a commutative Galois algebra with Galois group $G|_A \cong G$ and B' is a direct sum of projective group algebras.*

Proof. By Theorem 3.8 in [7], there exist central idempotents $\{E_j \mid j = 1, 2, ..., n$ for some integer $n\}$ such that $B = BE_0 \oplus (\oplus \sum_{j=1}^{n} BE_j)$ where BE_j is a central Galois algebra over CE_j with Galois group H_j contained in G for each $j = 1, 2, ..., n$ and BE_0 is a commutative Galois algebra over RE_0 with Galois group $G|_{BE_0} \cong G$. Since RE_j is a semi-local ring, CE_j is a semi-local ring; and so BE_j is a projective group algebra for each $j = 1, 2, ..., n$ by Theorem 3.4.

Acknowledgments

This paper was written under the support of a Caterpillar Fellowship at Bradley University. The authors would like to thank Caterpillar Inc. for the support.

References

1. F.R. DeMeyer, Some Notes on the General Galois Theory of Rings, *Osaka J. Math.*, **2**, 117 (1965).
2. F.R. DeMeyer, Galois Theory in Separable Algebras over Commutative Rings, *Illinois J. Math.*, **10**, 287 (1966).
3. F.R. DeMeyer, Projective Modules over Central Separable Algebras, *Canadian J. Math.*, **21**, 39 (1969).
4. F.R. DeMeyer and E. Ingraham, "Separable algebras over commutative rings", Volume 181, Springer Verlag, Berlin, Heidelberg, New York, 1971.
5. T. Kanzaki, On Galois Algebra over a Commutative Ring, *Osaka J. Math.*, **2**, 309 (1965).
6. K. Kishimoto and T. Nagahara, On G-extensions of a semi-connected ring. *Math. J. Okayama Univ.* **32**, 25 (1990).
7. G. Szeto and L. Xue, The Structure of Galois Algebras, *Journal of Algebra*, **237**(1), 238 (2001).

WEAK KRULL DIMENSION OVER COMMUTATIVE RINGS*

GAOHUA TANG

*Department of Mathematics and Computer Science,
Guangxi Teacher's College,
Nanning, 530001, P.R.China*

In this paper, the notion of weak Krull dimension over any commutative ring is introduced and the relations among Krull dimension, weak Krull dimension, weak global dimension, codimension and regularity of coherent local rings are studied. Particularly, the famous Serre's Theorem is partly generalized from Noetherian case to coherent case.

1. Introduction

Throughout this paper it is assumed that all rings are commutative and all modules are unitary.

In this paper, the notion of weak Krull dimension over any commutative ring is introduced and the relations among Krull dimension, weak Krull dimension, weak global dimension, codimension and regularity of coherent local rings are discussed.

It is well-known that Krull dimension is an important invariant in the study of Noetherian rings. For example, the three top theorems of commutative ring theory in order of importance, Krull dimension theorem(or Principal Ideal Theorem), Cohen's structure theorem for complete local rings and Serre's characterization of a regular Noetherian local ring[8], involve Krull dimension. We also know that global dimension and weak global dimension are two of the most important invariant in ring theory and homological algebra. By [9,Theorem 9.22], for a commutative Noetherian ring R, the global dimension of R is equal to the weak global dimension of R. Serre's theorem states that a Noetherian local ring R is regular if and only if $gl.dim R = k.dim R$. This grasps the essence of regular local rings and is

*This work is supported by guangxi natural science foundation(0221029), the support program for 100 young and middle-aged discipliary leaders in guangxi higher education institutions and scientific reserch foundation of guangxi educational committee.

also an important meeting-point of ideal theory and homological algebra. But in the study of coherent rings, the weak global dimension is more effective than the global dimension and it is difficult to use Krull dimension to study coherent rings. So we try to find an analogue of Krull dimension to study coherent rings with weak global dimension. In this paper, we find a new invariant, we call it weak Krull dimension, defined by finitely generated prime ideals(see section 2), which can be used to investigate coherent rings with codimension and weak global dimension. It is somewhat of using Krull dimension, codimension and global dimension to study Noetherian rings.

R is called a regular ring if every finitely generated ideal of R has finite projective dimension[5]. In the case R is a Noetherian ring, the notion of regularity given here coincides with that in [7].

Recall that the set of all prime ideals of R is called the spectrum of R, and written $specR$, the set of maximal ideals of R is called the maximal spectrum of R, and written $Max(R)$; the supremum of the length r, taken over all strictly decreasing chains $P_0 \supset P_1 \supset \cdots \supset P_r$ of prime ideals of R, is called the Krull dimension of R, and denoted $k.dimR$.

The set of all finitely generated prime ideals of R is called the finitely generated prime spectrum of R, and written $f.g.SpecR$; the supremum of the length r, taken over all strictly decreasing chains $P_0 \supset P_1 \supset \cdots \supset P_r$ of finitely generated prime ideals of R, is called the weak Krull dimension, and denoted $w.k.dimR$. Obviously, $w.k.dimR = k.dimR$ if R is a Noetherian. But the converse is not true.

In section 2, we introduce the notion of weak Krull dimension over any commutative rings and prove that for any two natural numbers $m \leq n$, there exists a coherent ring R and a non-coherent ring S such that $w.k.dimR = m$, $k.dimR = n$, $w.k.dimS = m$, and $k.dimS = n$.

In section 3, we study the regularity of coherent local rings. In [7,Theorem 60 and Theorem 69], Kaplansky proved that a Noetherian local ring R is regular if and only if the unique maximal ideal of R is generated by a regular R-sequence. In this section, we prove that a coherent local ring R with finitely generated maximal ideal m is regular if and only if m is generated by a regular R-sequence.

In section 4, we try to extend the Serre's theorem to the coherent case. We prove that if R is a regular coherent local ring with finitely generated maximal ideal then $w.k.dimR = w.gl.dimR$.

In this paper, we use J, $Spec(R)$, $f.g.Spec(R)$, $Max(R)$, $f.g.Max(R)$, $gl.dimR$, $w.gl.dimR$, $pd_R(M)$, $fd_R(M)$, $id_R(M)$, $codim_R(M)$, $FP\text{-}id_R(M)$ for the Jacobson radical, the prime spectrum, the finitely generated prime spectrum, the maximal spectrum, the finitely generated maximal spectrum, global dimension, weak global dimension of R, projective dimension, flat dimension, injective dimension, codimension, FP-injective dimension of R-module M , respectively.

2. Definitions and Examples

The set of all finitely generated prime ideals of a ring R is called the finitely generated prime spectrum of R, and written $f.g.SpecR$; The set of all finitely generated maximal ideals of R is called the finitely generated maximal spectrum of R, and written $f.g.Max(R)$.

Definition 2.1. Let R be a ring and X a subset of $SpecR$. The supremum of lengths r, taken over all strictly decreasing chains $P_0 \supset P_1 \supset \cdots \supset P_r$ of prime ideals of R in X, is called the Krull dimension of X, and denoted $k.dimX$; The supremum of lengths r, taken over all strictly decreasing chains $P_0 \supset P_1 \supset \cdots \supset P_r$ of finitely generated prime ideals of R in X, is called the weak Krull dimension of X, and denoted $w.k.dimX$. When $X = SpecR$, we denote $k.dimR = k.dimX$ and $w.k.dimR = w.k.dimX$, which are called Krull dimension of R and weak Krull dimension of R respectively. Clearly, $w.k.dimR = w.k.dim\{f.g.SpecR\} = k.dim\{f.g.SpecR\}$. If $f.g.SpecR = \emptyset$, we set $w.k.dimR = -1$.

We recall that for a prime ideal P of a ring R, the height of P is defined to be the supremum of the lengths r, taken over all strictly decreasing chains of prime ideals $P = P_0 \supset P_1 \supset \cdots \supset P_r$ starting from P, and denoted htP; Moreover, the supremum of the lengths r, taken over all strictly increasing chains of prime ideals $P = P_0 \subset P_1 \subset \cdots \subset P_r$ starting from P, is called the coheight of P and denoted $cohtP$. It follows from the definitions that $htP = k.dimR_P, cohtP = k.dimR/P$ and $htP + cohtP \leq k.dimR$. For an ideal I of R, the height of I is defined as:
$$htI = inf\{htP \mid I \subseteq P \in SpecR\}$$

Definition 2.2. Let R be a ring and X a subset of $SpecR$. For a prime ideal P in X, we define the height of P in X to be the supremum of the lengths r, taken over all strictly decreasing chains of prime ideals in X, $P = P_0 \supset P_1 \supset \cdots \supset P_r$ starting from P, and denoted ht_XP; the coheight of P in X to be the supremum of the lengths r, taken over all strictly increasing chains of prime ideals in X, $P = P_0 \subset P_1 \subset \cdots \subset P_r$ starting from P, and denoted $coht_XP$.

When $X = f.g.SpecR$ and $P \in f.g.SpecR$, $ht_X P$ and $coht_X P$ are called weak height of P and weak coheight of P respectively, denoted $w.htP$ and $w.cohtP$ respectively. It is clear that if $X = SpecR$ then $ht_X P = htP$ and $coht_X P = cohtP$.

It follows from the above definitions that
$w.htP \leq w.k.dimR_P, w.cohtP = w.k.dimR/P$ and $w.htP + w.cohtP \leq w.k.dimR$, for any $P \in f.g.SpecR$.

For an ideal I of R, we define the weak height of I in R as:
$$w.htI = inf\{w.htP \mid I \subseteq P \in f.g.SpecR\}$$
Here also we have the inequality
$$w.htI + w.k.dimR/I \leq w.k.dimR.$$

For a Noetherian ring R, it is obvious that $w.htI = htI$ for any ideal I of R and $w.k.dimR = k.dimR$. But, in general, the converse is not true.

Lemma 2.3. *Let D be a domain and $x_1, \cdots, x_n$ indeterminates over K, where K is the quotient field of D. Let $R = \{f \in K[x_1, \cdots, x_n] | f(0) \in D\}$, that is $R = D + K[x_1, \cdots, x_n]x_1 + \cdots + K[x_1, \cdots, x_n]x_n$.*

(1) R is Noetherian if and only if $D = K$;

(2) If D is not a field then R is coherent if and only if D is coherent and $n = 1$;

(3) $\forall P \in SpecD - \{0\}, R/PR \simeq D/P$ and therefore $PR \in SpecR$. For the sake of convenience, we also denote PR by P;

(4) If $Q \in SpecR$ and $Q \cap D \neq 0$, then $P(= Q \cap D)$ is a prime ideal of D and $PR = Q$;

(5) $SpecR = Y' \overset{.}{\cup} X$, where $Y' = \{P \in SpecR \mid P \cap D \neq 0\}$ and $X = \{P \in SpecR \mid P \cap D = 0\}$;

(6) Set $Y = Y' \cup \{0\}$. There exists a one-to-one order-preserving correspondence between Y and $SpecD$ by $P \to P \cap D$ for any $P \in Y$ and there exists a one-to-one order-preserving correspondence between X and $SpecK[x_1, \cdots, x_n]$;

(7) $X = X_1 \overset{.}{\cup} X_2$, where $X_1 = \{P \in X \mid \forall f(x) \in P, f(0) = 0\}$ and $X_2 = \{P \in X \mid \exists f(x) \in P, \text{ such that } f(0) \neq 0\}$;

(8) $\forall P \in X_1, \forall \alpha \in D - \{0\}$, we have $P \subset \alpha R$ and therefore $\forall P \in X_1, P' \in Y'$, we have $P \subset P'$;

(9) $\forall P \in X_1, P \neq 0$, P is infinitely generated;

(10) Set $P_n = \{f \in K[x_1, \cdots, x_n] \mid f(0) = 0\}$. Then P_n is the unique maximal element of X_1 and $ht_{X_1}P_n = htP_n = n$;

(11) $\forall P \in X_2, P' \in Y'$, we have $P \nsubseteq P'$ and $P' \nsubseteq P$;

(12) $k.dimY = k.dimD; \; w.k.dimY = w.k.dimD; \; k.dimX = k.dimX_1 = k.dimX_2 = n; \; w.k.dimX = w.k.dim(X_2 \cup \{0\}) = n; \; w.k.dimX_1 = 0;$

    (13) $k.dimR = k.dimD + n; \; w.k.dimR = max\{n, \, w.k.dimD\}.$

Theorem 2.4. *For any natural numbers $s \geq t \geq 1$, there exists a coherent local ring R with finitely generated maximal ideal m such that $w.k.dimR = t$, $k.dimR = s$ and m can be generated by a regular R-sequence of t elements.*

Theorem 2.5. *For any natural numbers $s > t \geq 2$, there exists a noncoherent ring R such that $w.k.dimR = t$, $k.dimR = s$.*

3. A Characterization of Regular Coherent Local Rings

Definition 3.1. Let R be a ring and M an R-module. The FP-injective dimension of M, denoted by $FP\text{-}id_R(M)$, is equal to the least integer $n \geq 0$ for which $Ext_R^{n+1}(P, M) = 0$ for every finitely presented R-module P. If no such n exists set $FP\text{-}id_R(M) = \infty$.

Lemma 3.2. *([6, Theorem 3 and Theorem 5]) If R is a coherent semilocal ring with Jacobson radical J and M is a finitely presented R-module, then*

    (1) The following statements are equivalent.

        (a) $pd_R M \leq n$;

        (b) $fd_R M \leq n$;

        (c) $Tor_{n+1}^{R}(M, R/J) = 0$;

        (d) $Ext_{n+1}^{R}(M, R/J) = 0$.

    (2) $w.gl.dimR = fd_R R/J = id_R R/J = FP\text{-}id_R R/J$.

Lemma 3.3. *([10, Theorem 2.6]) Let R be a coherent ring and M a finitely presented R-module. If m is a maximal ideal of R satisfying $fd_{R_m} M_m = fd_R M = n < \infty$ and $\alpha_1, \cdots, \alpha_s$ is a regular M-sequence in m, then*

$$pd_R(M/(\alpha_1, \cdots, \alpha_s)M) = pd_R M + s \; and$$
$$pd_{R_m}(M_m/(\alpha_1, \cdots, \alpha_s)M_m) = pd_R(M/(\alpha_1, \cdots, \alpha_s)M)$$

Lemma 3.4. *([10, Theorem 2.8]) Let R be a coherent ring and M a nonzero R-module and m a maximal ideal of R satisfying $pd_{R_m} M_m = pd_R M$.*

    (1) If M is finitely presented, then $pd_R M + m - codim_R M \leq w.gl.dimR_m$;

    (2) If M is finitely presented Noetherian R-module and $w.gl.dimR_m < \infty$, then $pd_R M + m - codim_R M = w.gl.dimR_m$.

Lemma 3.5. *Let R be a coherent local ring and M a finitely presented R-module. Then $codim_R M \leq w.gl.dim R$, particularly, $codim_R R \leq w.gl.dim R$.*

Theorem 3.6. *Let (R, m) be a coherent local ring with finitely generated maximal ideal m. Then the following statements are equivalent:*
  (a) R is regular;
  (b) $w.gl.dim R < \infty$;
  (c) $fd_R(R/m) < \infty$;
  (d) $id_R(R/m) < \infty$;
  (e) $FP\text{-}id_R(R/m) < \infty$;
  (f) $pure\text{-}dim R < \infty$.

Theorem 3.7. *Let (R, m) be a coherent local ring with finitely generated maximal ideal m. Then the following statements are equivalent:*
  (a) R is regular;
  (b) m is generated by a regular R-sequence.

Proof: $(b) \Rightarrow (a)$. Suppose m is generated by a regular R-sequence $\{\alpha_1, \cdots, \alpha_q\}$. Then R/m is a finitely presented R-module. By Lemma 3.2 and Lemma 3.3, we can get $w.gl.dim R = pd_R(R/(\alpha_1, \cdots, \alpha_q)R) = pd_R R + q = q < \infty$, which implies that R is regular by Theorem 3.6.

$(a) \Rightarrow (b)$. Since (R, m) is regular, by Theorem 3.6, we have $w.gl.dim R < \infty$. It follows from Lemma 3.5 that $codim_R R \leq w.gl.dim R < \infty$. Set $codim R = t$. We will prove the conclusion by induction on t.

If $t = 0$, then every element $\alpha \in m$ is a zero-divisor of R. Since R is a regular coherent local ring, it follows from [12,Corollary 5] that R is a GCD domain and thus $m = 0$ and the result is true when $t = 0$.

Now suppose $t > 0$. m/m^2 is a vector space over the field $K = R/m$. Since m is finitely generated, $dim_K(m/m^2) < \infty$. Assume $dim_K(m/m^2) = n$. Obviously $n \geq 1$.

For any $\alpha_1 \in m - m^2$, we use $\overline{\alpha}_1$ to denote the image of α_1 at the natural map $m \to m/m^2$. Clearly, $\overline{\alpha}_1 \neq 0$. So $\overline{\alpha}_1$ can be extended to a base $\{\overline{\alpha}_1, \overline{\alpha}_2, \cdots, \overline{\alpha}_n\}$ of m/m^2 over K. By [1,Exercise 1 at page 294], $\{\alpha_1, \alpha_2, \cdots, \alpha_n\}$ is a minimal set of generators of m. It follows from [12,Corollary 5] that R is a GCD domain and thus all of $\alpha_1, \cdots, \alpha_n$ are not zero divisors of R. Set $\overline{R} = R/\alpha_1 R$, then $\overline{R}$ is also a coherent local ring with unique maximal ideal $\overline{m} = m/\alpha_1 R$ and $\overline{m} = (\overline{\alpha}_2, \cdots, \overline{\alpha}_n)$. If $\overline{u}_2, \cdots, \overline{u}_p$ is a regular $\overline{R}$-sequence, then $\alpha_1, u_2, \cdots, u_p$, is obviously a regular R-sequence, where $u_i(2 \leq i \leq p)$ is a preimage of $\overline{u}_i$ at the natural map $\pi \colon R \to R/\alpha_1 R$.

Thus

$$codim_{\overline{R}}\overline{R} \le codim_R(R) - 1 = t - 1.$$

In order to apply the inductive hypothesis, we need to prove $\overline{R}$ is regular. It is sufficient to show that $w.gl.dim\overline{R} < \infty$, by Theorem 3.6.

Set $A = \alpha_1 R$, $B = \alpha_1 m + \alpha_2 R + \cdots + \alpha_n R$. It is easy to verify that

$$m = A + B, \; \alpha_1 m = A \cap B$$

So

$$m/\alpha_1 m = (A + B)/\alpha_1 m \simeq (A/\alpha_1 m) \bigoplus (B/\alpha_1 m)$$

and hence

$$\overline{m} = m/\alpha_1 R \simeq (m/\alpha_1 m)/(\alpha_1 R/\alpha_1 m)$$

$$\simeq (m/\alpha_1 m)/(A/\alpha_1 m) \simeq B/\alpha_1 m,$$

that is, $\overline{m}$ is isomorphic to a direct summand of $m/\alpha_1 m$ as an R-module, and so as an $\overline{R}$-module. Therefore $pd_{\overline{R}}(\overline{m}) \le pd_{\overline{R}}(m/\alpha_1 m)$. Since α_1 is not a zero divisor on m and R, by [5,Theorem 3.1.2] we have

$$pd_{\overline{R}}(\overline{m}) \le pd_{\overline{R}}(m/\alpha_1 m) = pd_{R/\alpha_1 R}(m/\alpha_1 m) = pd_R(m) < \infty$$

and

$$pd_{\overline{R}}(\overline{R}/\overline{m}) \le pd_{\overline{R}}(\overline{m}) + 1 < \infty$$

So by Lemma 3.2, we have

$$w.gl.dim\overline{R} = fd_{\overline{R}}(\overline{R}/\overline{m}) = pd_{\overline{R}}(\overline{R}/\overline{m}) < \infty.$$

By induction hypothesis, $\overline{m}$ can be generated by a regular $\overline{R}$-sequence $\{\overline{\beta}_2, \cdots, \overline{\beta}_q\}$. Thus $\overline{m} = \overline{\beta}_2 \overline{R} + \cdots + \overline{\beta}_q \overline{R}$ and $\alpha_1, \beta_2, \cdots, \beta_q$ is a regular R-sequence, where β_i is a preimage of $\overline{\beta}_i (2 \le i \le q)$ at the natural map π: $R \to \overline{R}$, and it is easy to verify that $m = \alpha_1 R + \beta_2 R + \cdots + \beta_n R$, which completes our proof. $\qquad \square$

Remark. Let R be a regular coherent local ring with finitely generated maximal ideal m. From the proof of Theorem 3.7, it is not difficult to see that every minimal set $\{\alpha_1, \cdots, \alpha_n\}$ of generators of m is a regular R-sequence and for any $\alpha \in m - m^2$, it can be extended to a regular R-sequence $\{\alpha, \alpha_2, \cdots, \alpha_n\}$ such that $m = \alpha R + \alpha_2 R + \cdots + \alpha_n R$ and $\alpha_i \in m - m^2$, $2 \le i \le n$.

Corollary 3.8. *Let R be a regular coherent local ring with finitely generated maximal ideal m. Then*

$$w.gl.dim R = codim_R R = q$$

222

where q is the length of a regular R-sequence $\{\alpha_1, \cdots, \alpha_q\}$ which generates m: $m = \alpha_1 R + \cdots + \alpha_q R$.

Corollary 3.9. *Let R be a regular coherent local ring with maximal ideal m and let $\alpha_1, \cdots, \alpha_q$ be a maximal regular R-sequence. Then the following statements are equivalent:*

    (a) $m = \alpha_1 R + \cdots + \alpha_q R$;
    (b) $w.gl.dim_R(R/(\alpha_1, \cdots, \alpha_q))R = 0$;
    (c) $w.gl.dim_R(R/(\alpha_1, \cdots, \alpha_q))R < \infty$.
    If any condition above holds, then $w.gl.dim R = codim_R(R) = q < \infty$.

Theorem 3.10. *For any natural numbers $s \geq t \geq 1$, there exists a regular coherent local ring R with finitely generated maximal ideal such that $w.gl.dim R = w.k.dim R = t$ and $k.dim R = s$.*

Proof: By Theorem 2.4, for any natural numbers $s \geq t \geq 1$, there exists a coherent local ring R with finitely generated maximal ideal m such that $w.k.dim R = t$, $k.dim R = s$ and m can be generated by a regular R-sequence $\{\alpha_1, \cdots, \alpha_t\}$. Thus by Theorem 3.7, R is regular and by Corollary 3.8, $w.gl.dim R = t$. $\qquad\qquad\square$

4. A Generalization of Serre's Theorem

Serre's Theorem states that if R is a Noetherian local ring, then R is regular if and only if $gl.dim R = k.dim R$[8,Theorem19.2]. In this section we try to extend this result to the case of coherent local rings. The main result of this section is that if R is a regular coherent local ring with finitely generated maximal ideal then $w.gl.dim R = w.k.dim R$ (Theorem 4.2).

Lemma 4.1. *If P is a finitely generated prime ideal of R which properly contained in m, then $w.gl.dim R_P < w.gl.dim R$.*

Proof: Since $w.gl.dim R_P \leq w.gl.dim R < \infty$, R_P is also a regular coherent local ring with unique finitely generated maximal ideal PR_P. Set $w.gl.dim R_P = t$. By Lemma 3.2, $pd_{R_P}(R_P/PR_P) = t < \infty$, which implies that
$Tor^t_{R_P}(R_P/PR_P, R_P/PR_P) \neq 0$ and therefore $Tor^t_R(R/P, R/P) \neq 0$.
So $pd_R(R/P) = fd_R(R/P) \geq t$. On the other hand, R/P is a finitely presented R-module, from Lemma 3.4, it follows that $pd_R R/P + codim_R(R/P) \leq w.gl.dim R$. Since P is properly contained in m, $codim_R(R/P) \geq 1$. Hence $pd_R(R/P) < w.gl.dim R$. Therefore $w.gl.dim R_P = pd_{R_P}(R_P/PR_P) = t \leq pd_R(R/P) < w.gl.dim R$. $\qquad\square$

Theorem 4.2. *Let R be a coherent local ring with finitely generated maximal ideal m. If R is regular then $w.gl.dimR = w.k.dimR$.*

Proof: From the proof of Theorem 3.7, we know that if m is generated by a regular R-sequence $\alpha_1, \cdots, \alpha_n$, then $m = (\alpha_1, \cdots, \alpha_n) \supset (\alpha_2, \cdots, \alpha_n) \supset (\alpha_3, \cdots, \alpha_n) \supset \cdots \supset (\alpha_n) \supset 0$ is a strictly decreasing chain of finitely generated prime ideals. Thus $w.k.dimR \geq n = w.gl.dimR$.

Now we need only to prove $w.k.dimR \leq w.gl.dimR$. We prove it by induction on n, where $n = w.gl.dimR$.

If $n = 0$, then R is a field. Clearly $w.k.dimR = 0 = w.gl.dimR$.

Now suppose $n > 0$ and the inequality $w.k.dimR \leq w.gl.dimR$ holds for any regular coherent local ring with finitely generated maximal ideal and $w.gl.dimR < n$. Assume that R is a regular coherent local ring with finitely generated maximal ideal m and $w.gl.dimR = n$. By Lemma 4.1, for any $P \in f.g.SpecR - m$, $w.gl.dimR_P < w.gl.dimR$. By the induction hypothesis, we have $w.k.dimR_P = w.gl.dimR_P$. Thus

$$
\begin{aligned}
w.k.dimR &= sup\{w.htP \mid P \in f.g.SpecR\} \\
&= sup\{w.htP \mid P \in f.g.SpecR,\ P \neq m\} + 1 \\
&\leq sup\{w.k.dimR_P \mid P \in f.g.SpecR,\ P \neq m\} + 1 \\
&= sup\{w.gl.dimR_P \mid P \in f.g.SpecR,\ P \neq m\} + 1 \\
&\leq w.gl.dimR.
\end{aligned}
$$

$\square$

Corollary 4.3. *Let R be a coherent ring . Then for any $P \in f.g.SpecR$, we have*

$$
w.htP \leq w.k.dimR_P \leq w.gl.dimR_P \leq pd_R R/P.
$$

Corollary 4.4. *Let R be a coherent ring . Then $w.k.dimR \leq w.gl.dimR$.*

Remark 1. The condition "coherent" in Theorem 4.2 is necessary. From the following theorem we can see that there exists a non-coherent local ring R with finitely generated maximal ideal such that $w.k.dimR < w.gl.dimR$.

Theorem 4.5. *For any natural numbers $s > t > 0$, there exists a non-coherent local ring R with finitely generated maximal ideal m such that $w.k.dimR = t,\ w.gl.dimR \geq s$.*

Remark 2. We conjecture that the converse of Theorem 4.2 is true.

References

1. Cheng, F.C., Homological Algebra, Guangxi Normal Univ. Press: Guilin, 1989.
2. Feng, K.Q., A First Course of Commutative Algebra, Higher Education Press:Beijing, 1985.
3. Gilmer, R., Multiplicative Ideal Theory, Marcel Dekker. INC. New York, 1972.
4. Gilmer, R., Prüfer domains and rings of integer-valued polynomials, J.Algebra, 1990, 129, 502-517.
5. Glaz, S., Commutative Coherent Rings, Lecture Notes in Math. 1371, Springer Verlag:Berlin Heidelberg, 1989.
6. Huang, Z.Y., Homological dimension over coherent semilocal rings II, Pitman Research Notes in Math. Series 1996, 346, 207-210.
7. Kaplansky, I., Commutative Rings, Univ. of Chicago Press:Chicago, 1974.
8. Matsumura, H., Commutative Ring Theory, Cambridge Univ. Press:Londeon, 1979.
9. Rotman, J.J., An Introduction to Homological Algebra, Academic Press, INC. London, 1979.
10. Tang,G.H.;Yin,X.B.;Tong,W.T., A Generalization of Auslander-Buchsbaum Theorem, (to appear)
11. Vasconcelos, W.V., The Rings of Dimension Two, Marcel Dekker. INC. New York And Basel, 1976.
12. Zhao, Y.C., On commutative indecomposable coherent regular rings, Comm. in Alg. 1992, 20(5), 1389-1394.

BAER PROPERTY OF MODULES AND APPLICATIONS

S. TARIQ RIZVI*

*Department of Mathematics
The Ohio State University
Lima, OH 45804-3576, USA
E-mail: rizvi.1@osu.edu*

COSMIN S. ROMAN

*Department of Mathematics
The Ohio State University
Lima, OH 45804-3576, USA
E-mail: cosmin@math.ohio-state.edu*

The notion of Baer rings has been of interest for several decades. Recently, the notion of Baer property was introduced in the module theoretical setting [26]. In this survey paper we discuss how this module theoretic concept fits in with the existing theory and present connections of Baer modules to extending modules. Some applications are presented, including a type decomposition for nonsingular extending modules.

1. Introduction and Preliminaries

The notion of Baer rings has its roots in functional analysis. For example, von Neumann algebras, such as the *-algebra of bounded operators on a Hilbert space containing the identity operator which are closed under the weak operator topology (also called W^*-algebras), possess a plethora of structures - algebraic, geometric and topological. For an algebraist, a boon is a rich supply of idempotents which these algebras have. In order to obtain an insight into the theory of von Neumann algebras, several authors started to axiomatize this theory, including I.M. Gel'fand, F.J. Murray, M.A. Naĭmark, von Neumann, C.E. Rickart and S.W.P. Steen. Algebraically, in any von Neumann algebra the right annihilator of any subset is generated as a right ideal by a projection (i.e. a self-adjoint idempotent

---

*Work partially supported by a research grant from The Ohio State University, Lima.

with respect to the involution *). Kaplansky [13], in 1951, defined the concept of abstract W^*-algebras, or AW^*-algebras, which took into account mainly the algebraic structure of von Neumann algebras (AW^*-algebras are Banach algebras with an involution such that $\|xx^*\| = \|x\|^2$ and which have the property that the right annihilator of any subset is generated by a projection). He also made the connection with von Neumann's study of continuous geometries, by noticing that the projection lattice of a "directly finite" AW^*-algebra is a continuous geometry [14]. Kaplansky in 1955 [15] defined the larger class of Baer *-rings by focusing on annihilators and projections of AW^*-algebras. A Baer *-ring is defined as a ring with involution in which the right annihilator of every subset (or left ideal) is a principal right ideal generated by a projection. The name honors Reinhold Baer, who studied this condition earlier in his book "Linear Algebra and Projective Geometry". Dropping the assumption of an involution in this definition, led Kaplansky to the concept of a Baer ring.

A ring is called *Baer* if the right annihilator of any left ideal (or any subset) is a right ideal generated by an idempotent. A number of interesting properties of Baer rings were shown by Kaplansky and this theory was further developed by several other mathematicians. Large classes of rings satisfy the Baer property (see Example 2.2). The theory of Baer rings has come to play an important role and major contributions to this theory have been made in recent years, providing a number of interesting results in the ring-theoretical setting. Some of the contributors include S.K. Berberian, G. F. Birkenmeier, A. W. Chatters, S. M. Khuri, J. Y. Kim, Y. Hirano, J. K. Park, A. Pollingher, K.G. Wolfson and A. Zaks, among others (see, for example, [31], [24], [20], [7], [5], [6], [3]).

For a given Baer ring R, a natural question that can be asked is: does the right module eR, for any $e^2 = e \in R$, have any kind of "Baer-ness"? More generally, what can be a suitable module theoretic analogue of a Baer ring? Connections of this analogue to the much studied concept of extending module are also of interest, in view of the well known connections of Baer rings to extending rings. We provide a suitable definition of a Baer module and show that this is a natural generalization of a Baer ring. The difficulties in our investigations of this new notion of modules arise due to the interplay of the base ring on one side of the module and the endomorphism ring on the other side. We show that many of the results known for Baer rings "lift" to the module case. However, due to the interplay mentioned above, the proofs are quite different and require new techniques and tools. In this paper, after providing the background in Section 1, and a selection of some basic results on Baer rings in Section 2, we provide the definition and a number of our results on Baer modules and their properties. In Section 4, connections between Baer modules and extending modules are established and in Section 5 we include results on endomorphism rings of Baer modules and show some applications of the theory. Open questions are provided at the end of the paper.

Throughout the paper, unless stated otherwise, ring properties are assumed to be on the right (e.g. right extending rings, right p.p. rings). All rings have an identity element and modules are unital right modules.

Recall that a ring R is called *(von Neumann) regular* if for every x in R there exists y in R such that $x = xyx$. A module is called *extending* (or *CS*) if every submodule is essential in a direct summand. A ring R is *right extending* if R_R is an extending module.

We denote the center of a ring R by $Z(R)$. The left annihilator of a set X in a ring (or module) Y will be denoted by $l_Y(X)$ and the right annihilator by $r_Y(X)$. $N \leq^e M$ denotes that the submodule N is essential in the module M; $N \leq^c M$ denotes that the submodule N is essentially closed in the module M. For a ring R, a right R-module M, $S = End_R(M)$ denotes the endomorphism ring of M.

2. Baer Rings

Integral domains form a large class of rings, having useful properties. One of the properties of domains is that right (or left) annihilator of a set is (trivially) a direct summand of R. A more general and larger class of rings satisfying this property is precisely the class of Baer rings.

Definition 2.1. ([15]) A ring R is called *Baer* if the right annihilator of every subset of R is of the form eR, where $e^2 = e \in R$ is an idempotent. Equivalently, a ring R is *Baer* if the left annihilator of every subset of R is of the form Re, where $e^2 = e \in R$ is an idempotent. Thus, the definition is left-right symmetric.

Baer rings are ubiquitous in literature, as is evident from the few examples below.

Example 2.2. (1) Any domain (with a unit); (2) the ring of linear transformations on a vector space over a division ring; more generally, the ring of endomorphism of any semisimple module; (3) the ring of all bounded operators on a Hilbert space is a Baer ring (suitable subrings are also Baer), von Neumann algebras; (4) considering a pair of dual vector spaces V, W

over a division ring, then R is Baer, where R is the ring of "continuous" linear transformations on V, i.e. linear transformations that have an adjoint on W, granted that the pair V, W is splittable as defined in [23] (as a special case, this property applies when both V, W have countable dimension); (5) the direct product of any family of Baer rings is a Baer ring; (6) any right self-injective, regular ring, or any right semihereditary right noetherian ring.

If we restrict the above condition to annihilators of single elements, we obtain an even more general concept.

Definition 2.3. A ring R is called *right Rickart* if the right annihilator of each element in R is of the form eR, where $e^2 = e \in R$ is an idempotent.

The class of right Rickart rings properly contains the class of Baer rings (see Example 7.4 in [18]). However, if the ring has no infinite set of orthogonal idempotents, the two concepts coincide by a result of Small ([18]).

Unlike Baer rings, the concept of Rickart rings is not left-right symmetric. On the other hand, such rings enjoy certain homological properties: a ring R is right Rickart if and only if it is right p.p. (i.e. each principal right ideal is projective) as a right R-module. Thus any right semihereditary ring is a right Rickart ring.

Some basic properties of Baer rings include:

Theorem 2.4. *(Theorems 4, 6, 7, 9 [15]; 3.7, [2])*

> *(1) If R is a Baer ring and e is an idempotent in R, then eRe is a Baer ring.*
>
> *(2) In any Baer ring the annihilator of any central subset is a (ring) direct summand.*
>
> *(3) The center of a Baer ring is a Baer ring.*
>
> *(4) Let x be an element in a Baer ring. In the boolean algebra of central idempotents there exists a smallest v satisfying $vx = x$.*

The smallest central idempotent, obtained in theorem above, is called the *central cover* of x, and is denoted by $C(x)$.

Example 2.5. Let $R = M_n(\mathbb{Z})$ and $1 \leq k \leq n$. The central idempotents are matrices that have either 0 or 1 on the diagonal. The central cover of an element x is the central idempotent matrix v with fewest count of 1 on its diagonal, subject to $vx = x$. For instance, if $\forall\, k,\, 1 \leq k \leq n$, there exists a column of x so that in kth position there is a nonzero integer, then $C(x) = I_n$.

If the ring is reduced (i.e. with no nonzero nilpotent elements) and satisfies a polynomial identity (PI ring), the center of the ring is Baer if and only if the ring is Baer ([1]).

In general, the Baer property does not always extend to matrix rings or polynomial extension. For example, let $R = M_2(\mathbb{Z})$; then R is a Baer ring, but $R[x]$ is not Baer ([1]; [12]). However, for matrix ring extensions we have the following result.

Theorem 2.6. *([15]) For a commutative integral domain R, the following are equivalent:*
(a) R is Prüfer
(b) $M_n(R)$ is a Baer ring, for $n > 1$.

Furthermore, when the ring is reduced, the Baer property is carried over to the polynomial rings, namely for a reduced ring R, $R[X]$ is a Baer ring if and only if R is a Baer ring ([1]).

The Baer property also provides an insight on the structure of ideals of a ring; for example:

Proposition 2.7. *([30]) A regular ring is Baer if and only if its lattice of principal right ideals is complete.*

It is easy to see from its definition that any Baer ring is right (and left) nonsingular. In the presence of nonsingularity, large classes of rings satisfy the Baer property.

Proposition 2.8. *([2]) Let R be a right nonsingular ring.*

(1) If R is a right self-injective then R is a regular Baer ring;
(2) The maximal ring of quotients of R is regular, right self-injective, hence Baer.

The first statement can, in fact, be generalized to (right) continuous rings (see [21]).

Since idempotents play an important role in a Baer ring, countability of the set of idempotents, in such a ring, yields interesting consequences.

Proposition 2.9. *(Theorem 1, [29]) If R is a right Rickart ring with no infinite set of orthogonal idempotents, then R is a Baer ring.*

Theorem 2.10. *(1.9. [8]) A regular Baer ring with countably many idempotents is compressible (i.e., for each idempotent e of R, $Z(eRe) = eZ(R)$).*

Theorem 2.11. *(Theorem 2, [17]) If R is a Baer ring with only countable many idempotents, then R has no infinite set of orthogonal idempotents.*

In the case when R is regular, more can be said.

Theorem 2.12. *(Theorem 3, [17]; 1.5, [8]) Let R be a regular Baer ring with only countable many idempotents. Then R is semisimple artinian.*

Corollary 2.13. *(Theorem 1, [25]; 1.6, [8]; Corollary 4, [17]) Any countable regular Baer ring is semisimple artinian.*

We now focus our attention on special idempotents, which determine a decomposition of a Baer ring, and allow for a type theory for Baer rings to develop.

Definition 2.14. An idempotent e in a Baer ring is called *faithful* if 0 is the only central idempotent orthogonal to e. Equivalently, e is faithful if the smallest central idempotent v in S satisfying $ve = e$ is 1 (i.e. $C(e) = 1$; recall that such a central cover always exists in a Baer ring).

Definition 2.15. A Baer ring R is called *abelian* if all idempotents of R are central (i.e. commute with any other element of R).

Definition 2.16. A ring is *finite* (also known as *directly finite*) if $xy = 1$ implies $yx = 1$, for x, y elements of the ring. An idempotent e in the ring R is *finite* if eRe is finite. A ring (respectively an idempotent) is *infinite* if it is not finite. An idempotent is called *purely infinite* if eRe contains no nonzero central, finite idempotent.

Using the above, Kaplansky defined various types of Baer rings (based on the study in [22]), as follows.

Definition 2.17. ([15]) A Baer ring is of *type I* if it has a faithful abelian idempotent. A Baer ring is of *type II* if it has a faithful finite idempotent, but no non-zero abelian idempotents. It is of *type III* if it has no nonzero finite idempotents. A Baer ring is *purely infinite* if it has no nonzero central finite idempotents.

Kaplansky [15] proved that any Baer ring can be uniquely decomposed as a direct sum of ring direct summands of these three, main, types. Also, by further decomposing the type I and type II summands, respectively, into a direct sum of a directly finite and a purely infinite part, he further refined this decomposition to a total of five types. Thus, a Baer ring decomposes uniquely into a sum of five components, as described below.

A simple artinian ring is a Baer ring of type I, finite, while the endomorphism ring of any infinite-dimensional vector space is a Baer ring of type I, infinite. Examples of type II and type III can be found in [10] (e.g. Example 10.7 and Example 10.11).

Theorem 2.18. *([15]) A Baer ring decomposes uniquely into a ring direct sum of Baer rings of types: I and directly finite (I_f); I and purely infinite (I_∞); II and directly finite (II_f); II and purely infinite (II_∞); III.*

3. Baer Modules

In this section we introduce the notion of the Baer property of rings in the general module theoretic setting. Most of the results presented in sections 3, 4 and 5 are drawn from [26], [28] and [27]. The details of proofs and discussions can be found in those references. For proofs of results 3.4 through 3.12, see [26]. One of the motivations for our study in this module theoretic setting is to answer the question: Given a Baer ring R, what kind of Baer property does the right R-module eR have (where $e^2 = e \in R$ is an idempotent)? Another motivation follows from a result of Chatters and Khuri, which characterizes Baer rings in terms of nonsingular extending rings. More precisely, can we obtain a characterization of nonsingular extending modules in terms of a Baer property of modules?

Recall that, for any right R-module M we let $S = End_R(M)$.

Definition 3.1. ([26]) M is called a *Baer module* if $\forall N \leq M, l_S(N) = Se$, with $e^2 = e \in S$. Equivalently, $\forall I \leq {}_SS, r_M(I) = eM$ where $e^2 = e \in S$.

Example 3.2. Every Baer ring is a Baer module over itself. All semisimple modules are obviously Baer modules. $\mathbb{Z}^n$ is a Baer $\mathbb{Z}$-module, $\forall n \in \mathbb{N}$. More examples will be evident later.

Definition 3.3. A module M is said to have the *summand intersection property (SIP)* if the intersection of any two direct summands of M is a direct summand. M is said to have the *strong summand intersection property (SSIP)* if the intersection of any family of direct summands of M is a direct summand.

232

The following useful characterization shows the connections of Baer modules to modules with SSIP.

Theorem 3.4. *A module M is Baer if and only if M has the strong summand intersection property and $Ker(\varphi) \leq^{\oplus} M$, $\forall\ \varphi \in S$.*

An obvious consequence of Theorem 3.4 is the following description of indecomposable Baer modules.

Theorem 3.5. *M is an indecomposable Baer module if and only if $\forall\ 0 \neq \varphi \in End(M)$, φ is a monomorphism.*

Next we show that the Baer property is inherited by direct summands, consequently answering one of the questions posed above.

Theorem 3.6. *Let M be a Baer module. Then every direct summand N of M is also a Baer module.*

This yields the following interesting consequence, showing that all right ideals eR have the module Baer property if R is a Baer ring and $e^2 = e \in R$.

Corollary 3.7. *Let R be a Baer ring, and let $e^2 = e \in R$ be any idempotent of R. Then $M = eR$ is an R-module which is Baer.*

It is of interest to know which abelian groups are Baer. For the finitely generated abelian groups we obtain:

Proposition 3.8. *A finitely generated $\mathbb{Z}$-module M is Baer if and only if M is either semisimple or torsion-free.*

We remark that if a Baer module M can be decomposed into a finite direct sum of indecomposable summands, then it can be shown that any other arbitrary direct sum decomposition of M is finite.

Next we show that the direct sum of Baer modules need not be Baer.

Example 3.9. The modules $\mathbb{Z}$ and $\mathbb{Z}_p$, where p is a prime integer, are Baer $\mathbb{Z}$-modules (the former is a domain, hence a Baer ring; the latter is a simple module). On the other hand, the module $M = \mathbb{Z} \oplus \mathbb{Z}_p$ is not Baer (the kernel of the endomorphism $\varphi(m, \hat{n}) = \hat{m}$ is not a direct summand of M).

The question when direct sums of Baer modules are Baer poses a challenging problem, in view of Example 3.9. We can however obtain the inheritance of the Baer module property to direct sums of Baer modules in some instances. In view of Theorem 4.8 in the following section, this may help provide a new approach to solving the question of when are (arbitrary) direct sums of extending modules extending.

Example 3.9 shows that a direct sum of Baer modules cannot be Baer if the maps between any pair of Baer modules do not behave "properly". The results below stress this idea.

Definition 3.10. We say that Baer modules M and N are *relatively Baer* to each other if $\forall\ \varphi \in Hom(M,N)$ and $\forall\ \psi \in Hom(N,M)$, $Ker\varphi \leq^{\oplus} M$ and $Ker\psi \leq^{\oplus} N$.

It is easy to see that, if $M \oplus N$ is a Baer module, then M and N are relatively Baer. Also, every Baer module is relatively Baer to itself. To investigate direct sums of Baer modules, we connect the various pieces by the relative Baer property.

The following results provide some instances of when direct sums of Baer modules inherit the property.

Theorem 3.11. *Let $\{M_i\}_{i\leq n}$ be a class of Baer modules, where $n \in \mathbb{N}$. For any $i \neq j$, M_i and M_j are relative Baer and relative injective. Then, $\bigoplus_{i\leq n} M_i$ is a Baer module.*

The following result provides a characterization in a special case, which will prove useful in our investigation on types for Baer modules.

Proposition 3.12. *Let $M = \bigoplus_{i\in\mathcal{I}} M_i$ ($\mathcal{I}$ an index set) and let $M_i \trianglelefteq M$, $\forall\ i \in \mathcal{I}$. Then $\bigoplus_{i\in\mathcal{I}} M_i$ is a Baer module if and only if M_i is a Baer module, $\forall\ i \in \mathcal{I}$.*

Proposition 3.13. *Let M be a finite direct sum of copies of some finite rank, torsion-free module whose endomorphism ring is a PID. Then M is Baer.*

A natural question to ask is for what classes of rings, every R-module is Baer? Our next result completely characterizes this class of rings.

Theorem 3.14. *Let R be a ring. The following are equivalent:*
1. *Every (right) R-module is Baer*
2. *Every injective right R module is Baer*
3. *R is semisimple artinian.*

We mention that Definition 3.1 can be modified to define a right Rickart module, as a module theoretic analogue to right Rickart rings discussed in

Section 2. This notion will be investigated in depth in a sequel to this paper.

Definition 3.15. M is called a *right Rickart module* if $\forall\ \varphi \in S$, $r_M(\varphi) = Ker\varphi = eM$ where $e^2 = e \in S$.

It is easy to see that every Baer module is a right Rickart module, and that R_R is a right Rickart module if R is a right Rickart ring. This also shows that the two module classes do not coincide, in general.

4. Connections to Extending Property

An interesting result of Chatters and Khuri connects the concept of a Baer ring to that of an extending ring. This useful result had no analogue in the module theoretic setting. In this section we show that similar connections exist between the newly defined concept of a Baer module and the existing notion of an extending module.

Recall that a ring R is called *right cononsingular* if $\forall I \leq R_R$, $rI \neq 0$ for all $0 \neq r \in R$ implies $I \leq^e R$ (for example, any right extending ring).

Theorem 4.1. *(Theorem 2.1, [7]) Let R be any ring. Then R is a right nonsingular right extending rings if and only if R is a right cononsingular Baer ring.*

Corollary 4.2. *(2.2, [4]) The following conditions are equivalent:*
(i) R is right nonsingular and right extending;
(ii) R is right extending and a Baer ring;
(iii) R is right extending and right p.p.;
(iv) R is right nonsingular and every principal right ideal of R is extending.

Since the concept of a Baer module depends on its endomorphism ring, we introduce concepts of nonsingularity and cononsingularity that take this characteristic in account. This new nonsingularity generalizes the usual concept of nonsingularity, for the case of modules, and coincides with it when $M_R = R_R$.

Definition 4.3. We say a module M is $\mathcal{K}$-*nonsingular* if, for all $\varphi \in S$, $r_M(\varphi) = Ker\varphi \leq^e M$ implies $\varphi = 0$.

Definition 4.4. A module M is called $\mathcal{K}$-*cononsingular* if, for all $N \leq M$, $l_S(N) = 0$ implies $N \leq^e M$ (equivalently, $\varphi(N) \neq 0$ for all $0 \neq \varphi \in S$ implies $N \leq^e M$).

Example 4.5. Every semisimple module is $\mathcal{K}$-nonsingular. Any uniform (or extending) module is $\mathcal{K}$-cononsingular.

Recall that a module M is called *polyform* (or *non-M-singular*) if, for any $K \leq M$ and $f : K \to M$, $Kerf \leq^c M$.

Proposition 4.6. *([26]) Every non-M-singular module M is $\mathcal{K}$-nonsingular. In particular, every nonsingular module in $\mathcal{K}$-nonsingular.*

That $\mathcal{K}$-nonsingularity is distinct from nonsingularity is shown in the following example.

Example 4.7. $\mathbb{Z}_p = \mathbb{Z}/p\mathbb{Z}$ is a singular module (hence not nonsingular). Since it is a simple module, its endomorphisms are either zero or isomorphisms, hence $\mathbb{Z}_p$ is $\mathcal{K}$-nonsingular.

Next, we provide a module-theoretic version of the Chatters-Khuri result. This result will prove to be useful in applications of this theory.

Theorem 4.8. *([26]) A module M is extending and $\mathcal{K}$-nonsingular if and only if M is Baer and $\mathcal{K}$-cononsingular.*

The proof of this theorem is based on four different lemmas, which are of interest on their own ([26]).

We can use the above result to provide a rich source of Baer modules and, at the same time, exhibit a strong connection to extending modules.

Corollary 4.9. *Let M be an extending module. Then $M/Z_2(M)$ is a Baer module, where $Z_2(M)$ is the second singular submodule of M.*

Corollary 4.10. *(see [6] and Theorem 4.8) If R_R is extending, then every nonsingular cyclic R-module M is extending, hence M is a Baer module.*

5. Applications

In this section, the focus is on the endomorphism ring of a Baer module and its properties. We use our results to show some interesting applications of the theory of Baer modules, including a type decomposition theory for Baer modules. As a consequence of Theorem 4.8, we obtain a type theory also for nonsingular extending modules, generalizing the type decomposition for nonsingular injective modules, provided in [10], [11].

Our first result connects the Baer property of a module to that of its endomorphism ring.

236

Theorem 5.1. *([26]) Let M be a Baer module. Then $S = End(M)$ is a Baer ring.*

A similar connection exists for the case of right Rickart modules and their endomorphism rings.

Theorem 5.2. *([27]) Let M be a Rickart module. Then $S = End(M)$ is a right Rickart ring.*

Proposition 5.3. *([28]) Let M be an extending module such that its endomorphism ring S is a regular ring. Then M is a Baer module, and subsequently S is a Baer ring.*

Example 5.4. Let $M = \mathbb{Z}_{p^\infty}$, considered as a $\mathbb{Z}$-module. Then it is well-known that $End_\mathbb{Z}(M)$ is the ring of p-adic integers (Example 3, page 216 in [9]). Since the ring of p-adic integers is a commutative domain, it is a Baer ring. However $M = \mathbb{Z}_{p^\infty}$ is not a Baer module.

Definition 5.5. A module M is called *retractable* if $Hom(M, N) \neq 0$, $\forall$ $0 \neq N \leq M$ (or, equivalently, $\exists\, 0 \neq \varphi \in S$ with $Im(\varphi) \subseteq N$).

It is easy to see that free modules are retractable.

Proposition 5.6. *([26]) Let M be retractable. Then M is Baer if and only if S is a Baer ring.*

Example 5.7. (Example 3.4 in [16]) Let K be a subfield of complex numbers $\mathbb{C}$. Let R be the ring $\begin{bmatrix} K & 0 \\ \mathbb{C} & \mathbb{C} \end{bmatrix}$. The R is left nonsingular left extending ring. Consider the module $M = Re$ where $e = \begin{pmatrix} 1 & 0 \\ 0 & 0 \end{pmatrix}$. Then M is projective, extending and nonsingular (as it is summand of R) hence is Baer. But M is not retractable, since the endomorphism ring of M, which is isomorphic to K, consists of isomorphisms and the zero endomorphism; on the other hand, M is not simple, and so by retractability it should have endomorphisms which are not onto.

As Example 5.7 shows, a Baer module does not satisfy the condition of retractability, in general. However, a more general concept is implied by the Baer property of a module.

Definition 5.8. A module M is called *quasi-retractable* if $\forall\, I \leq {}_SS$ with $0 \neq r_M(I)$, $Hom(M, r_M(I)) \neq 0$ (or, equivalently, if $r_M(I) \neq 0$ then $r_S(I) \neq 0$, $\forall\, I \leq {}_SS$).

Example 5.7 is also an example of a quasi-retractable module that is not retractable.

We use this notion to characterize a Baer module.

Theorem 5.9. *([28],[27]) A module M is a Baer module if and only if its endomorphism ring S is a Baer ring and M is quasi-retractable.*

It is known that finite matrix rings over a commutative integral domain are Baer if the domain is Prüfer ([15]). Given the strong connection between the endomorphism ring and the module provided by Theorem 5.9, we obtain a characterization of when finitely generated free modules over an integral domain are Baer modules. We remind the reader that this characterization also relates to the more general question of when is a direct sum of Baer modules a Baer module (a domain is a Baer ring, hence a Baer module over itself).

Theorem 5.10. *([28],[27]) Let R be a commutative integral domain. Then any finite rank (> 1) free module M over R is Baer if and only if R is a Prüfer domain.*

Our next result shows that if every (finitely generated) free R-module is Baer, then R is forced to be (semi-) hereditary.

Theorem 5.11. *([28],[27]) For a ring R, if any (finitely generated) free module over R is a Baer module, implies that R is right (semi-) hereditary.*

We use Theorem 5.1 and Theorem 2.11 to obtain the following decomposition of a Baer module.

Proposition 5.12. *([28]) If M is a Baer module, with only countably many direct summands, then M is a finite direct sum of indecomposable summands.*

Proposition 5.13. *([28],[27]) Let M be a module with a semisimple artinian endomorphism ring S. Then M is a Baer module.*

We remark that, in the hypothesis of Proposition 5.13, the module itself may not be semisimple artinian, and hence it might not be retractable (for example, take $\mathbb{Q}_{\mathbb{Z}}$). On the other hand, if the module is retractable, M is semisimple artinian (a straightforward proof, using Wedderburn-Artin's Theorem).

The next result is a corrected version of Theorem 4.10 in [26], where a typographic omission has been removed. It can be viewed as a module

238

theoretic version of Theorem 2.12, showing instances when a Baer module is semisimple artinian.

Theorem 5.14. *([26]) Let M be a Baer module with only countably many direct summands. Then M is semisimple artinian if any of the following conditions hold:*

> *(i) M is retractable and S is a regular ring;*
> *(ii) every cyclic submodule of M is a direct summand of M; or*
> *(iii) $\forall\, m \in M$, $\exists\, f \in Hom(M, R_R)$ such that $m = mfm$ (Zelmanowitz [32] calls such a module a regular module).*

Since the endomorphism ring of a Baer module is a Baer ring, it is natural to ask whether the decomposition theory obtained by Kaplansky for Baer rings can be extended to the module theoretic setting. Goodearl and Boyle ([10], [11]) have provided a similar decomposition theory for nonsingular injective modules (which are Baer modules, by Theorem 4.8). For the detailed proofs of these next results (5.16 through 5.23) we refer the reader to [27].

We start with a number of definitions which generalize the corresponding definitions for idempotents and rings. We will study properties of these classes of modules, in the light of similar results of Goodearl and Boyle.

Definition 5.15. A module M is called *abelian* if all idempotent endomorphisms are central (i.e. commute with any endomorphism). An idempotent endomorphism e is called abelian if eM is an abelian module.

Proposition 5.16. *([28],[27]) For any Baer module M, the following conditions are equivalent:*

> *(1) M is abelian;*
> *(2) all direct summands of M are fully invariant;*
> *(3) isomorphic summands of M are equal;*
> *(4) if N_1, N_2 are summands of M and $N_1 \cap N_2 = 0$ then $Hom(N_1, N_2) = 0$.*

Proposition 5.17. *([28],[27]) Let M be a Baer module.*

> *(1) If $N \leq^{\oplus} M$, and M is abelian, then N is an abelian Baer module;*
> *(2) Let M_i, $i \in \mathcal{I}$ be a family of modules. Then $\bigoplus_{i \in \mathcal{I}} M_i$ is abelian Baer module if and only if each M_i is abelian Baer module and $Hom(M_i, M_j) = 0$, $\forall\, i \neq j$, $i, j \in \mathcal{I}$.*

Definition 5.18. Recall that a ring R is called *directly finite* if $xy = 1 \Rightarrow yx = 1$, $\forall\ x, y \in R$. A module M is called directly finite if $S = End(M)$ is a directly finite ring. An idempotent endomorphism e is called directly finite if eM is a directly finite module. A module that is not directly finite will be called *directly infinite*.

Recall that a module M is directly finite if and only if M is not isomorphic to any proper direct summand of itself ([10]).

Proposition 5.19. *([28],[27]) Let M be a Baer module. Then the following hold:*

(1) if M is an abelian module, then M is directly finite;

(2) $N \leq^{\oplus} M$ and M directly finite; then N is directly finite Baer module;

(3) Let (M_i), $i \in \mathcal{I}$ a family of modules with $Hom(M_i, M_j) = 0$ $\forall$ $i \neq j$, $i, j \in \mathcal{I}$ ($\mathcal{I}$ an index set); then $\bigoplus_{i \in \mathcal{I}} M_i$ is directly finite Baer module if and only if M_i is a directly finite Baer module, $\forall$ $i \in \mathcal{I}$.

Definition 5.20. We call a Baer module M of type (T) if $S = End(M)$ is of type (T) (where T is one of the five types described above: I_f; I_∞; II_f; II_∞; III).

Example 5.21. Let R be a Baer ring, let e be an idempotent of R. The Baer module $M = eR$ (Corollary 3.7)is of type (T) if R is of type (T), where (T) is one of I_f, I_∞, II_f, II_∞, III. For examples of Baer rings of types (T), see [10], [15]. Furthermore, any nonsingular injective module (hence a Baer module, by Theorem 4.8) of type (T) provides another such example ([10], [11]).

Theorem 5.22. *([28],[27]) A Baer module decomposes uniquely into a sum of fully invariant summands of types I_f; I_∞; II_f; II_∞; III.*

As a consequence to Theorem 5.22, since every nonsingular extending module is Baer (Theorem 4.8), we get:

Corollary 5.23. *A nonsingular extending module decomposes uniquely into a sum of fully invariant summands of types I_f; I_∞; II_f; II_∞; III.*

In view of Theorem 4.8, the preceding results in Theorem 5.22 and Corollary 5.23 may be helpful in providing another approach to the question of when is the direct sum of extending modules extending, by restricting the discussion to the types described (for example, in the presence of $\mathcal{K}$-nonsingularity).

Open Questions

(1) Find necessary and sufficient conditions for a (finite, countable, arbitrary) direct sum of Baer modules to be a Baer module.

(2) Completely characterize abelian groups, and modules over PID, that are Baer.

(3) Find conditions for a Baer module to be decomposed into a direct sum of indecomposables.

(4) Find internal characterizations of Baer modules that are abelian, or of types I_f, I_∞, II_f, II_∞, III.

(5) Characterize Baer modules which are not nonsingular.

(6) Find conditions for transfer of properties from and to the endomorphism ring of a Baer module.

References

1. Armendariz, E. P., A Note On Extensions Of Baer And p.p.-Rings, J. Australian Math. Soc. **1974** *18*, 470–473

2. Berberian, S. K., *Baer Rings*, manuscript, 1988

3. Birkenmeier, G. F.; Heatherly, H. E.; Kim J. Y.; Park, J. K., Triangular Matrix Representations, J. Algebra **2000**, *230*, 558–595

4. Birkenmeier, G. F.; Kim J. Y.; Park, J. K., When Is The CS Condition Hereditary?, Comm. Alg. **1999**, *27(8)*, 3875–3885

5. Birkenmeier, G. F.; Müller, B. J.; Rizvi, S. T., Modules In Which Every Fully Invariant Submodule Is Essential In A Direct Summand, Comm. Algebra **2002**, *30*, 1395–1415

6. Birkenmeier, G. F.; Park, J. K.; Rizvi, S. T., Modules With Fully Invariant Submodules Essential In Fully Invariant Summands, Comm. Algebra **2002**, *30*, 1833–1852

7. Chatters, A. W.; Khuri, S. M., Endomorphism Rings Of Modules Over Nonsingular CS Rings, J. London Math. Soc. **1980**, *21* (2), 434–444

8. Cho, I.-H.; Kim, J. Y.; Lim, J. I.; Park, D. Y., Compressibility And Annihilator Conditions, J. Korean Math. Soc. **1988**, *25(2)*, 303–308

9. Fuchs, L, *Infinite Abelian Groups*, Pure And Applied Mathematics Series, Vol. 1, Academic Press, 1970

10. Goodearl, K. R., *Von Neumann Regular Rings*, 2nd edition; Krieger Publishing Company, 1991

11. Goodearl, K. R.; Boyle, A. K., *Dimension Theory for Nonsingular Injective Modules*, Memoirs of the American Mathematical Society, number 177; American Mathematical Society, 1976

12. Jøndrup, S., p.p.-Rings And Finitely Generated Flat Ideals, Proc. Amer. Math. Soc. **1971**, *28*, 431–435

13. Kaplansky, I., Projections In Banach Algebras, Ann. of Math. **1951**, *53(2))*, 235–249

14. Kaplansky, I., Any Orthocomplemented Complete Modular Lattice Is A Continuous Geometry, Ann. of Math. **1955**, *61(2))*, 524–541

15. Kaplansky, I., *Rings Of Operators*, Mathematics Lecture Note Series; W. A. Benjamin: New York, 1968

16. Khuri, S. M., Endomorphism Rings Of Nonsingular Modules, Ann. Sc. Math. Québec **1980**, *IV(2)*, 145–152

17. Kim, J. Y.; Park, J. K., When Is A Regular Ring A Semisimple Artinian Ring?, Math. Japonica **1997**, *45* (2), 311–313

18. Lam, T. Y., *Lectures On Modules And Rings*, GTM 189; Springer Verlag: Berlin-Heidelberg-New York, 1999

19. Maeda, S., On The Lattice Of Projections Of A Baer ∗-Ring, J. Sci. Hiroshima Univ. Ser. A **1958**, *22*, 75–88

20. Mewborn, A. C., Regular Rings And Baer Rings, Math. Z. **1971**, *121*, 211–219

21. Mohamed, S. H.; Müller, B. J., *Continuous and Discrete Modules*, London Mathematical Society Lecture Note Series; Cambridge University Press: Cambridge, 1990

22. Murray, F. J.; von Neumann, J., On Rings Of Operators, Ann. of Math. **1936**, *37*, 116–229

23. Ornstein, D., Dual Vector Spaces, Ann. of Math. **1959**, *69(2)*, 520–534

24. Pollingher, A.; Zaks, A., On Baer And Quasi-Baer Rings **1970**, *37*, 127–138

25. Rangaswany, K. M., Regular And Baer Rings, Proc. Amer. Math. Soc. **1974**, *42(2)*, 254–358

26. Rizvi, S. T.; Roman, C. S., Baer And Quasi-Baer Modules, Comm. Alg. **2004**, *32(1)*, 103–123

27. Rizvi, S. T.; Roman, C. S., Endomorphism rings of Baer Modules and Type Theory, preprint

28. Roman, C. S., *Baer And Quasi-Baer Modules*, Ph.D. Thesis **2004**

29. Small, L. W., Semihereditary Rings, Bull. Amer. Math. Soc. **1967**, *73*, 656–658

30. Stenström, B., *Rings Of Quotients*, GTM 217; Springer Verlag: Berlin-Heidelberg-New York, 1975

31. Wolfson, K. G., Baer Rings Of Endomorphisms, Math. Annalen **1961**, *143*, 19–28

32. Zelmanowitz, J., Regular Modules, Trans. Am. Math. Soc. **1972**, *163*, 341–355

PROPERTIES OF GRADED FORMAL TRIANGULAR MATRIX RINGS

YAO WANG AND Y.L REN

Department of Mathematics, Anshan Normal University,
Anshan, Liaoning 114005, China

School of Mathematical Sciences, Nankai University,
Tianjin, 300071, China
E-mail: wangyao@mail.asnc.edu.cn

Given two M-graded rings $R = \bigoplus_{x \in M} R_x, A = \bigoplus_{x \in M} A_x$ and one M-graded bimodule $V =_R V_A = \bigoplus_{x \in M} V_x$ we can obtain a graded formal triangular matrix ring $T = \begin{pmatrix} R & V \\ 0 & A \end{pmatrix} = \bigoplus_{x \in M} \begin{pmatrix} R_x & V_x \\ 0 & A_x \end{pmatrix}$. In this paper we carry out a systematic study of various graded ring theoretic properties of graded formal triangular matrix rings. Some definitive results are obtained on these rings concerning properties such as being respectively graded unit regular, graded Von Neumann regular, graded weakly regular, graded left (right) strongly regular, graded semilocal, graded semiperfect, graded left (right) perfect, graded semiprimary and weakly graded direct finiteness.

1. Introduction

All the rings considered will be associative rings with identity, all the modules considered will be unital modules and every monoid M considered be left (or right) cancellative with identity, i.e. for all $x, y, z \in M, zx = zy$ (or $xz = yz$) implies $x = y$. Let $R = \bigoplus_{x \in M} R_x$ and $A = \bigoplus_{x \in M} A_x$ be two given M-graded rings and $V =_R V_A = \bigoplus_{x \in M} V_x$ be a $(R, A) - M-$ graded bimodule. The M-graded formal triangular matrix ring $T = \begin{pmatrix} R & V \\ 0 & A \end{pmatrix} = \bigoplus_{x \in M} \begin{pmatrix} R_x & V_x \\ 0 & A_x \end{pmatrix}$ has as its x-component formal matrix $\begin{pmatrix} R_x & V_x \\ 0 & A_x \end{pmatrix}$ for any $x \in M$ and as its homogeneous elements of degree x formal matrices

$\begin{pmatrix} r_x & v_x \\ 0 & a_x \end{pmatrix}$ where $r_x \in R_x, a_x \in A_x$ and $v_x \in V_x$, with addition defined co-ordinatewise and multiplication given by

$$\begin{pmatrix} r_x & v_x \\ 0 & a_x \end{pmatrix} \begin{pmatrix} r_y & v_y \\ 0 & a_y \end{pmatrix} = \begin{pmatrix} r_x r_y & r_x v_y + v_x a_y \\ 0 & a_x a_y \end{pmatrix}.$$

Let $R^\alpha = \underset{x \in M}{\oplus} R_x^\alpha (\alpha \in W)$ be a family of M-graded rings, and R denote the set of all functions

$$\sigma : \quad W \longmapsto \{R^\alpha | \alpha \in W\}$$
$$\alpha \longmapsto \sigma(\alpha) \in R^\alpha.$$

In R, define the addition and the multiplication as follows:

$$(\sigma + \tau)(\alpha) = \sigma(\alpha) + \tau(\alpha), \qquad \sigma\tau(\alpha) = \sigma(\alpha)\tau(\alpha),$$

for every $\alpha \in W$ and $\sigma, \tau \in R$. Then R is an associative ring.

For any $x \in M$, put $R_x = \{\sigma \in R | \sigma(\alpha) \in R_x^\alpha, \forall \alpha \in W\}$. Then every R_x is a subgroup of $(R, +)$. In particular, R_e is a subring of R. Denote $gr \prod_{\alpha \in W} R^\alpha = \sum_{x \in M} R_x = \underset{x \in M}{\oplus} R_x$. Since $\sigma\tau(\alpha) = \sigma(\alpha)\tau(\alpha) \in R_x^\alpha R_y^\alpha \subseteq R_{xy}^\alpha$ for any $\sigma \in R_x, \tau \in R_y$ and $\alpha \in W, \sigma\tau \in R_{xy}$. It shows that $gr \prod_{\alpha \in W} R^\alpha$ is a graded subring of R. $gr \prod_{\alpha \in W} R^\alpha$ is called the graded direct product of the M-graded ring family R^α $(\alpha \in W)$. It is easy to see that $\sigma = \sum_{x \in M} \sigma_x$ where $\sigma_x \in R_x$ and $x \in M$, for any $\sigma \in gr \prod_{\alpha \in W} R^\alpha$. Moreover, $(\sigma(\alpha))_x = \sigma_x(\alpha)$.

Also let A be a graded subring of $R = gr \prod_{\alpha \in W} R^\alpha$. For any $\alpha \in W$, set

$$\theta_\alpha : \quad A \longrightarrow R^\alpha$$
$$\sigma \longmapsto \theta(\sigma) = \sigma(\alpha) = \sum_{x \in M} \sigma_x(\alpha) = \sum_{x \in M} (\sigma(\alpha))_x.$$

Then θ_α is a graded ring homomorphism. If $\theta_\alpha(A) = R^\alpha$ for all $\alpha \in W$, then we call A a graded subdirect product of $\{R^\alpha | \alpha \in W\}$. Similar to the case of associative rings we can prove the following:

Proposition 1.1. Let I_α $(\alpha \in W)$ be a set of graded ideals of M-graded ring A. Then A is a graded subdirect product of graded quotient rings $B_\alpha = A/I_\alpha$ $(\alpha \in W)$ if and only if $\underset{\alpha \in W}{\cap} I_\alpha = 0$.

244

2. Graded Regularity

In this section, we take $M = G$, any group.

Let $R = \underset{g \in G}{\oplus} R_g$ be a G-graded ring. Recall that R is said to be graded Von Neumann regular if, for any $g \in G$ and any $a \in R_g$, there exists $r \in R_{g^{-1}}$ such that $a = ara$, R is said to be graded weakly regular if for any $g \in G$ and any $a \in R_g$ there exist $r' \in R_e$ and $r'' \in R_{g^{-1}}$ (or $r' \in R_{g^{-1}}, r'' \in R_e$) such that $a = ar'ar''$, and R is said to be graded strongly regular if for any $g \in G$ and any $a \in R_g$ there exists $a' \in R_{g^{-1}}$ such that $a = a^2 a'$. Moreover, R is said to be graded unit regular if for any $g \in G$ and any $a \in R_g$ there exists a graded invertible homogeneous element $u \in R_{g^{-1}}$ such that $a = aua$.

Theorem 2.1. *The following are equivalent:*

(1) G-graded formal triangular matrix ring $T = \begin{pmatrix} R & V \\ 0 & A \end{pmatrix} = \underset{g \in G}{\oplus} \begin{pmatrix} R_g & V_g \\ 0 & A_g \end{pmatrix}$ *is graded weakly regular (graded Von Neumann regular, graded unit regular, graded strongly regular, respectively).*

(2) $R = \underset{g \in G}{\oplus} R_g, A = \underset{g \in G}{\oplus} A_g$ are both graded weakly regular (graded Von Neumann regular, graded unit regular, graded strongly regular, respectively) and $V = 0$.

Proof. $(1) \Longrightarrow (2)$. Let $T = \begin{pmatrix} R & V \\ 0 & A \end{pmatrix} = \underset{g \in G}{\oplus} \begin{pmatrix} R_g & V_g \\ 0 & A_g \end{pmatrix}$ be a graded weakly regular ring. For any $\in G$ and $a \in R_g$, $\begin{pmatrix} a & 0 \\ 0 & 0 \end{pmatrix} \in T_g$, by the hypothesis, we may assume that there exist $\begin{pmatrix} r' & v' \\ 0 & a' \end{pmatrix} \in T_e$ and $\begin{pmatrix} r'' & v'' \\ 0 & a'' \end{pmatrix} \in T_{g^{-1}}$ such that

$$\begin{pmatrix} a & 0 \\ 0 & 0 \end{pmatrix} = \begin{pmatrix} a & 0 \\ 0 & 0 \end{pmatrix} \begin{pmatrix} r' & v' \\ 0 & a' \end{pmatrix} \begin{pmatrix} a & 0 \\ 0 & 0 \end{pmatrix} \begin{pmatrix} r'' & v'' \\ 0 & a'' \end{pmatrix}.$$

It follows that $\begin{pmatrix} a & 0 \\ 0 & 0 \end{pmatrix} = \begin{pmatrix} ar'ar'' & 0 \\ 0 & 0 \end{pmatrix}$. Hence $a = ar'ar''$. This shows that $R = \underset{g \in G}{\oplus} R_g$ is a graded weakly regular ring. Similarly, the $A = \underset{g \in G}{\oplus} A_g$ is also graded weakly regular.

Moreover, for any $g \in G$ and any $v_g \in V_g$, $\begin{pmatrix} 0 & v_g \\ 0 & 0 \end{pmatrix} \in T_g$, according to the graded weakly regularity of $T = \underset{g \in G}{\oplus} T_g$, we can asume that there exist

$$\begin{pmatrix} r' & v' \\ 0 & a' \end{pmatrix} \in T_e \text{ and } \begin{pmatrix} r'' & v'' \\ 0 & a'' \end{pmatrix} \in T_{g^{-1}} \text{ such that}$$

$$\begin{pmatrix} 0 & v_g \\ 0 & 0 \end{pmatrix} = \begin{pmatrix} 0 & v_g \\ 0 & 0 \end{pmatrix} \begin{pmatrix} r' & v' \\ 0 & a' \end{pmatrix} \begin{pmatrix} 0 & v_g \\ 0 & 0 \end{pmatrix} \begin{pmatrix} r'' & v'' \\ 0 & a'' \end{pmatrix}.$$

It implies that $\begin{pmatrix} 0 & v_g \\ 0 & 0 \end{pmatrix} = \begin{pmatrix} 0 & 0 \\ 0 & 0 \end{pmatrix}$. Hence $v_g = 0$. This proves that $V = 0$.

$(2) \Longrightarrow (1)$. Let $R = \underset{g \in G}{\oplus} R_g$ and $A = \underset{g \in G}{\oplus} A_g$ are both graded weakly regular, and $V = 0$. Then $T = \begin{pmatrix} R & V \\ 0 & A \end{pmatrix} = \begin{pmatrix} R & 0 \\ 0 & A \end{pmatrix} = \underset{g \in G}{\oplus} \begin{pmatrix} R_g & 0 \\ 0 & A_g \end{pmatrix}$.

For any $g \in G$ and $r \in R_g, a \in A_g$, by the hypothesis we may assume that there exist $r' \in R_e, s' \in A_e$ and $r'' \in R_{g^{-1}}, s'' \in A_{g^{-1}}$ such that $r = rr'rr''$ and $a = aa'aa''$. From this we have that

$$\begin{pmatrix} r & 0 \\ 0 & a \end{pmatrix} \begin{pmatrix} r' & 0 \\ 0 & a' \end{pmatrix} \begin{pmatrix} r & 0 \\ 0 & a \end{pmatrix} \begin{pmatrix} r'' & 0 \\ 0 & a'' \end{pmatrix} = \begin{pmatrix} rr'rr'' & 0 \\ 0 & aa'aa'' \end{pmatrix} = \begin{pmatrix} r & 0 \\ 0 & a \end{pmatrix}$$

This shows that $\begin{pmatrix} R & 0 \\ 0 & A \end{pmatrix}$ is graded weakly regular.

The other conclusions can be proved similarly.

Definition 2.1. A G-graded ring $R = \underset{g \in G}{\oplus} R_g$ is said to be graded strongly π-regular if for any $g \in G$ and $r \in R_g$ there exist a $s \in R_{g^{-1}}$ and a positive integer n such that $r^n = r^{n+1}s$.

Let $I = \underset{g \in G}{\oplus} I_g$ be a graded ideal of $R = \underset{g \in G}{\oplus} R_g$. Since $R/I = \underset{g \in G}{\oplus} (R_g + I/I)$, then $R = \underset{g \in G}{\oplus} R_g$ is graded strongly π-regular implies R/I is too.

Proposition 2.1. The following are equivalent:

(1) $R = \underset{g \in G}{\oplus} R_g$ is graded strongly π-regular.

(2) There exist graded ideals $I_1, I_2, \cdots, I_n$ $(n \geq 2)$ of $R = \underset{g \in G}{\oplus} R_g$ with $I_1 I_2 \cdots I_n = 0$ such that $R/I_1, R/I_2, \cdots, R/I_n$ are all graded strongly π-regular.

Proof. $(1) \Longrightarrow (2)$ is trival.

$(2) \Longrightarrow (1)$. When $n = 2$, for any $g \in G$ and any $r \in R_g$. Then there exist $s, t \in R_{g^{-1}}$ and two positive integers m, l such that $r^l = r^{l+1}s$ (mod I_1), $r^m = r^{m+1}t$ (mod I_2). We may assume that $m \leq l$. By $r^m = r^{m+1}t + j_1$ $(j_1 \in I_2)$ we have $r^l = r^{l+1}t + j$ $(j \in I_2)$, by $r^l = r^{l+1}s + i$ $(i \in I_1)$ we have $r^l = r^{l+2}s^2 + i_1 = r^{l+3}s^3 + i_2 = \cdots = r^{2l+1}s' + i'$, where $s' = s^{l+1} \in R_{g^{-l-1}}$

and $i_1, i_2, \cdots, i' \in I_1$. Similarly, $r^l = r^{l+1}t + j$ $(j \in I_2)$ implies that $r^l = r^{2l+1}t' \pmod{I_2}$ where $t' = t^{l+1} \in R_{g-l-1}$. It follows that $r^l - r^{2l+1}s' \in I_1, r^l - r^{2l+1}t' \in I_2$. Hence $(r^l - r^{2l+1}s')(r^l - r^{2l+1}t') \in I_1 I_2 = 0$. Therefore $r^{2l} = r^{2l+1}(r^l r' + s' r^l - s' r^{2l+1} t') \in r^{2l+1} R_{g-1}$. This shows that $R = \underset{g \in G}{\oplus} R_g$ is graded strongly π-regular.

Assume that the result follows whenever $n \le k$ $(k \ge 2)$. Let $n = k+1$ and $J = I_1 I_2 \cdots I_k$. Then $J I_{k+1} = 0$ and $I_1/J_1, I_2/J, \cdots, I_k/J$ are all graded ideals of R/J. Since $R/J/I_i/J \cong_{gr} R/I_i$ is graded strongly π-regular for all $1 \le i \le k$ and $I_1/J \cdot I_2/J \cdots I_k/J = 0, R/J$ is graded strongly π-regular. So R/J and R/I_{k+1} are both graded strongly π-regular. Since $J I_{k+1} = 0$, by the proceeding, $R = \underset{g \in G}{\oplus} R_g$ is graded strongly π-regular.

Corollary 2.2. Every finite graded subdirect product of graded strongly π-regular rings is graded strongly π-regular.

Proof. Let $R = \underset{g \in G}{\oplus} R_g$ and $A = \underset{g \in G}{\oplus} A_g$ be graded strongly π-regular, S the graded subdirect product of R and A. Then there exist two graded ideals K, L of S with $K \cap L = 0$ such that S/K and S/L are both graded strongly π-regular. Since $KL \subseteq K \cap L = 0, S$ is graded strongly π-regular by proposition 2.1.

Theorem 2.3. *The following are equivalent:*

(1) $T = \begin{pmatrix} R & V \\ 0 & A \end{pmatrix} = \underset{g \in G}{\oplus} \begin{pmatrix} R_g & V_g \\ 0 & A_g \end{pmatrix}$ *is graded strongly π-regular.*

(2) $R = \underset{g \in G}{\oplus} R_g$ *and* $A = \underset{g \in G}{\oplus} A_g$ *are graded strongly π-regular.*

Proof. $(2) \Longrightarrow (1)$. Assume that $R = \underset{g \in G}{\oplus} R_g$ and $A = \underset{g \in G}{\oplus} A_g$ are graded strongly π-regular. Let

$$I = \underset{g \in G}{\oplus} \begin{pmatrix} 0 & V_g \\ 0 & A_g \end{pmatrix}, \qquad J = \underset{g \in G}{\oplus} \begin{pmatrix} R_g & V_g \\ 0 & 0 \end{pmatrix}.$$

Then I and J are both graded ideals of T with $IJ = 0$. Since $T/I \cong_{gr} R$ and $T/J \cong_{gr} A$, we know that T/I and T/K are both graded strongly π-regular. By virtue of proposition 2.1, T is graded strongly π-regular.

$(1) \Longrightarrow (2)$. Assume that T is graded strongly π-regular. Let I and J be graded ideals of T as above. Since $R \cong_{gr} T/I$ and $A \cong_{gr} T/J$, it is easy to verify that R and A are graded strongly π-regular.

3. Graded Jacobson Radical

In the following two sections, we consider M-graded rings.

Recall that the graded Jacobson radical of any M-graded ring $B = \underset{x \in M}{\oplus} B_x$ is $J_G(B) = \cap\{I | I \text{ is maximal graded left ideal of } B\}$.

Proposition 3.1. Let $T = \begin{pmatrix} R & V \\ 0 & A \end{pmatrix} = \underset{x \in M}{\oplus} \begin{pmatrix} R_x & V_x \\ 0 & A_x \end{pmatrix}$ be a graded formal triangular matrix ring. Then every maximal graded left ideal of T is given by $\begin{pmatrix} I & V \\ 0 & J \end{pmatrix} = \underset{x \in M}{\oplus} \begin{pmatrix} I_x & V_x \\ 0 & J_x \end{pmatrix}$ where either $I = R = \underset{x \in M}{\oplus} R_x$ and $J = \underset{x \in M}{\oplus} J_x$ is a maximal graded left ideal of A or $I = \underset{x \in M}{\oplus} I_x$ is a maximal graded left ideal of R and $J = A = \underset{x \in M}{\oplus} A_x$.

Proof. Let $\begin{pmatrix} I & U \\ 0 & J \end{pmatrix} = \underset{x \in M}{\oplus} \begin{pmatrix} I_x & U_x \\ 0 & J_x \end{pmatrix}$ be a maximal graded left ideal of T. Then $I = \underset{x \in M}{\oplus} I_x$ and $J = \underset{x \in M}{\oplus} J_x$ are both graded left ideals with $R_y U_x + V_y J_x \subseteq U_{yx}$ for any $x, y \in M$. If $J \neq A$, then choosing a maximal graded left ideal of A with $J' \supseteq J$ but $J' \neq J$. We see that $\begin{pmatrix} R & V \\ 0 & J' \end{pmatrix}$ is a graded left ideal of T with $\begin{pmatrix} I & U \\ 0 & J \end{pmatrix} \subset \begin{pmatrix} R & V \\ 0 & J' \end{pmatrix}$. The maximality of $\begin{pmatrix} I & U \\ 0 & J' \end{pmatrix}$ yields $I = R, U = V$ and $J' = J$. If on the other hand $J = A$ then from $R_x U_e + V_x A_e \subseteq U_x$ for any $x \in M$ we see that $V_x A_e \subseteq U_x$. Since $1_A \in A_e$, it implies that $U_x = V_x, \forall x \in M$. This proves that $U = V$. The maximality of $\begin{pmatrix} I & U \\ 0 & J \end{pmatrix} = \begin{pmatrix} I & V \\ 0 & A \end{pmatrix}$ now implies that I is a maximal graded left ideal. Thus any maximal graded left ideal of T has to be either $\begin{pmatrix} R & V \\ 0 & J \end{pmatrix}$ with J a maximal graded left ideal of A or $\begin{pmatrix} I & V \\ 0 & A \end{pmatrix}$ with I a maximal graded left ideal of R. Conversely, graded left ideals of the above form are clearly maximal graded left ideals of T.

Corollary 3.2. $J_G(T) = \begin{pmatrix} J_G(R) & V \\ 0 & J_G(A) \end{pmatrix}$.

Let $B = \underset{x \in M}{\oplus} B_x$ be any M-graded ring. Recall that B is said to be graded semisimple if B is the direct sum of finite minimal graded left ideals of B. B is said to be graded semilocal if $B/J_G(B)$ is graded semisimple. B is said to be graded semiperfect if idempotents mod $J_G(B)$ can be lifted and $B/J_G(B)$ is graded semisimple. B is said to be graded left (resp., right) perfect if $B/J_G(B)$ is graded semisimple and $J_G(B)$ is graded left (resp., right) T-nilpotent, i.e. for any sequence of homogeneous elements $\{a_1, a_2, a_3, \cdots\}$ of $J_G(B)$, there exists an integer $n \geq 1$ such that $a_1 a_2 \cdots a_n = 0$ (resp.,

248

$a_n \cdots a_2 a_1 = 0$). B is said to be graded semiprimary if $J_G(B)$ is nilpotent and $B/J_G(B)$ is graded semisimple.

Corollary 3.3. The following hold:

(1) The mapping

$$\varphi : \quad T/J_G(T) \to R/J_G(R) \times_{gr} A/J_G(A)$$
$$\begin{pmatrix} r_x & v_x \\ 0 & a_x \end{pmatrix} + J_G(T) \longmapsto (r_x + J_G(R), a_x + J_G(A))$$

is a graded ring isomorphism.

(2) The homogeneous idempotents of $T/J_G(T)$ can be lifted to T if and only if the homogenous idempotents of $R/J_G(R)$ can be lifted to R and the homogenous idempotents of $A/J_G(A)$ can be lifted to A.

(3) T is graded semilocal if and only if R and A are both graded semilocal.

(4) T is graded semiperfect if and only if R and A are both graded semiperfect.

(5) T is graded left (resp., right) perfect if and only if R and A are both graded left (resp., right) graded perfect.

(6) T is graded semiprimary if and only if R and A are both graded semiprimary.

Proof. (1) and (2) are immediate consequences of corollary 3.2, and (3), (4) and (5) are immediate consequences of (1) and (2). Now we prove (6).

Let R and A be graded semiperfect. By (3), we only need to show that $J_G(T)$ is nilpotent. Since $J_G(R), J_G(A)$ are nilpotent, there exists a positive integer k such that $J_G(R)^k = J_G(A)^k = 0$. Let $I_1 = \begin{pmatrix} J_G(R) & V \\ 0 & 0 \end{pmatrix}$ and $I_2 = \begin{pmatrix} 0 & V \\ 0 & J_G(A) \end{pmatrix}$. Then I_1 and I_2 are graded ideals of T with $I_1^{k+1} = I_2^{k+1} = 0$. But $J_G(T) = I_1 + I_2, I_2 I_1 = 0$ and $I_1 I_2 \subseteq \begin{pmatrix} 0 & V \\ 0 & 0 \end{pmatrix}$. Hence $J_G(T)^{k+2} = 0$. Conversely, the conclusion is easy to see from (3) and Corollary 3.2.

4. Graded Direct Finiteness

Definition 4.1. Let $T = \underset{x \in M}{\oplus} T_x$ be a M-graded ring with identity 1_T. If $ts = 1_T$ implies $st = 1_T$ for all $t, s \in T_e$, then T is called a weakly graded directly finite ring.

Proposition 4.1. Finite graded subdirect product of weakly graded directly finite rings is weakly graded directly finite.

Proof. Let $R = \bigoplus_{x \in M} R_x$ and $A = \bigoplus_{x \in M} A_x$ be weakly graded directly finite, S the graded subdirect product of R and A. Then there exist graded epimorphism $\varphi_1 : S \to R$ and $\varphi_2 : S \to A$ such that $\ker \varphi_1 \cap \ker \varphi_2 = 0$. Let $R \times_{gr} A = (\bigoplus_{x \in M} R_x) \times_{gr} (\bigoplus_{x \in M} A_x)$ denote the graded direct product of R and A. We construct a graded ring homomorphism $\psi : S \to R \times_{gr} A$ given by $\psi(s_x) = (\varphi_1(s_x), \varphi_2(s_x))$ for any $s_x \in S_x$. It is easy to verify that $\ker \psi = \ker \varphi_1 \cap \ker \varphi_2 = 0$. So S is graded isomorphic to a graded subring of $R \times_{gr} A$.

Given $(t_1, t_2) \in (R \times_{gr} A)_e$ and $(s_1, s_2) \in (R \times_{gr} A)_e$. If $(t_1, t_2)(s_1, s_2) = (1_R, 1_A)$, then $t_1 s_1 = 1_R$ and $t_2 s_2 = 1_A$. Since R and A are both weakly graded directly finite, we have $s_1 t_1 = 1_R$ and $s_2 t_2 = 1_A$. So $(s_1, s_2)(t_1, t_2) = (1_A, 1_R)$. Thus $R \times_{gr} A$ is weakly graded directly finite. By virtue of [5, Corollary 2], we know that S is weakly graded directly finite.

Proposition 4.2. The following are equivalent:

(1) $R = \bigoplus_{x \in M} R_x$ is weakly graded directly finite.

(2) There exist graded ideals $I_1, I_2, \cdots, I_n$ of R with $I_1 I_2 \cdots I_n = 0$ such that $R/I_1, R/I_2, \cdots, R/I_n$ are weakly graded directly finite.

Proof. $(1) \Longrightarrow (2)$. Obviously, the result is valid when $n = 1$. Assume that the result follows whenever $n \leq k$ $(k \geq 1)$. Let $n = k + 1$ and $I = I_1 I_2 \cdots I_k, J = I_{k+1}$. The $IJ = 0$ and $I_1/I, I_2/I, \cdots, I_k/I$ are all graded ideals of R/I. Since $R/I/I_i/I \cong_{gr} R/I_i$ is weakly graded directly finite for all $1 \leq i \leq k$ and $\prod_{i \leq k} I_i/I = 0$, by the assumption above, R/I is weakly graded directly finite. Thus R/I and R/J are weakly graded directly finite. Given any $t = \sum_{x \in M} t_x \in I \cap I_{k+1}, r, s \in R_e$. Then $(r t_e s)^2 \in II_{k+1} = 0, (1 - r t_e s)(1 + r t_e s) = 1$. By virtue of [4, Lemma 4.3], $t_e \in J(R_e)$. This shows that $t = \sum_{x \in M} t_x$ is a weakly left quasi-regular element (see [2]). By [2, Theorem 2.3], $I \cap J \subseteq J_G(R)$. If $I \cap J = 0$, then R is graded isomorphic to graded subdirect product of R/I and R/J. From proposition 4.1, we know that R is weakly graded directly finite. If $I \cap J \neq 0$, then $I/I \cap J$ and $J/I \cap J$ are graded ideals of $R/I \cap J$ with $(I/I \cap J)(J/I \cap J) = 0$. Since $R/I \cap J/I/I \cap J \cong_{gr} R/I$ and $R/I \cap J/J/I \cap J \cong_{gr} R/J$ are weakly graded directly finite, by the discussion above, $R/I \cap J$ is weakly graded directly finite. For any $t, s \in R_e$, if $ts = 1$ in R_e, then $\bar{t}\bar{s} = \bar{1}$ in $R_e/I_e \cap J_e$. So $\bar{s}\bar{t} = \bar{1} \in R_e/I_e \cap J_e$. Since $I \cap J \subseteq J_G(R), st - 1 \in (J_G(R))_e$. By the

250

structure of $J_G(R)$ given by [2], st is invertible in R_e. From $ts = 1$ we have $st = s(ts)t = (st)(st)$. So $st = 1$. Thus R is weakly graded directly finite.

$(2) \Longrightarrow (1)$. Take $n = 1$ and $I_1 = 0$, the result follows.

Theorem 4.3. *The following are equivalent:*

(1) $T = \begin{pmatrix} R & V \\ 0 & A \end{pmatrix} = \bigoplus_{x \in M} \begin{pmatrix} R_x & V_x \\ 0 & A_x \end{pmatrix}$ *is weakly graded directly finite.*

(2) $R = \bigoplus_{x \in M} R_x$ *and* $A = \bigoplus_{x \in M} A_x$ *are weakly graded directly finite.*

Proof. $(1) \Longrightarrow (2)$. Assume that T is weakly graded directly finite. Given $t, s \in R_e$ such that $ts = 1_R$. Then there exist $\begin{pmatrix} t & -tv \\ 0 & 1_A \end{pmatrix}, \begin{pmatrix} s & v \\ 0 & 1_A \end{pmatrix} \in T_e$ for any $v \in V_e$ such that $\begin{pmatrix} t & -tv \\ 0 & 1_A \end{pmatrix} \begin{pmatrix} s & v \\ 0 & 1_A \end{pmatrix} = \begin{pmatrix} 1_R & 0 \\ 0 & 1_A \end{pmatrix}$. Since T is weakly graded directly finite, we have $\begin{pmatrix} s & v \\ 0 & 1_A \end{pmatrix} \begin{pmatrix} t & -tv \\ 0 & 1_A \end{pmatrix} = \begin{pmatrix} 1_R & 0 \\ 0 & 1_A \end{pmatrix}$. So $st = 1_R$, hence $R = \bigoplus_{x \in M} R_x$ is weakly graded directly finite. Similarly, we can prove A is weakly graded directly finite.

$(2) \Longrightarrow (1)$. Let $R = \bigoplus_{x \in M} R_x$ and $A = \bigoplus_{x \in M} A_x$ are weakly graded directly finite. Let $I = \bigoplus_{x \in M} \begin{pmatrix} 0 & V_x \\ 0 & A_x \end{pmatrix}, J = \bigoplus_{x \in M} \begin{pmatrix} R_x & V_x \\ 0 & 0 \end{pmatrix}$. Then I and J are both graded ideals of T with $IJ = 0$. Given any $t \in I \cap J, r, s \in T_e$. Then $(rt_e s)^2 \in IJ = 0$. So $1 - rt_e s \in U(T_e)$, and then $t_e \in J(T_e)$. Thus $I \cap J \subseteq J_G(T)$. Obviously, $T/I \cong_{gr} R$ and $T/J \cong_{gr} A$. So R and A are weakly graded directly finite. By virtue of proposition 4.2, T is weakly graded directly finite.

References

1. G. Karpilovsky, The Jacobson Radical of Classical Rings, New York: John Wiley &   Sons Inc, 1991.
2. Y. Wang, Graded Jacobson Radical of Graded Rings, Act Math. Sinica, **41**(2)(1998), 347–354.
3. A. Haghany and K. Varadarajan, Study of Formal Triangular Matrix Rings, Comm. Algebra, **27**(11)(1999), 5507–5525.
4. T.Y Lam, A First Course in Noncommutative Rings, New York: Spring-Verleg, 1996.
5. R.E. Hartwing and J. Luh, On Finite Regular Rings, Pacific J. Math, **69**(1997), 73–95.

POWER-SUBSTITUTION AND EXCHANGE RINGS

JIAQUN WEI

Department of Mathematics, Nanjing Normal University,
Nanjing 210097, P.R.China
E-mail: weijiaqun@njnu.edu.cn

We introduce the unit power-substitution property for rings and give some basic results. Some characterizations of exchange rings with the power-substitution property are also proved in this paper.

1. Introduction

All rings in this paper are associative with identity. Let $M_n(R)$ be the $n \times n$ matrices over a ring R. Denote by I_n the identity matrix in $M_n(R)$. Recall that a ring R is said to have the right (left) power-substitution property if for any $ax + b = 1$ in R, there exist a positive integer n and $Q \in M_n(R)$ such that $aI_n + bQ$ $(xI_n + Qb)$ is a unit in $M_n(R)$. By [7], this definition is left-right symmetric. The power-substitution property was introduced by Goodearl [6] to study power cancellation of groups and modules. Goodearl in [6] also proved, among other things, that the power-substitution property of rings is preserved under taking corners and fractions. However, it is not Morita invariant as shown in [5]. The power-substitution property has also been studied by several other authors, see for instance [4, 11, 8].

The purpose of this paper is to investigate some characterizations of the power-substitution property over exchange rings and to study the unit power-substitution property. In the second section we deduce some necessary and sufficient conditions for the exchange ring having the power-substitution property. For example, Theorem 2.5 says, an exchange ring R has the power-substitution property if and only if whenever $a_1 R + \cdots + a_m R = R$, there exist a positive integer n and W_i's $(1 \le i \le m)$ such that W_i's are unit-regular in $M_n(R)$ and that $a_1 W_1 + \cdots + a_m W_m = I_n$. In the third section, we introduce the unit power-substitution property and give some basic results. We show that all algebraic algebras over a field have the unit power-substitution property. A natural problem is to consider whether the unit power-substitution property is left-right symmetric. Under some conditions we also give an affirmative answer to this question.

Throughout this paper, $U(R)$ always denotes the set of all units in the ring R. Recall that R is an exchange ring if for every right R-module A and any two decompositions $A = M' \oplus N = \oplus_{i \in I} A_i$, where $M' \simeq R$ and the index set I is finite, there exist submodules $A'_i \subseteq A_i$ such that $A = M' \oplus (\oplus_{i \in I} A'_i)$. The class of exchange rings is very large, which includes local rings, semiperfect rings, semiregular rings and others. A ring R is said to have the right (left) stable range one if for any $ax + b = 1$ in R there exists $u \in R$ such that $a + bu$ $(x + ub)$ is a unit in R. If moreover u is a unit in R then R is said to have the right (left) unit stable rang one. It is well known that the (unit) stable range one is left-right symmetric.

2. The power-substitution property

As shown in [2, 11, 5], some results on stable range one could be generalized to the power-substitution property. Following this idea we will give some characterizations of the power-substitution property over exchange rings in this section. For simpleness we introduce the following notion.

Definition 2.1. Let R be a ring. R is said to have the power stable range one, denoted $psr(R) \leq 1$, if R has the power-substitution property.

Recall a ring R is said to be directly finite if all one-sided inverses in R are two-sided. Our first result shows that all rings having the power-substitution property are directly finite.

Lemma 2.2. *Let R be a ring such that $psr(R) \leq 1$. Then R is directly finite.*

Proof. Let x be a one-sided inverse in R. Assume that $xy = 1$ for some $y \in R$. Since $psr(R) \leq 1$, there exist a positive integer n and $Q \in M_n(R)$ such that $xI_n + 0Q = xI_n$ is a unit in $M_n(R)$. It follows that x is two-sided inverse. Now assume that $yx = 1$ for some $y \in R$. By the symmetric property of the power-substitution, there exist a positive integer n and $Q \in M_n(R)$ such that $xI_n + Q0 = xI_n$ is a unit in $M_n(R)$. It also follows that x is two-sided inverse. Hence R is directly finite.

Lemma 2.3. *The following are equivalent for a ring R.*
  (1) $psr(R) \leq 1$.

(2) Whenever $aR + bR = R$, there exists $Q \in M_n(R)$ such that $aI_n + bQ$ is a unit in $M_n(R)$.

Proof. $(1) \Rightarrow (2)$. Let $ax + by = 1$ for some $x, y \in R$. Since $psr(R) \le 1$, we have $aI_n + byP$ is a unit in $M_n(R)$, where $P \in M_n(R)$. Now $aI_n + bQ$ is a unit in $M_n(R)$, where $Q = yP \in M_n(R)$.

$(2) \Rightarrow (1)$. For any $ax + b = 1$, we have $aR + bR = R$. It follows that there exists $Q \in M_n(R)$ such that $aI_n + bQ$ is a unit in $M_n(R)$ by assumptions. Hence $psr(R) \le 1$.

Corollary 2.4. *Let R be an exchange ring. The following are equivalent:*
(1) $psr(R) \le 1$.
(2) Whenever $aR + eR = R$, where $e = e^2$, there exist a positive integer n and $Q \in M_n(R)$ such that $aI_n + eQ$ is a unit in $M_n(R)$.

Proof. $(1) \Rightarrow (2)$. By Lemma 2.3.

$(2) \Rightarrow (1)$. For any $aR + bR = R$, there is an idempotent $e = e^2 \in bR$ such that $aR + eR = R$ since R is exchange [10]. Then, by assumptions, there exist a positive integer n and $Q \in M_n(R)$ such that $aI_n + eQ$ is a unit in $M_n(R)$. Let $e = br$. Now $aI_n + bP$ is a unit in $M_n(R)$, where $P = rQ$. Hence $psr(R) \le 1$.

Theorem 2.5. *Let R be exchange ring. The following are equivalent:*
(1) $psr(R) \le 1$.
(2) Whenever $aR + bR = R$, there exist a positive integer n and W_1, W_2 such that W_1, W_2 are unit-regular in $M_n(R)$ and that $aW_1 + bW_2 = I_n$.
(3) Whenever $a_1 R + \cdots + a_m R = R$, there exist a positive integer n and W_i's $(1 \le i \le m)$ such that W_i's are unit-regular in $M_n(R)$ and that $a_1 W_1 + \cdots + a_m W_m = I_n$.

Proof. $(1) \Rightarrow (3)$. Let $a_1 R + \cdots + a_m R = R$. Then there are orthogonal idempotents $e_i \in a_i R$ $(1 \le i \le m)$ such that $e_1 + \cdots + e_m = 1$, since R is an exchange ring. Let $e_i \in a_i x_i$ and set $w_i = x_i e_i$ for each $1 \le i \le m$. Then $a_1 w_1 + \cdots + a_m w_m = e_1^2 + \cdots + e_m^2 = e_1 + \cdots + e_m = 1$. Since $psr(R) \le 1$, there exist positive integers n_i's such that each $w_i I_{n_i}$ is unit-regular in $M_{n_i}(R)$ for each $1 \le i \le m$ by [11, Theorem 3.1]. Let $n = n_1 \cdots n_m$. Then each $w_i I_n$ is unit-regular in $M_n(R)$. Set $W_i = w_i I_n$. Combining arguments above, we see that $a_1 W_1 + \cdots + a_m W_m = I_n$, where each W_i is unit-regular in $M_n(R)$.

$(3) \Rightarrow (2)$ is clear.

$(2) \Rightarrow (1)$. Assume $aR + bR = R$. By assumptions, there exist a positive integer n and W_1, W_2 such that W_1, W_2 are unit-regular in $M_n(R)$ and that $aW_1 + bW_2 = I_n$. Let $W_1 = W_1 V W_1$, where V is a unit in $M_n(R)$. Then $W_1 = UE$, where $U = V^{-1}$ and $E = VW_1 = E^2 \in M_n(R)$. Now

$$I_n - E = (I_n - E)(aW_1 + bW_2) = (I_n - E)aW_1 + (I_n - E)bW_2$$
$$= (I_n - E)aUE + (I_n - E)bW_2.$$

Consequently, we have that

$$W_1 + U(I_n - E)bW_2 = UE + U(I_n - E)bW_2 = U(E + (I_n - E)bW_2)$$
$$= U(I_n - (I_n - E)aUE).$$

Note that $(I_n - (I_n - E)aUE)(I_n + (I_n - E)aUE) = I_n$ and U is a unit in $M_n(R)$, so $W_1 + U(I_n - E)bW_2$ is a unit in $M_n(R)$. By [7, Lemma 3.1], there exists some $X \in M_n(R)$ such that $aI_n + bW_2X$ is a unit in $M_n(R)$. Hence $psr(R) \leq 1$.

Proposition 2.6. *Let R be an exchange ring. The following are equivalent:*

(1) $psr(R) \leq 1$.

(2) Whenever $aR = bR$, there exist a positive integer n and Q such that Q is a unit in $M_n(R)$ and that $aI_n = bQ$.

(3) Whenever $aR = bR$, there exist a positive integer n and W such that W is unit-regular in $M_n(R)$ and that $aI_n = bW$.

Proof. $(1) \Rightarrow (2)$. Assume $aR = bR$. Then $at = b$ and $a = br$ for some $t, r \in R$. It follows that $brt = at = b$ and $b(1 - rt) = 0$. Now from $rt + 1 - rt = 1$ we deduce that $rI_n + (1 - rt)P = Q$ is a unit in $M_n(R)$ for some positive integer n and $P \in M_n(R)$, since $psr(R) \leq 1$. Then $aI_n = brI_n = brI_n + b(1 - rt)P = b(rI_n + (1 - rt)P) = bQ$. This shows that (2) holds.

$(2) \Rightarrow (3)$. Note that every unit is obviously unit-regular, so the conclusion follows by setting $W = Q$.

$(3) \Rightarrow (1)$. By [11, Theorem 3.1], we need only to show that for any regular element $x \in R$, there exists a positive integer n such that xI_n is unit-regular in $M_n(R)$. Assume that $x = xyx$ for some $y \in R$. Then we have that $xR = (xy)R$. By assumptions, there exist a positive integer n and W such that W is unit-regular in $M_n(R)$ and $xI_n = (xy)W$. Now from the fact that $xy + 1 - xy = 1$, we deduce that

$$W = W - (xy)W + (xy)W = (I_n - (xy)I_n)W + xI_n.$$

Then

$$I_n = I_n + Wy - Wy + (xI_n - (xy)I_n)$$
$$= Wy + I_n - (xy)I_n + (xy)Wy - Wy$$

$$= Wy + (I_n - (xy)I_n)(I_n - Wy).$$

Assume that $W = WVW$, where V is a unit in $M_n(R)$. Let $E = WV$ and $U = V^{-1}$. Then $E^2 = E$ and U is a unit in $M_n(R)$. Now

$$I_n - E = (Wy + (I_n - (xy)I_n)(I_n - Wy))(I_n - E)$$
$$= EUy(I_n - E) + (I_n - (xy)I_n)(I_n - Wy)(I_n - E).$$

It follows that

$$W + (I_n - (xy)I_n)(I_n - Wy)(I_n - E)U$$
$$= EU + (I_n - (xy)I_n)(I_n - Wy)(I_n - E)U$$
$$= (E + (I_n - (xy)I_n)(I_n - Wy)(I_n - E))U$$
$$= (I_n - EUy(I_n - E))U.$$

Note also that $(I_n - EUy(I_n - E))(I_n + EUy(I_n - E)) = I_n$ and that U is a unit in $M_n(R)$, so $W + (I_n - (xy)I_n)(I_n - Wy)(I_n - E)U$ is a unit in $M_n(R)$. Therefore, we have that

$$xI_n = (xy)W = (xy)W + xy(I_n - (xy)I_n)(I_n - Wy)(I_n - E)U$$
$$= (xy)(W + (I_n - (xy)I_n)(I_n - Wy)(I_n - E)U).$$

Set $P = W + (I_n - (xy)I_n)(I_n - Wy)(I_n - E)U$. Then $xI_n = (xy)P$. It follows that $xP^{-1} = (xy)I_n$. Since $xy = (xy)^2$ we have that $xP^{-1} = xP^{-1}xP^{-1}$. Hence $xI_n = xP^{-1}x$ and xI_n is unit-regular in $M_n(R)$.

Proposition 2.7. *Let R be an exchange ring. The following are equivalent:*

(1) $psr(R) \leq 1$.

(2) Whenever x is a regular element in R, there exist a positive integer n and $U, E \in M_n(R)$ such that U is a unit and E is an idempotent and $xI_n = EU$.

(3) Whenever x is a regular element in R, there exist a positive integer n and $U, E \in M_n(R)$ such that U is a unit and E is an idempotent and $xI_n = UE$.

Proof. (1) $\Rightarrow$ (2). Assume that $x = xyx$. From the fact $xy = (1 - xy) = 1$ we derive that $xI_n + (1 - xy)Q = U$ is a unit in $M_n(R)$ for some positive integer n and $Q \in M_n(R)$. Then

$$xI_n = (xy)(xI_n) = (xy)(xI_n) + (xy)(1 - xy)Q$$
$$= (xy)(xI_n + (1 - xy)Q) = (xy)U.$$

Setting $E = xyI_n$, we have that $E^2 = E$ and $xI_n = EU$ as desired.

(2) $\Rightarrow$ (1). Again we show that for any regular element $x \in R$ there exists a positive integer n such that xI_n is unit-regular in $M_n(R)$. Assume that $x = xyx$. By assumptions, there exist a unit U and an idempotent E in $M_n(R)$ for some positive integer n such that $xI_n = EU$. Since $xy + (1 - xy) = 1$, we have that

$$I_n - E = (xyI_n + (I_n - xyI_n))(I_n - E)$$

256

$$= EUy(I_n - E) + (I_n - xyI_n))(I_n - E).$$

It follows that

$$xI_n + (I_n - xyI_n))(I_n - E)U = EU + (I_n - xyI_n))(I_n - E)U$$
$$= (E + (I_n - xyI_n))(I_n - E))U = (I_n - EUy(I_n - E))U$$

is a unit in $M_n(R)$, since $(I_n - EUy(I_n - E))$ and U are units in $M_n(R)$. By [7,Lemma 3.1], we have that $yI_n + Q(1 - xy) = V$ is a unit in $M_n(R)$ for some $Q \in M_n(R)$. Therefore, we obtain that

$$xI_n = (xyx)I_n = xyxI_n + xQ(1 - xy)x$$
$$= x(yI_n + Q(1 - xy))x = xVx.$$

Hence, xI_n is unit-regular in $M_n(R)$.

(1) $\Leftrightarrow$ (3). The conclusion follows from the fact that the power-substitution property is left-right symmetric and the proof of (1) $\Leftrightarrow$ (2).

Proposition 2.8. *Let R be an exchange ring. The following are equivalent:*

(1) $psr(R) \leq 1$.

(2) Whenever x is a regular element in R, there exist a positive integer n and Q such that Q is a unit in $M_n(R)$ and Qx is an idempotent in $M_n(R)$.

(3) Whenever x is a regular element in R, there exist a positive integer n and Q such that Q is a unit in $M_n(R)$ and xQ is an idempotent in $M_n(R)$.

Proof. (1) $\Rightarrow$ (2). By [11,Theorem 3.1], if $x \in R$ is regular, then xI_n is unit-regular in $M_n(R)$ for some positive integer n. Assume that $xI_n = xQx$, where $Q \in M_n(R)$ is a unit. Then $Qx = Q(xQx)$ is an idempotent in $M_n(R)$.

(2) $\Rightarrow$ (1). Assume that $x = xyx$. By assumptions, Qx is an idempotent for for some positive integer n and some unit $Q \in M_n(R)$. Let $E = Qx$. Then $xI_n = UE$. From $xy + (1 - xy) = 1$, we derive that $Ey + Q(1 - xy) = Q$. Then we have that $E(y + Q(1 - xy)) = EQ$. Note that $EQxy = Qxy$, so that

$$Q = EQ + (Q - EQ) + (EQxy - Qxy) = EQ + (I_n - E)Q(1 - xy)$$
$$= E(yI_n + Q(1 - xy)) + (I_n - E)Q(1 - xy).$$

Let $U = Q^{-1}$. Then we have that

$$(I_n - E)Q(1 - xy) = (I_n - E)Q(1 - xy) - (I_n - E)Q((1 - xy)x)Q$$
$$= (I_n - E)Q(1 - xy)(I_n - xQ)$$
$$= (I_n - E)Q(1 - xy)(UQ - UQxQ - UQxy + UQxQxy)$$
$$= (I_n - E)Q(1 - xy)U(I_n - E)Q(1 - xy) = FQ(1 - xy),$$

where $F = (I_n - E)Q(1 - xy)U(I_n - E)$. By the above process, $Q = E(yI_n + Q(1 - xy)) + FQ(1 - xy)$. Note that $E = E^2$ and that

$$F^2 = (I_n - E)Q(1 - xy)U(I_n - E)(I_n - E)Q(1 - xy)U(I_n - E)$$

$$= ((I_n - E)Q(1 - xy)U(I_n - E)Q(1 - xy))U(I_n - E)$$
$$= (I_n - E)Q(1 - xy)U(I_n - E = F.$$

Obviously, $EF = FE = 0$ since $E(I_n - E) = 0 = (I_n - E)E$. Hence, we have that

$$Q = E(yI_n + Q(1 - xy)) + FQ(1 - xy)$$
$$= (E + F)[E(yI_n + Q(1 - xy)) + FQ(1 - xy)]$$
$$= (E + F)Q = [E + (I_n - E)Q(1 - xy)U(I_n - E)]Q$$
$$= [E(I_n - EQ(1 - xy)U(I_n - E)) + Q(1 - xy)U(I_n - E)]Q$$
$$= Q[x(I_n - EQ(1 - xy)U(I_n - E)) + (1 - xy)U(I_n - E)]Q.$$

Since $[I_n + EQ(1 - xy)U(I_n - E)][I_n - EQ(1 - xy)U(I_n - E)] = I_n$, we have that

$$Q = Q[x(I_n - EQ(1 - xy)U(I_n - E)) + (1 - xy)U(I_n - E)$$
$$[(I_n + EQ(1 - xy)U(I_n - E))(I_n - EQ(1 - xy)U(I_n - E))]]Q$$
$$= Q[xI_n + (1 - xy)U(I_n - E)(I_n + EQ(1 - xy)U(I_n - E))]$$
$$(I_n - EQ(1 - xy)U(I_n - E))Q.$$

It follows that $xI_n + (1 - xy)U(I_n - E)(I_n + EQ(1 - xy)U(I_n - E))$ is a unit in $M_n(R)$. By [7, Lemma 3.1], we have that $yI_n + P(1 - xy) = W$ is a unit in $M_n(R)$. Therefore, $xI_n = xyxI_n = xyxI_n + xP(1 - xy)x = x(yI_n + P(1 - xy))x$ is unit-regular. Hence $psr(R) \leq 1$.

(1) $\Leftrightarrow$ (3). By the left-right symmetric property of the power-substitution and the proof of (1) $\Leftrightarrow$ (2).

3. The unit power-substitution property

Goodearl and Menal [9] studied rings satisfying the unit stable range one. From then rings satisfying the unit stable range one have been studied by many authors. Taken into account the relation between rings satisfying the stable range one and rings satisfying the power-substitution property, it is suitable for us to introduce the following notions.

Definition 3.1. Let R be a ring. R is said to have right (left, resp.) unit power-substitution property or right (left, resp.) unit power stable range one, denoted by $rupsr(R) \leq 1$ ($lupsr(R) \leq 1$, resp.), if whenever $a, b, x \in R$ satisfy $ax + b = 1$, there exist a positive integer n and a unit $Q \in M_n(R)$ such that $aI_n + bQ$ ($xI_n + Qb$, resp.) is a unit in $M_n(R)$.

If $rupsr(R) \leq 1$ and $lupsr(R) \leq 1$, then we simply denote $upsr(R) \leq 1$. For example, this is the case if R is a commutative ring.

Of course, if R has the unit stable range one then R also has the unit power stable range one. The converse is in general false, see for instance

258

Example 3.10.

Proposition 3.2. *Let K be a two-sided ideal of R.*

(1) If $rupsr(R) \leq 1$ then $rupsr(R/K) \leq 1$.

(2) If $K \subseteq J(R)$ and $rupsr(R/K) \leq 1$ then $rupsr(R) \leq 1$.

Proof. (1). It's trivial.

(2). Since $K \subseteq J(R)$, every unit $\bar{Q}$ in $M_n(R/K)$ lifts to a unit Q in $M_n(R)$ for every positive integer n. From this fact we easily deduce the conclusion.

Proposition 3.3. *Let R be a commutative ring. Then the following are equivalent:*

(1) $upsr(R) \leq 1$.

(2) Whenever $a, b, x \in R$ satisfy $ax + b = 1$, there are $y \in U(R)$ and $z \in R$ such that $a^n + b^n y + abz \in U(R)$.

Proof. (1) $\Rightarrow$ (2). Assume that $upsr(R) \leq 1$. Then given $ax + b = 1$ in R, there exist a positive integer n and $Q \in U(M_n(R))$ such that $aI_n + bQ \in U(M_n(R))$. Hence, $\det(aI_n + bQ) = q \in U(R)$. It follows that $a^n + b^n y + abz \in U(R)$ for some $z \in R$, where $y = \det(Q) \in U(R)$.

(2) $\Rightarrow$ (1). Assume that $ax + b = 1$ and that $a^n + b^n y + abz \in U(R)$, where $y \in U(R)$. Then $a^n + b^n y + abz = a^n + b^n y + ab(ax + b)^{n-2}z = a^n + a_1 a^{n-1} b + \cdots + a_{n-1} ab^{n-1} + yb^n$. Set

$$
Q = \begin{pmatrix}
a_1 & -a_2 & a_3 & \cdots & (-1)^{n+1}y \\
1 & 0 & 0 & \cdots & 0 \\
0 & 1 & 0 & \cdots & 0 \\
& & \vdots & & \\
0 & 0 & \cdots & 1 & 0
\end{pmatrix}.
$$

By [6, Lemma 3.1], $\det(aI_n + bQ) = a^n + b^n y + abz$. Hence $aI_n + bQ \in U(M_n(R))$. Note $\det(Q) = (-1)^{n+1}y \in U(R)$. Therefore $upsr(R) \leq 1$.

Remark 3.4. By [6], $psr(\mathbf{Z}) \leq 1$. However, it is easy to check that $\mathbf{Z}$ does not have the unit power-substitution property by the previous proposition.

Proposition 3.5. *Let R be a ring. If for any $x, y \in R$, there exist a positive integer n and a unit $Q \in M_n(R)$ such that $xI_n - Q$ and $yI_n - Q^{-1}$ are units in $M_n(R)$, then $rupsr(R) \leq 1$.*

Proof. Let $ax + b = 1$. By assumptions, for $a, x \in R$, there exist a positive integer n and a unit $Q \in M_n(R)$ such that $xI_n - Q = V$ and $yI_n - Q^{-1} = W$ are units in $M_n(R)$. Therefore, we have that

$$aV + bI_n = a(xI_n - Q) + bI_n = axI_n - aQ + bI_n$$
$$= (ax + b)I_n - aQ = I_n - aQ = I_n - (W + Q^{-1})Q$$
$$= I_n - WQ - I_n = WQ \in U(M_n(R)).$$

It follows that $aI_n + bV^{-1} = -WQV^{-1} \in U(M_n(R))$. Hence $rupsr(R) \leq 1$.

The following results show that there are many rings satisfying the right unit power-substitution property.

Proposition 3.6. *Let R be an algebraic algebra over a field F. Then $rupsr(R) \leq 1$.*

Proof. This is essentially in the proof of [9, Theorem 3.6], where it was shown that for any $x, y \in R$, there exists a unit $Q \in M_n(R)$ for some positive integer n such that $xI_n - Q$ and $yI_n - Q^{-1}$ are both units in $M_n(R)$. Hence, by applying the previous theorem we derive the conclusion.

Corollary 3.7. *Let R be an algebra over a field F. If all $M_n(R)$ are algebraic over F, then $rupsr(A) \leq 1$, where $A = End_R N$ for some finitely generated R-module N.*

Proof. By [9], the endomorphism ring of every finitely generated R-module is algebraic over F. Now apply the previous proposition.

A ring R is said to have many units if for any $x, y \in R$, there exists a unit $q \in R$ such that $x - q$ and $y - q^{-1}$ are units in R. By Proposition 3.5, we easily obtain the following corollary.

Corollary 3.8. *Let R be a ring. If there exists some positive integer n such that $M_n(R)$ has many units, then $rupsr(R) \leq 1$.*

A generalization of the previous result is the following.

Proposition 3.9. *Let R be a ring. If there exists some positive integer n such that $M_n(R)$ has unit stable range one, then $rupsr(R) = lupsr(R) \leq 1$.*

Proof. Let $ax + b = 1$. By assumptions, there exists some positive integer n such that $M_n(R)$ has the unit stable range one. Hence, we have $Q \in U(M_n(R))$ such that $aI_n + bQ \in U(M_n(R))$ following from the fact $aI_n xI_n + bI_n = I_n$. Therefore $rupsr(R) \leq 1$. Note that the unit stable range one condition for a ring R is left-right symmetric, so we also have that $lupsr(R) \leq 1$.

Example 3.10. Let $R = (\mathbf{Z}/2\mathbf{Z})$. It is easy to check that $usr(R) \leq 1$ for every $n \geq 2$. By the previous proposition, $upsr(R) \leq 1$. On the other hand, it is easy to see that R has not the unit stable range one.

From Example 3.10 we also see that $usr(M_n(R)) \leq 1 \not\Rightarrow usr(R) \leq 1$. However, for the unit power-substitution property we have the following.

Proposition 3.11. *Let R be a ring such that $rupsr(M_n(R)) \leq 1$ for some positive integer n. Then $rupsr(R) \leq 1$.*

Proof. Let $ax + b = 1$. Since $rupsr(M_n(R)) \leq 1$, we have $(aI_n)(I_n)_m + (bI_n)Q = U \in U(M_m(M_n(R)))$ for some m and $Q \in U(M_m(M_n(R)))$. That is, $aI_{nm} + bQ \in U(M_{nm}(R))$ and $Q \in U(M_{nm}(R))$. Hence $rupsr(R) \leq 1$ by the definition.

Lemma 3.12. *Let R be a ring satisfying one of the following conditions:*
   (∗) For every positive integer n, $P = (p_{ij})_{1 \leq i,j \leq n}$ is a unit in $M_n(R)$ implies there is some i, $1 \leq i \leq n$, such that at least one of p_{ii}, $\sum_{j=1}^{n} p_{ij}$, $\sum_{j=1}^{n} p_{ji}$ is a unit in R.
   Then the following are equivalent:
   (1) $rupsr(R) \leq 1$.
   (2) Whenever $aR + bR = R$, there exist a positive integer n and a unit $Q \in M_n(R)$ such that $aI_n + bQ$ is a unit in $M_n(R)$.

Proof. $(1) \Rightarrow (2)$. Assume that $ax + by = 1$. Since $rupsr(R) \leq 1$, we have $aI_t + byV = P \in U(M_t(R))$ for some positive integer t and $V \in U(M_t(R))$. Then $aP^{-1} + byVP^{-1} = I_t$. Let $P^{-1} = (p_{ij})_{1 \leq i,j \leq n}$. Then, there is some i, $1 \leq i \leq n$, such that at least one of p_{ii}, $\sum_{j=1}^{n} p_{ij}$, $\sum_{j=1}^{n} p_{ji}$ is a unit in R. Take one unit among them, denoted by x. Then we have that $ax + bz = 1$ for some $z \in R$. Since $rupsr(R) \leq 1$, there exist a positive integer n and a unit $W \in M_n(R)$ such that $bI_n + axW$ is a unit in $M_n(R)$. Therefore, $aI_n + bQ$ is a unit in $M_n(R)$, where $Q = W^{-1}x^{-1} \in U(M_n(R))$.
   $(2) \Rightarrow (1)$ is obvious.

We end this paper by considering a natural problem: is the unit power-substitution property for rings left-right symmetric? The first result shows that the answer is affirmative for rings satisfying assumptions in Lemma 3.12.

Proposition 3.13. *Assume that R satisfies the same condition $(*)$ in Lemma 3.12. Then the following are equivalent:*

(1) $rupsr(R) \leq 1$.

(2) For any $x, y \in R$, there exist a positive integer n and a unit $Q \in M_n(R)$ such that $I_n + x(yI_n - Q)$ is a unit in $M_n(R)$.

(3) $lupsr(R) \leq 1$.

Proof. $(1) \Rightarrow (2)$. For any $x, y \in R$ we have $(1 + xy) + (-x)y = 1$. By Lemma 3.12, $(1 + xy)I_n + (-x)Q$ is a unit in $M_n(R)$ for some positive integer n and $Q \in U(M_n(R))$. That is, $I_n + x(yI_n - Q)$ is a unit in $M_n(R)$.

$(2) \Rightarrow (1)$. Let $ax + b = 1$. By assumptions, for $-a, x \in R$ there exist a positive integer n and a unit $Q \in M_n(R)$ such that $I_n + (-a)(xI_n - Q) = P$ is a unit in $M_n(R)$. Note $P = I_n + (-a)(xI_n - Q) = aQ + bI_n$, so $aI_n + bQ^{-1} = PQ^{-1}$ is a unit in $M_n(R)$. Hence $rupsr(R) \leq 1$.

$(3) \Rightarrow (2)$. For any $x, y \in R$ we have $(1 + yx) + y(-x) = 1$. By Lemma 3.12, $(1 + yx)I_n + Q(-x) = I_n + (yI_n - Q)x$ is a unit in $M_n(R)$ for some positive integer n and $Q \in U(M_n(R))$. From the fact that for any c, d in a ring $1 - cd$ is a unit if and only if $1 - dc$ is a unit, we deduce that $I_n + x(yI_n - Q)$ is a unit in $M_n(R)$.

$(2) \Rightarrow (3)$. Let $ax + b = 1$. By assumptions, for $-x, a \in R$ there exist a positive integer n and a unit $Q \in M_n(R)$ such that $I_n + (-x)(aI_n - Q)$ is a unit in $M_n(R)$. Since for any c, d in a ring $1 - cd$ is a unit if and only if $1 - dc$ is a unit, we have $I_n + (aI_n - Q)(-x) = bI_n + Qx = P$ is also a unit in $M_n(R)$. Then $xI_n + Q^{-1}b = Q^{-1}P$ is a unit in $M_n(R)$, too. Hence $lupsr(R) \leq 1$.

Proposition 3.14. *Let R be a ring satisfying that $rupsr(M_n(R)) \leq 1$ for all integers $n \geq 2$. Then*

(1) $rupsr(R) \leq 1$.

(2) Whenever $aR + bR = R$, there exist a positive integer n and a unit $Q \in M_n(R)$ such that $aI_n + bQ$ is a unit in $M_n(R)$.

(3) For any $x, y \in R$, there exist a positive integer n and a unit $Q \in M_n(R)$ such that $I_n + x(yI_n - Q)$ is a unit in $M_n(R)$.

(4) $lupsr(R) \leq 1$.

Proof. (1) follows from Proposition 3.11.

(2). Assume that $ax + by = 1$. Since $rupsr(R) \leq 1$, we have that $aI_t + byV = P \in U(M_t(R))$ for some positive integer t and $V \in U(M_t(R))$. Then $aP^{-1} + byVP^{-1} = I_t$. By assumptions, $rupsr(M_t(R)) \leq 1$, hence there exist some positive integer m and $W \in U(M_m(M_t(R)))$ such that

262

$bI_{tm} + (aP^{-1})W \in U(M_m(M_t(R)))$. It follows that $aI_n + bQ \in U(M_n(R))$, where $n = tm$ and $Q = W^{-1}P \in U(M_n(R))$.

(3) and (4) follow from proofs similar to (1) $\Rightarrow$ (2) and (2) $\Rightarrow$ (3) in Proposition 3.13.

References

1. P. Ara, Strongly π-regular rings have stable range one, *Proc. AMS.* 124 (1996), 3293-3298.
2. H. Chen, Rings with stable range conditions, *Comm. in Algebra* 26 (1998), 3653-3668.
3. H. Chen, On stable range conditions, *Comm. in Algebra* 28 (2000), 3913-3924.
4. H. Chen, Power-substitution, exchange rings and unit π-regularity, *Comm. in Algebra* 28 (2000), 5123-5233.
5. R.Camps and P.Menal, The power substitution property for rings of continious functions, *J. Algebra* 161 (1995), 480-503.
6. K.R.Goodearl, Power-cancellation of groups and modules, *Pacific J. Math.* 64 (1976), 387-411.
7. K.R.Goodearl, Cancellation of low-rank vector bundles, *Pacific J. Math.* 113 (1984), 289-302.
8. R.Guralnik, Power-cancellation of modules, *Pacific J. Math.* 124 (1986), 131-144.
9. K.R.Goodearl and P. Menal, Stable range one for rings with many units, *J. Pure. Appl. Alg.* 54 (1988), 261-287.
10. W.K.Nicholson, Lifting idempotents and exchange rings, *Trans. AMS.* 229 (1977), 269-278.
11. T. Wu, The power-substitution condition of endomorphism rings of quasi-projective modules, *Comm. in Algebra* 28 (2000), 407-418.
12. H.P. Yu, Stabke range one for exchange rings, *J. Pure Appl. Alg.* 98 (1995), 105-109.

FROM GALOIS FIELD EXTENSIONS TO GALOIS COMODULES

ROBERT WISBAUER

Department of Mathematics, HHU, 40225 Düsseldorf, Germany

e-mail: wisbauer@math.uni-duesseldorf.de

web site: http://math.uni-duesseldorf.de/~wisbauer

Given a finite automorphism group G of a field extension $E \supset K$, E can be considered as module over the group algebra $K[G]$. Moreover, E can also be viewed as a comodule over the bialgebra $K[G]^*$ and here a canonical isomorphism involving the subfield fixed under the action of G arises. This isomorphism and its consequences were extended and studied for group actions on commutative rings, for actions of Hopf algebras on noncommutative algebras, then for corings with grouplike elements and eventually to comodules over corings. The purpose of this note is to report about this development and to give the reader some idea about the notions and results involved in this theory (without claiming to be comprehensive).

1. Preliminaries

To begin with we recall the algebraic structures for which Galois type conditions are applied. We follow the notation in [9]. Throughout R will denote a commutative associative ring with unit.

1.1. Algebras and modules. A, or more precisely $(A, \mu, 1_A)$, stands for an associative R-algebra with multiplication $\mu : A \otimes_R A \to A$ and unit 1_A. Right A-modules are defined as R-modules M with an action $\varrho_M : M \otimes_R A \to M$.

For the category of right A-modules we write $\mathbf{M}_A$ and denote the morphisms between $M, N \in \mathbf{M}_A$ by $\mathrm{Hom}_A(M, N)$. It is well known that A is a projective generator in $\mathbf{M}_A$.

1.2. Coalgebras and comodules. An R-*coalgebra* is a triple (C, Δ, ε) where C is an R-module, $\Delta : C \to C \otimes_R C$ is the coproduct and $\varepsilon : C \to R$

is the counit. Right C-comodules are R-modules M with a coaction $\varrho^M : M \to M \otimes_R C$.

The category of right C-comodules is denoted by $\mathbf{M}^C$ and the morphisms between $M, N \in \mathbf{M}^C$ are written as $\mathrm{Hom}^C(M, N)$. As a right comodule, C is a subgenerator in $\mathbf{M}^C$, that is, every right C-comodule is a subcomodule of a C-generated comodule. Note that $\mathbf{M}^C$ need not have projectives even if R is a field.

Left (co)modules and their categories are defined and denoted in an obvious way.

1.3. Bialgebras and Hopf modules. An R-*bialgebra* is a quintuple $(B, \Delta, \varepsilon, \mu, 1_B)$ where (B, Δ, ε) is an R-coalgebra and $(B, \mu, 1_B)$ is an R-algebra such that Δ is an algebra morphism (equivalently μ is a coalgebra morphism).

An R-module M that is a right B-module by $\varrho_M : M \otimes_R B \to M$ and a right B-comodule by $\varrho^M : M \to M \otimes_R B$ is called a *right B-Hopf module* provided for any $m \in M$ and $b \in B$, $\varrho^M(mb) = \varrho^M(m)\Delta(b)$. The category of all right B-Hopf modules is denoted by $\mathbf{M}_B^B$. The module $B \otimes_R B$ allows for a right B-Hopf module structure and with this it is a subgenerator in $\mathbf{M}_B^B$. For $M \in \mathbf{M}_B^B$ the *coinvariants* are defined as

$$M^{coB} = \{m \in M \mid \varrho^M(m) = m \otimes_R 1_B\} \simeq \mathrm{Hom}_B^B(A, M).$$

An R-bialgebra B is called a *Hopf algebra* if there is an antipode, that is, an R-linear map $S : B \to B$ which is the inverse of the identity of B with respect to the convolution product in $\mathrm{End}_R(B)$ (see also 2.5).

For any R-algebra A which is finitely generated and projective as R-module, the dual $A^* = \mathrm{Hom}_R(A, R)$ can be considered as an R-coalgebra with natural comultiplication and counit. Here we are interested in the following special case.

1.4. Group algebras and their dual. Let G be a finite group of order $n \in \mathbb{N}$ and $R[G]$ the group algebra, that is, $R[G]$ is a free R-module with basis the group elements $\{g_1, \ldots, g_n\}$ and the product given by the group multiplication. Furthermore, $R[G]$ is an R-coalgebra with coproduct induced by $\Delta(g) = g \otimes g$ and counit $\varepsilon(g) = 1_R$, for $g \in G$. With these structures $R[G]$ is an R-bialgebra, and even a Hopf algebra with antipode S induced by $S(g) = g^{-1}$ for $g \in G$.

The R-dual $R[G]^* = \mathrm{Hom}_R(R[G], R)$ is also a Hopf algebra. The multiplication of $f, g \in R[G]^*$ is given by $f * g(x) = f(x)g(x)$ for $x \in G$. To describe the coalgebra structure let $\{p_g\}_{g \in G} \subset R[G]^*$ be the dual basis to $\{g\}_{g \in G}$. Then coproduct and counit are defined by

$$\Delta(p_g) = \sum_{kh=g} p_k \otimes p_h, \quad \varepsilon(P_g) = \delta_{1,g}.$$

The antipode S of $R[G]^*$ is induced by $S(p_g) = p_{g^{-1}}$ for $g \in G$.

1.5. Comodule algebras and relative Hopf modules. Let B be an R-bialgebra. An R-algebra A is called *right B-comodule algebra* if A is a right B-comodule by $\varrho^A : A \to A \otimes_R B$ such that ϱ^A is an algebra morphisms.

A *right (A, B)-Hopf module* is an R-module M which is a right A-module and a right B-comodule by $\varrho^M : M \to M \otimes_R B$ such that for all $m \in M$ and $a \in A$, $\varrho^M(ma) = \varrho^M(m)\varrho^A(a)$. The category of these modules is denoted by $\mathbf{M}_A^H$ and it has $A \otimes_R H$ as a subgenerator. For $M \in \mathbf{M}_A^H$ the coinvariants are defined as

$$M^{coB} = \{m \in M \mid \varrho^M(m) = m \otimes_R 1_B\} \simeq \mathrm{Hom}_A^B(A, M).$$

Note that in the above construction the right (A, B)-Hopf modules may be replaced by the category $\mathbf{M}(B)_A^D$ of *right (A, D)-Hopf modules* where D is a right B-module coalgebra and the objects are right D-comodules which are also right A-modules satisfying some compatibility condition. Then $A \otimes_R D$ is a subgenerator $\mathbf{M}(B)_A^D$ (see [13], [18]).

Under weak (projectivity) conditions, for all the structures considered above the related (co)module categories can be understood as module categories over some algebra subgenerated by a suitable module. We refer to [24] for more details. All this settings are subsumed as special cases of

1.6. Corings and comodules. An A-coring is a triple $(\mathcal{C}, \underline{\Delta}, \underline{\varepsilon})$ where $\mathcal{C}$ is an (A, A)-bimodule with coproduct $\underline{\Delta} : \mathcal{C} \to \mathcal{C} \otimes_A \mathcal{C}$ and counit $\underline{\varepsilon} : \mathcal{C} \to A$. Associated to this there are the right and left dual rings $\mathcal{C}^* = \mathrm{Hom}_A(\mathcal{C}, A)$ and $^*\mathcal{C} = {}_A\mathrm{Hom}(\mathcal{C}, A)$ with the convolution products.

A *right $\mathcal{C}$-comodule* is a right A-module M together with an A-linear $\mathcal{C}$-coaction $\varrho^M : M \to M \otimes_A \mathcal{C}$. These comodules form a category which we denote by $\mathbf{M}^{\mathcal{C}}$. It is an additive category with coproducts and cokernels, and $\mathcal{C}$ is a subgenerator in it. The functor $- \otimes_A \mathcal{C} : \mathbf{M}_A \to \mathbf{M}^{\mathcal{C}}$ is right adjoint to the forgetful functor by the isomorphisms, for $M \in \mathbf{M}^{\mathcal{C}}$ and $X \in \mathbf{M}_A$,

$$\mathrm{Hom}^{\mathcal{C}}(M, X \otimes_A \mathcal{C}) \to \mathrm{Hom}_A(M, X), \ f \mapsto (I_X \otimes \underline{\varepsilon}) \circ f,$$

with inverse map $\quad h \mapsto (h \otimes I_{\mathcal{C}}) \circ \varrho^M$.

Notice that for any monomorphism (injective map) $f : X \to Y$ in $\mathbf{M}_A$, the colinear map $f \otimes I_{\mathcal{C}} : X \otimes_A \mathcal{C} \to Y \otimes_A \mathcal{C}$ is a monomorphism in $\mathbf{M}^{\mathcal{C}}$ but need not be injective. In case $_A\mathcal{C}$ is flat, monomorphisms in $\mathbf{M}^{\mathcal{C}}$ are injective maps and in this case $\mathbf{M}^{\mathcal{C}}$ is a Grothendieck category (see 18.14 in [9]).

Any right $\mathcal{C}$-comodule (M, ϱ^M) allows for a left $^*\mathcal{C}$-module structure by putting $f{\rightharpoonup}m = (I_M \otimes f) \circ \varrho^M(m)$, for any $f \in {}^*\mathcal{C}$, $m \in M$. This yields a faithful functor $\Phi : \mathbf{M}^{\mathcal{C}} \to {}_{*_{\mathcal{C}}}\mathbf{M}$ which is a full embedding if and only if the map

$$\alpha_K : K \otimes_A \mathcal{C} \to \mathrm{Hom}_A(^*\mathcal{C}, K), \quad n \otimes c \mapsto [f \mapsto nf(c)],$$

is injective for any $K \in \mathbf{M}_A$. This is called the *left α-condition* on $\mathcal{C}$ and it holds if and only if $_A\mathcal{C}$ is locally projective. In this case $\mathbf{M}^{\mathcal{C}}$ can be identified with $\sigma[_{*_{\mathcal{C}}}\mathcal{C}]$, the full subcategory of $_{*_{\mathcal{C}}}\mathbf{M}$ whose objects are subgenerated by $\mathcal{C}$.

1.7. A as a $\mathcal{C}$-comodule. An element g of an A-coring $\mathcal{C}$ is called a *grouplike element* if $\underline{\Delta}(g) = g \otimes g$ and $\underline{\varepsilon}(g) = 1_A$. Such a grouplike element g exists if and only if A is a right or left $\mathcal{C}$-comodule, by the coactions

$$\varrho^A : A \to \mathcal{C}, \ a \mapsto ga, \quad {}^A\varrho : A \to \mathcal{C}, \ a \mapsto ag.$$

Write A_g or $_gA$ to consider A with the right or left comodule structure induced by g. Given an A-coring $\mathcal{C}$ with a grouplike element g and $M \in \mathbf{M}^{\mathcal{C}}$, the *$g$-coinvariants* of M are defined as the R-module

$$M_g^{co\mathcal{C}} = \{m \in M \mid \varrho^M(m) = m \otimes g\} = \mathrm{Ke}\,(\varrho^M - (- \otimes g)),$$

and there is an isomorphism

$$\mathrm{Hom}^{\mathcal{C}}(A_g, M) \to M_g^{co\mathcal{C}}, \quad f \mapsto f(1_A).$$

The bijectivity of this map is clear by the fact that any A-linear map with source A is uniquely determined by the image of 1_A. As special cases we have the coinvariants

(1) $\operatorname{End}^{\mathcal{C}}(A_g) \simeq A_g^{co\mathcal{C}} = \{a \in A_g \mid ga = ag\}$, the centraliser of g in A.

(2) For any $X \in \mathbf{M}_A$, $(X \otimes_A \mathcal{C})^{co\mathcal{C}} \simeq \operatorname{Hom}^{\mathcal{C}}(A_g, X \otimes_A \mathcal{C}) \simeq X$, and for $X = A$,

$$\mathcal{C}^{co\mathcal{C}} \simeq \operatorname{Hom}^{\mathcal{C}}(A_g, \mathcal{C}) \simeq \operatorname{Hom}_A(A_g, A) \simeq A,$$

which is a left A- and right $\operatorname{End}^{\mathcal{C}}(A_g)$-morphism.

Given any right B-module M, $M \otimes_B A$ is a right $\mathcal{C}$-comodule via the coaction

$$\varrho^{M \otimes_B A} : M \otimes_B A \to M \otimes_B A \otimes_A \mathcal{C} \cong M \otimes_B \mathcal{C}, \quad m \otimes a \mapsto m \otimes ga.$$

This yields a functor $- \otimes_B A : \mathbf{M}_B \to \mathbf{M}^{\mathcal{C}}$. Right adjoint to this is the *g-coinvariants functor* $\operatorname{Hom}^{\mathcal{C}}(A_g, -) : \mathbf{M}^{\mathcal{C}} \to \mathbf{M}_B$.

For $N \in \mathbf{M}_B$ the unit of the adjunction is given by

$$N \to (N \otimes_B A)^{co\mathcal{C}}, \quad n \mapsto n \otimes 1_A,$$

and for $M \in \mathbf{M}^{\mathcal{C}}$, the counit reads

$$M^{co\mathcal{C}} \otimes_B A \to M, \quad m \otimes a \mapsto ma.$$

1.8. Coring of a projective module. For R-algebras A, B, let P be a (B, A)-bimodule that is finitely generated and projective as a right A-module. Let $p_1, \ldots, p_n \in P$ and $\pi_1, \ldots, \pi_n \in P^* = \operatorname{Hom}_A(P, A)$ be a dual basis for P_A. Then the (B, B)-bimodule $P \otimes_A P^*$ is an algebra by the isomorphism

$$P \otimes_A P^* \to \operatorname{End}_A(P), \quad p \otimes f \mapsto [q \mapsto pf(q)],$$

and the (A, A)-bimodule $P^* \otimes_B P$ is an A-coring with coproduct and counit

$$\underline{\Delta} : P^* \otimes_B P \to (P^* \otimes_B P) \otimes_A (P^* \otimes_B P), \quad f \otimes p \mapsto \sum_i f \otimes p_i \otimes \pi_i \otimes p,$$

$$\underline{\varepsilon} : P^* \otimes_B P \to A, \quad f \otimes p \mapsto f(p).$$

As a special case, for the (A, A)-bimodule $P = A^n$, $n \in \mathbb{N}$, $P^* \otimes_A P$ can be identified with the $n \times n$-matrices $M_n(A)$ over A, endowed with an A-coring structure (matrix coring).

268

1.9. The Sweedler coring. Given an R-algebra morphism $\phi : B \to A$, the tensor product $\mathcal{C} = A \otimes_B A$ is an A-coring with coproduct

$$\underline{\Delta} : \mathcal{C} \to \mathcal{C} \otimes_A \mathcal{C} \simeq A \otimes_B A \otimes_B A, \quad a \otimes a' \mapsto a \otimes 1_A \otimes a',$$

and counit $\underline{\varepsilon}(a \otimes a') = aa'$. $\mathcal{C}$ is called the *Sweedler A-coring* associated to the algebra (or ring) morphism $\phi : B \to A$. Clearly $1_A \otimes 1_A$ is a grouplike element in $\mathcal{C}$.

Since A is finitely generated and projective as right A-module, in view of 1.7 this is a special case of 1.8.

1.10. Entwining structures. Given an R-algebra A and an R-coalgebra C one may think about compatibility conditions between these two structures. This led to the notion of a (right-right) *entwining structure* which is given by an *entwining map*, that is, an R-module map $\psi : C \otimes_R A \to A \otimes_R C$ satisfying the conditions

  (1) $\psi \circ (I_C \otimes \mu) = (\mu \otimes I_C) \circ (I_A \otimes \psi) \circ (\psi \otimes I_A)$,

  (2) $(I_A \otimes \Delta) \circ \psi = (\psi \otimes I_C) \circ (I_C \otimes \psi) \circ (\Delta \otimes I_A)$,

  (3) $\psi \circ (I_C \otimes \iota) = \iota \otimes I_C$,

  (4) $(I_A \otimes \varepsilon) \circ \psi = \varepsilon \otimes I_A$.

Associated to any entwining structure (A, C, ψ) is the category of (right-right) (A, C, ψ)-*entwined modules* denoted by $\mathbf{M}_A^C(\psi)$. An object $M \in \mathbf{M}_A^C(\psi)$ is a right A-module with multiplication ϱ_M and a right C-comodule with coaction ϱ^M satisfying

$$\varrho^M \circ \varrho_M = (\varrho_M \otimes I_C) \circ (I_M \otimes \psi) \circ (\varrho^M \otimes I_A),$$

and morphisms in $\mathbf{M}_A^C(\psi)$ are maps which are right A-module as well as right C-comodule morphisms.

Entwining structures were introduced in [7] in the context of gauge theory on noncommutatice spaces. It then turned out that they are instances of corings since - with the data given above - $A \otimes_R C$ is an A-coring with (A, A)-bimodule struture

$$b(a' \otimes c)a = ba'\psi(c \otimes a), \text{ for } a, a', b \in A, \, c \in C,$$

coproduct $\underline{\Delta} = I_A \otimes \Delta$ and counit $\underline{\varepsilon} = I_A \otimes \varepsilon$ (see 32.6 in [9]). With this correspondence the category $\mathbf{M}_A^C(\psi)$ can be identified with the comodule category $\mathbf{M}^{A \otimes_R C}$.

1.11. Bialgebras and corings (see 33.1 in [9]). Let $(B, \Delta_B, \varepsilon_B)$ be an R-bialgebra. Then $B \otimes_R B$ is a B-coring by the coproduct $\underline{\Delta} : I_B \otimes \Delta_B$, the counit $\underline{\varepsilon} = I_B \otimes \varepsilon_B$, and the (B, B)-bimodule structure

$$a(c \otimes d)b = (ac \otimes d)\Delta_B(b) \text{ where } a, b, c, d \in B.$$

With this structure the right B-Hopf modules can be identified with the right $B \otimes_R B$-comodules, that is, $\mathbf{M}_B^B = \mathbf{M}^{B \otimes_R B}$. Clearly $1_B \otimes 1_B$ is a grouplike element in $B \otimes_R B$ and the ring of $B \otimes_R B$-covariants of B is isomorphic to R.

1.12. Comodule algebras and corings (see 33.2 in [9]). Let $(B, \Delta_B, \varepsilon_B)$ be an R-bialgebra. Then for a right B-comodule algebra A, $A \otimes_R B$ is an A-coring with coproduct $\underline{\Delta} = I_A \otimes \Delta_B$, counit $\underline{\varepsilon} = I_A \otimes \varepsilon_B$, and (A, A)-bimodule structure

$$a(c \otimes b)d = (ac \otimes b)\varrho^A(d), \text{ for } a, c, d \in A \text{ and } b \in B.$$

Here the right relative (A, B)-Hopf modules are just the right $A \otimes_R B$-comodules, that is, $\mathbf{M}_A^B = \mathbf{M}^{A \otimes_R B}$.

1.13. Cointegrals. An (A, A)-bilinear map $\delta : \mathcal{C} \otimes_A \mathcal{C} \to \mathcal{C}$ is called a *cointegral in* $\mathcal{C}$ if

$$(I_{\mathcal{C}} \otimes \delta) \circ (\Delta \otimes I_{\mathcal{C}}) = (\delta \otimes I_{\mathcal{C}}) \circ (I_{\mathcal{C}} \otimes \Delta).$$

Cointegrals are characterised by the fact that for any $M \in \mathbf{M}^{\mathcal{C}}$, the map

$$(I_M \otimes \delta) \circ (\varrho^M \otimes I_{\mathcal{C}}) : M \otimes_A \mathcal{C} \to M$$

is a comodule morphism (or by the corresponding property for left $\mathcal{C}$-comodules).

In [10], Section 5, these maps are related to the counit for the adjoint pair of functors $- \otimes_A \mathcal{C}$ and the forgetful functor. For R-coalgebras C over a commutative ring R with C_R locally projective, a cointegral is precisely a C^*-balanced R-linear map $C \otimes_R C \to R$ (e.g., 6.4 in [9]).

Recall some properties of relative injectivity from [27], Section 2:

1.14. Relative injectivity. Let $M \in \mathbf{M}^{\mathcal{C}}$ and $S = \operatorname{End}^{\mathcal{C}}(M)$.

M is $(\mathcal{C}, A)$-*injective* provided the structure map $\varrho^M : M \to M \otimes_A \mathcal{C}$ is split by a $\mathcal{C}$-morphism $\lambda : M \otimes_A \mathcal{C} \to M$.

M is called *strongly (C, A)-injective* if this λ is C-colinear and S-linear. Given a subring $B \subseteq S$, M is said to be *B-strongly (C, A)-injective* if λ is C-colinear and B-linear.

M is called *fully (C, A)-injective* if there is a cointegral $\delta_M : C \otimes_A C \to C$ such that ϱ^M is split by $(I_M \otimes \delta_M) \circ (\varrho^M \otimes I_C)$.

The notions for left C-comodules are defined symmetrically.

For R-coalgebras C, B-strongly (C, R)-injective comodules are named *B-equivariantly C-injective* (see Definition 5.1 in [20]).

1.15. Fully (C, A)-injective comodules. *Let $M \in \mathbf{M}^C$ with $S = \mathrm{End}^C(M)$.*

(1) M is fully (C, A)-injective if and only if

$$(I_M \otimes \widetilde{\delta}_M) \circ \varrho^M = I_M \quad \text{where} \quad \widetilde{\delta}_M = \delta_M \circ \Delta : C \to A.$$

(2) C is a fully (C, A)-injective right (left) comodule if and only if C is a coseparable coring.

(3) Let M be fully (C, A)-injective. Then:

(i) Every comodule in $\sigma[M]$ is fully (C, A)-injective.

(ii) If M is a subgenerator in $\mathbf{M}^C$ then C is a coseparable coring.

(iii) For any subring $B \subset S$ and $X \in \mathbf{M}_B$, $X \otimes_B M$ is fully (C, A)-injective.

(iv) If M_A is finitely generated and projective, then M^* is a fully (C, A)-injective left C-comodule.

2. Galois extensions and comodules

Classical Galois theory studies the action of a finite automorphism group G on a field E and then considers E as extension of the subfield of the elements which are left unchanged by the action of G. This can be understood as a comodule situation (compare [19], Chapter 8).

2.1. Galois field extension. Let G be a finite automorphism group of a field extension $E \supset K$ and let $F = E^G$ be the fixed field of G. Thus the group algebra $K[G]$ acts on E and so its dual, the Hopf algebra $H = \mathrm{Hom}_K(K[G], K) = K[G]^*$ coacts on E.

To describe this let $G = \{g_1, \ldots, g_n\}$ and choose $\{b_1, \ldots, b_n\} \subset E$ as a basis of the F-vectorspace E. Denote by $\{p_1, \ldots, p_n\} \subset K[G]^*$ the dual basis to $\{g_1, \ldots, g_n\} \subset K[G]$. Then E is a right $K[G]^*$-comodule by the coaction

$$\varrho^E : E \to E \otimes_K K[G]^*, \quad a \mapsto \sum_{i=1}^n (g_i \cdot a) \otimes p_i,$$

and we can define the *Galois map*

$$\gamma : E \otimes_F E \to E \otimes_K K[G]^*, \quad a \otimes b \mapsto \sum_{i=1}^n a(g_i \cdot b) \otimes p_i.$$

For any $w = \sum_j a_j \otimes b_j \in \mathrm{Ke}\,\gamma$, we have $\sum_{j,i} a_j(g_i \cdot b_j) \otimes p_i = 0$ and by the independence of the $p_1, \ldots, p_n$, $\sum_j a_j(g_i \cdot b_j) = 0$ for all i. Now Dedekind's lemma on the independence of automorphisms implies that all $a_j = 0$ and thus $w = 0$. This shows that γ is injective and for dimension reasons it is in fact bijective.

Notice that the coinvariants of the $K[G]^*$-comodule E are

$$\{a \in E \mid \sum_{i=1}^n (g_i \cdot a) \otimes p_i = a \otimes \varepsilon\} = E^G,$$

since for each such $a \in E$ and $g_i \in G$, $g_i \cdot a = (g_i \cdot a)p_i(g_i) = a\varepsilon(g_i) = a$.

The definition of Hopf Galois extensions goes back to Chase-Harrison-Rosenberg [11] where the classical Galois theory of fields was extended to groups acting on commutative rings. This was generalised in Chase-Sweedler [12] to coactions of Hopf algebras on commutative R-algebras and then, in Kreimer-Takeuchi [17], to coactions on noncommutative R-algebras.

2.2. Comodule algebras. Let H be a Hopf R-algebra and A a right H-comodule algebra with structure map $\varrho^A : A \to A \otimes_R H$ and $B = A^{coH}$. Then $B \subset A$ is called *right H-Galois* if the following map is bijective:

$$\gamma : A \otimes_B A \to A \otimes H, \quad a \otimes b \mapsto (a \otimes 1)\varrho^A(b).$$

For examples and more information about such extensions we refer to [19], Section 8. Further investigation on such structures were done in particular by Doi, Takeuchi and Schneider [14], [15], [21], [22], [23].

Generalising results about the action of an affine algebraic group scheme on an affine scheme the following theorem was proved in [21]. This shows (again) that H-Galois extensions are closely related to modules inducing equivalences.

Schneider's Theorem. *Let H be a Hopf algebra over a field R with bijective antipode. Then for a right H-comodule algebra A and $B = A^{coH}$ the following are equivalent:*

(a) $B \subset A$ is a H-Galois extension and A is faithfully flat as a left B-module;

(b) $B \subset A$ is a H-Galois extension and A is faithfully flat as a right B-module;

(c) $- \otimes_B A : \mathbf{M}_B \to \mathbf{M}_A^H$ is an equivalence;

(d) $A \otimes_B - : {}_B\mathbf{M} \to {}_A\mathbf{M}^H$ is an equivalence.

Notice that the above theorem shows a left right symmetry which will not be maintained in (most of) the subsequent generalisations.

As mentioned in 1.5, the (A, H)-Hopf modules can be generalised to (A, D)-Hopf modules where D is a right H-module coalgebra yielding the category $\mathbf{M}(H)_A^D$. If there is a *grouplike element* $x \in D$, then A is in $\mathbf{M}(H)_A^D$ and for any $M \in \mathbf{M}(H)_A^D$ coinvariants can be defined as $\mathrm{Hom}_A^D(A, M)$. Then $B = \mathrm{Hom}_A^D(A, A)$ is a subring of A and the inclusion $B \hookrightarrow A$ is called a *right Hopf-Galois extension* provided the canonical map

$$A \otimes_B A \to A \otimes_R D, \quad a \otimes b \mapsto (a \otimes x)\varrho^A(b)$$

is bijective. For this setting an extension of Schneider's Theorem is proved by Menini and Zuccoli (see Theorem 3.29 in [18]).

2.3. Coalgebra-Galois extensions. Let C be an R-coalgebra and A an R-algebra and a right C-comodule with coaction $\varrho^A : A \to A \otimes_R C$. Define the *coinvariants* of A as

$$B = \{b \in A \mid \text{for all } a \in A, \ \varrho^A(ba) = b\varrho^A(a)\}.$$

The extension $B \hookrightarrow A$ is called a *coalgebra-Galois extension* (or a *C-Galois extension*) if the following left A-module, right C-comodule map is bijective:

$$\gamma : A \otimes_B A \to A \otimes_R C, \quad a \otimes a' \mapsto a\varrho^A(a').$$

Notice that here the definition of covariants does not require the existence of a grouplike element in C and thus coalgebra-Galois extensions are defined for arbitrary coalgebras. This notion was introduced in [6], following their appearance as generalised principal bundles in [7]. The main geometric motivation for this was the need for principal bundles with coalgebras playing the role of a structure group. The main result Theorem 2.7 in [6] shows how coalgebra Galois extensions are related to entwining structures.

Theorem. *Let R be a field and A a C-Galois extension of B (as defined above). Then there exists a unique entwining map $\psi : C \otimes_R A \to A \otimes_R C$ such that $A \in \mathbf{M}_A^C(\psi)$ with structure map ϱ^A.*

2.4. Galois corings. Let $\mathcal{C}$ be an A-coring with a grouplike element g and $B = A_g^{co\mathcal{C}}$. Following Definition 5.3 in [4], $(\mathcal{C}, g)$ is called a *Galois coring* if the canonical map

$$\chi : A \otimes_S A \to \mathcal{C}, \quad a \otimes a' \mapsto aga',$$

is an isomorphism (of corings). It was pointed out in [26] that this can be seen as the evaluation map

$$\mathrm{Hom}^{\mathcal{C}}(A_g, \mathcal{C}) \otimes_S A \to \mathcal{C}, \quad f \otimes a \mapsto f(a).$$

The following assertions are equivalent (4.6 in [26]):

(a) $(\mathcal{C}, g)$ is a Galois coring;

(b) for every $(\mathcal{C}, A)$-injective comodule $N \in \mathbf{M}^{\mathcal{C}}$, the evaluation

$$\mathrm{Hom}^{\mathcal{C}}(A_g, N) \otimes_B A \to N, \quad f \otimes a \mapsto f(a),$$

is an isomorphism.

Notice that here the canonical isomorphism can be extended to related isomorphisms for the class of all relative injective comodules.

The following is a one-sided generalization of Schneider's theorem (see 4.8 in [26]).

The Galois Coring Structure Theorem.

(1) The following are equivalent:

    (a) $(\mathcal{C}, g)$ is a Galois coring and $_BA$ is flat;

    (b) $_A\mathcal{C}$ is flat and A_g is a generator in $\mathbf{M}^{\mathcal{C}}$.

(2) The following are equivalent:

   (a) $(\mathcal{C}, g)$ is a Galois coring and $_B A$ is faithfully flat;

   (b) $_A \mathcal{C}$ is flat and A_g is a projective generator in $\mathbf{M}^{\mathcal{C}}$;

   (c) $_A \mathcal{C}$ is flat and $\mathrm{Hom}^{\mathcal{C}}(A_g, -) : \mathbf{M}^{\mathcal{C}} \to \mathbf{M}_B$ is an equivalence with inverse $- \otimes_B A : \mathbf{M}_B \to \mathbf{M}^{\mathcal{C}}$ (cf. 1.7).

If the base ring A is injective as right A-module, then $\mathcal{C}$ is injective as right $\mathcal{C}$-comodule and thus (see 4.9 in [26]) we obtain the

Corollary. *Assume A to be a right self-injective ring and let $\mathcal{C}$ be an A-coring with grouplike element g.*

(1) The following are equivalent:

   (a) $(\mathcal{C}, g)$ is a Galois coring;

   (b) for every injective comodule $N \in \mathbf{M}^{\mathcal{C}}$, the evaluation

$$\mathrm{Hom}^{\mathcal{C}}(A_g, N) \otimes_B A \to N, \quad f \otimes a \mapsto f(a),$$

   is an isomorphism.

(2) The following are equivalent:

   (a) $(\mathcal{C}, g)$ is a Galois coring and $_B A$ is (faithfully) flat;

   (b) $_B A$ is (faithfully) flat and for every injective comodule $N \in \mathbf{M}^{\mathcal{C}}$, the following evaluation map is an isomorphism:

$$\mathrm{Hom}^{\mathcal{C}}(A_g, N) \otimes_B A \to N, \quad f \otimes a \mapsto f(a).$$

2.5. Hopf algebras. Given an R-bialgebra B, by definition the B-coring $B \otimes_R B$ is Galois provided the canonical map

$$\gamma : B \otimes_R B \to B \otimes_R B, \quad a \otimes b \mapsto (a \otimes 1)\Delta(b)$$

is an isomorphism. Since bijectivity of this map is equivalent to the existence of an antipode (see 15.2 in [9]) we have:

For a bialgebra B the following are equivalent:

(a) $B \otimes_R B$ is a Galois B-coring;

(b) B is a Hopf algebra (has an antipode);

(c) $\mathrm{Hom}_B^B(B, -) : \mathbf{M}_B^B \to \mathbf{M}_R$ is an equivalence (with inverse $- \otimes_R B$).

If (any of) these conditions hold, B is a projective generator in $\mathbf{M}_B^B$.

The notion of Galois corings was extended to comodules by El Kaoutit and Gómez-Torrecillas in [16], where to any bimodule $_SP_A$ with P_A finitely generated and projective, a coring $P^* \otimes_S P$ was associated (see 1.8) and it was shown that the map

$$\varphi : \mathrm{Hom}_A(P, A) \otimes_S P \simeq \mathrm{Hom}^C(P, C) \otimes_S P \to C$$

is a coring morphism provided P is also a right C-comodule and $S = \mathrm{End}^C(P)$.

In [9], 18.25, such comodules P are termed *Galois comodules* provided φ is bijective, and it is proved in [9], 18.26, that this condition implies that the functors $\mathrm{Hom}_A(P, -) \otimes_S P$ and $- \otimes_A C$ from $\mathbf{M}_A$ to $\mathbf{M}^C$ are isomorphic.

2.6. Galois comodules. Let P be a right C-comodule such that P_A is finitely generated and projective and let $S = \mathrm{End}^C(P)$. Then P is called a *Galois comodule* if the evaluation map

$$\mathrm{Hom}^C(P, C) \otimes_S P \to C, \ f \otimes m \mapsto f(m),$$

is an isomorphism of right C-comodules.

Considering $P^* \otimes_S P$ as an A-coring (via 1.8), the following are equivalent:

(a) P is a Galois comodule;

(b) there is a (coring) isomorphism

$$P^* \otimes_S P \to C, \quad \xi \otimes m \mapsto \sum (\xi \otimes I_C) \varrho^P(m);$$

(c) for every (C, A)-injective comodule $N \in \mathbf{M}^C$, the evaluation

$$\mathrm{Hom}^C(P, N) \otimes_S P \to N, \quad f \otimes m \mapsto f(m),$$

is a (comodule) isomorphism;

(d) for every right A-module X, the map

$$\mathrm{Hom}_A(P, X) \otimes_S P \to X \otimes_A C, \quad g \otimes m \mapsto (g \otimes I_C) \varrho^P(m),$$

is a (comodule) isomorphism.

The next theorem - partially proved in [16] - shows which additional conditions on a Galois comodule are sufficient to make it a (projective) generator in $\mathbf{M}^C$ (see 18.27 in [9]).

The Galois comodule structure theorem.

(1) The following are equivalent:

> *(a) P is a Galois comodule and $_S P$ is flat;*
>
> *(b) $_A \mathcal{C}$ is flat and P is a generator in $\mathbf{M}^\mathcal{C}$.*

(2) The following are equivalent:

> *(a) M is a Galois comodule and $_S P$ is faithfully flat;*
>
> *(b) $_A \mathcal{C}$ is flat and P is a projective generator in $\mathbf{M}^\mathcal{C}$;*
>
> *(c) $_A \mathcal{C}$ is flat and $\mathrm{Hom}^\mathcal{C}(P, -) : \mathbf{M}^\mathcal{C} \to \mathbf{M}_S$ is an equivalence with the inverse $- \otimes_S P : \mathbf{M}_S \to \mathbf{M}^\mathcal{C}$.*

These Galois comodules are further investigated in Brzeziński [5] and their relevance for descent theory, vector bundles, and non-commutative geometry is pointed out there. In particular *principal comodules* are considered, that is, Galois comodules in the above sense which are projective as modules over their endomorphism rings. Related questions are, for example, also considered by Caenepeel, De Groot and Vercruysse in [10].

3. General Galois comodules

Recall that for a Galois $\mathcal{C}$-comodule P in the sense of 2.6 (where P_A is finitely generated and projective) the functors $- \otimes_A \mathcal{C}$ and $\mathrm{Hom}_A(P, -) \otimes_S P$ are isomorphic. In [27] it is suggested to take this property as definition without further condition on the A-module structure of P.

Throughout this section let $\mathcal{C}$ be an A-coring, $P \in \mathbf{M}^\mathcal{C}$ and $S = \mathrm{End}^\mathcal{C}(P)$, $T = \mathrm{End}_A(P)$.

3.1. Galois comodules. We call P a *Galois comodule* if

$$- \otimes_A \mathcal{C} \simeq \mathrm{Hom}_A(P, -) \otimes_S P \text{ as functors} : \mathbf{M}_A \to \mathbf{M}^\mathcal{C}.$$

The following are equivalent ([27], 2.1):

(a) P is a Galois comodule;

(b) $\mathrm{Hom}_A(P, -) \otimes_S P$ is right adjoint to the forgetful functor $\mathbf{M}^\mathcal{C} \to \mathbf{M}_A$, that is, for $K \in \mathbf{M}_A$ and $M \in \mathbf{M}^\mathcal{C}$, there is a (bifunctorial) isomorphism

$$\mathrm{Hom}^\mathcal{C}(M, \mathrm{Hom}_A(P, K) \otimes_S P) \to \mathrm{Hom}_A(M, K);$$

(c) for any $K \in \mathbf{M}_A$ there is a functorial isomorphism of comodules

$$\operatorname{Hom}_A(P, K) \otimes_S P \to K \otimes_A \mathcal{C}, \ g \otimes p \mapsto (g \otimes I_{\mathcal{C}})\varrho^P(p);$$

(d) for every $(\mathcal{C}, A)$-injective $N \in \mathbf{M}^{\mathcal{C}}$,

$$\operatorname{Hom}^{\mathcal{C}}(P, N) \otimes_S P \to N, \ f \otimes p \mapsto f(p),$$

is an isomorphism (in $\mathbf{M}^{\mathcal{C}}$).

These comodules have good properties (see 2.2 in [27]):

3.2. Isomorphisms for Galois comodules. *Let $P \in \mathbf{M}^{\mathcal{C}}$ be a Galois comodule.*

(1) For any $(\mathcal{C}, A)$-injective $N \in \mathbf{M}^{\mathcal{C}}$, there is a canonical isomorphism

$$\operatorname{Hom}^{\mathcal{C}}(P, N) \to \operatorname{Hom}^{\mathcal{C}}(P, \operatorname{Hom}^{\mathcal{C}}(P, N) \otimes_S P).$$

(2) For any $K \in \mathbf{M}_A$, there is a canonical isomorphism

$$\operatorname{Hom}_A(P, K) \to \operatorname{Hom}^{\mathcal{C}}(P, \operatorname{Hom}_A(P, K) \otimes_S P).$$

(3) There are right $\mathcal{C}$-comodule isomorphisms

$$\operatorname{Hom}^{\mathcal{C}}(P, \mathcal{C}) \otimes_S P \simeq \mathcal{C} \simeq \operatorname{Hom}_A(P, A) \otimes_S P.$$

(4) There is a T-linear isomorphism

$$T \otimes_S P \to P \otimes_A \mathcal{C}, \quad t \otimes p \mapsto (t \otimes I_{\mathcal{C}})\varrho^P(p).$$

(5) For any $K \in \mathbf{M}_A$ and index set Λ,

$$\operatorname{Hom}^{\mathcal{C}}(P, (K \otimes_A \mathcal{C})^{\Lambda}) \otimes_S P \simeq \operatorname{Hom}_A(P, K)^{\Lambda} \otimes_S P \simeq K^{\Lambda} \otimes_A \mathcal{C}.$$

It is clear from the definition that $(\mathcal{C}, A)$-injective modules are of particular interest in this setting (see 2.3 in [27]):

3.3. $(\mathcal{C}, A)$-injective modules. *Let P be a Galois comodule.*

(1) For $N \in \mathbf{M}^{\mathcal{C}}$ the following are equivalent:

  (a) N is $(\mathcal{C}, A)$-injective;

  (b) $\operatorname{Hom}^{\mathcal{C}}(P, \varrho^N) : \operatorname{Hom}^{\mathcal{C}}(P, N) \to \operatorname{Hom}^{\mathcal{C}}(P, N \otimes_A \mathcal{C})$ is a coretraction in $\mathbf{M}_S$.

(2) For P the following are equivalent:

  (a) P is $(\mathcal{C}, A)$-injective;

(b) the inclusion $S \hookrightarrow T$ is split by a right S-linear map.

(3) For P the following are equivalent:

(a) P is strongly (C, A)-injective;

(b) the inclusion $S \hookrightarrow T$ is split by an (S, S)-bilinear map.

(4) For P the following are equivalent:

(a) P is fully (C, A)-injective;

(b) C is a coseparable A-coring.

Notice that so far we did not make any assumptions neither on the A-module nor on the S-module structure of P. Of course special properties of this type influence the behaviour of Galois comodules. For the S-module structure we get (see 4.8 in [27]):

3.4. Module properties of $_S P$. *Let $P \in \mathbf{M}^C$ be a Galois comodule.*

(1) If $_S P$ is finitely generated, then $_A C$ is finitely generated.

(2) If $_S P$ is finitely presented, then $_A C$ is finitely presented.

(3) If $_S P$ is projective, then $_A C$ is projective.

(4) If $_T P$ is finitely generated and $_S P$ is locally projective, then $_A C$ is locally projective.

(5) If $_S P$ is flat, then $_A C$ is flat and P is a generator in $\mathbf{M}^C$.

(6) If $_S P$ is faithfully flat, then $_A C$ is flat and P is a projective generator in $\mathbf{M}^C$.

If $_A C$ is flat as an A-module then $\mathbf{M}^C$ is a Grothendieck category (see 18.14 in [9]) and the endomorphism ring of any semisimple right C-comodule is a (von Neumann) regular ring. This implies part of the next proposition (see 4.11 in [27]).

3.5. Semisimple Galois comodules. *Assume $_A C$ to be flat. For a semisimple right C-comodule P, the following are equivalent:*

(a) P is a Galois comodule;

(b) P is a generator in $\mathbf{M}^C$;

(c) $\mu_C : \mathrm{Hom}^C(P, C) \otimes_S P \to C$ is surjective.

In this case C is a right semisimple coring (and $_A C$ is projective).

Recall that $P^* \otimes_S P$ has a coring structure provided P_A is finitely generated and projective (see 1.8). Moreover, $P^* = \mathrm{Hom}_A(P, A)$ is a left $\mathcal{C}$-comodule canonically and we have a left-right symmetry for Galois comodules (see 5.3 in [27]):

3.6. Galois comodules with P_A f.g. projective. *Assume P_A to be finitely generated and projective. Then the following are equivalent:*

(a) P is a Galois right $\mathcal{C}$-comodule;

(b) $\mathrm{Hom}^{\mathcal{C}}(P, \mathcal{C}) \otimes_R P \simeq \mathcal{C}$ as right $\mathcal{C}$-comodule;

(c) P^* is a Galois left $\mathcal{C}$-comodule;

(d) ${}^{\mathcal{C}}\mathrm{Hom}(P^*, \mathcal{C}) \otimes P^* \simeq \mathcal{C}$ as left $\mathcal{C}$-comodule;

(e) $P^* \otimes_S P \simeq \mathcal{C}$ as A-corings.

In case A is a $\mathcal{C}$-comodule, that is, there is a grouplike element $g \in \mathcal{C}$, and $S = \mathrm{End}^{\mathcal{C}}(A)$, it is a Galois (right) comodule $((\mathcal{C}, g)$ is a Galois coring) if and only if the map

$$A \otimes_S A \to \mathcal{C}, \quad a \otimes a' \mapsto aga',$$

is an isomorphism. Under the given conditions, $A \otimes_S A$ has a canonical coring structure (Sweedler coring, 1.9) and the map is a coring isomorphisms (see 28.18 in [9]).

At various places we have observed a nice behaviour of strongly $(\mathcal{C}, A)$-injective comodules. For Galois comodules this property is symmetric in the following sense - an observation also proved in [5], Theorem 7.2.

3.7. Strongly $(\mathcal{C}, A)$-injective Galois comodules. *Let P be a Galois comodule with P_A finitely generated and projective. Then the following are equivalent:*

(a) P is strongly $(\mathcal{C}, A)$-injective;

(b) P^* is strongly $(\mathcal{C}, A)$-injective;

(c) the inclusion $S \hookrightarrow T$ is split by an (S, S)-bilinear map.

Proof. This follows from 3.3 and symmetry.

Finally we consider various conditions which imply that a Galois comodule induces an equivalence (see 5.7 in [27]).

3.8. Equivalences. *Let $P \in \mathbf{M}^C$ be a Galois comodule with P_A finitely generated and projective. Then*

$$\mathrm{Hom}^C(P, -) : \mathbf{M}^C \to \mathbf{M}_S$$

is an equivalence with inverse functor $- \otimes_S P$ provided that

(i) P is strongly (C, A)-injective, or

(ii) P^* is (C, A)-injective and $_S P$ is flat, or

(iii) P^* is coflat and $_S P$ is flat, or

(iv) C is a coseparable coring.

3.9. Remarks. (1) Entwining structures can be considered as corings and hence the assertions in 3.3 may be compared with Lemma 4.1 and Remarks 4.2 and 5.3 in Schauenburg and Schneider [20].

(2) *Weak Galois corings* are considered in [25], 2.4. For such corings the action of A on C is not required to be unital.

(3) For a deeper study of weak entwining and weak coalgebra-Galois extensions the reader may consult Brzeziński, Turner and Wrightson [8].

(4) For recent investigation of the Galois theory for Hopf algebroids we refer to Böhm [1].

Acknowledgement. The author appreciates useful remarks on the manuscript by Tomasz Brzeziński.

References

1. Böhm, G., *Galois theory for Hopf algebroids*, arXiv:math.RA/0409513 (2004)
2. Brzeziński, T., *On modules associated to coalgebra-Galois extensions*, J. Algebra 215, 290–317 (1999)
3. Brzeziński, T., *Coalgebra-Galois extensions from the extension theory point of view*, in Hopf Algebras and Quantum Groups, Caenepeel and van Oystaeyen (Eds.), LN PAM 209, Marcel Dekker, New York (2000)
4. Brzeziński, T., *The structure of corings. Induction functors, Maschke-type theorem, and Frobenius and Galois-type properties*, Algebras Rep. Theory 5, 389–410 (2002)
5. Brzeziński, T., *Galois Comodules*, arXiv:math.RA/0312159v3 (2004)
6. Brzeziński, T., Hajac, P.M., *Coalgebra extensions and algebra coextensions of Galois type*, Comm. Algebra 27, 1347–1367 (1999)
7. Brzeziński, T., Majid, S., *Coalgebra bundles*, Comm. Math. Phys. 191, 467–492 (1998)

8. Brzeziński, T., Turner, R.B, and Wrightson, A.P., *The structure of weak coalgebra-Galois extensions*, to appear in Comm. Algebra

9. Brzeziński, T., Wisbauer, R., *Corings and comodules*, London Math. Soc. LNS 309, Cambridge University Press (2003)

10. Caenepeel, S., De Groot. E., Vercruysse, J., *Galois theory for comatrix corings: Descent theory, Morita theory, Frobenius and separability properties*, arXiv:math.RA/0406436 (2004)

11. Chase, S.U., Harrison, D.K., and Rosenberg, A., *Galois theory and cohomology of commutative rings*, AMS Memoirs 52 (1962)

12. Chase, S.U., Sweedler, M.E., *Hopf algebras and Galois theory*, Lect. Notes in Math. 97, Springer Verlag (1969)

13. Doi, Y., *Unifying Hopf modules*, J. Algebra 153, 373–385 (1992)

14. Doi, Y., Takeuchi, M., *Cleft comodule algebras for a bialgebra*, Comm. Algebra 14, 801–817 (1986)

15. Doi, Y., Takeuchi, M., *Hopf-Galois extensions of algebras, the Miyashitu-Ulbrich action, and Azumaya algebras*, J. Algebra 121, 488–516 (1989)

16. El Kaoutit, L., Gómez-Torrecillas, J., *Comatrix corings: Galois corings, descent theory, and a structure theorem for cosemisimple corings*, Math. Z. 244, 887–906 (2003)

17. Kreimer, H.F. and Takeuchi, M., *Hopf algebras and Galois extensions of an algebra*, Indiana Univ. Math. J. 30, 675–692 (1981)

18. Menini, C., Zuccoli, M., *Equivalence theorems and Hopf-Galois extensions*, J. Algebra 194, 245–274 (1997)

19. Montgomery, S., *Hopf Algebras and Their Actions on Rings*, Reg. Conf. Series in Math., CBMS 82, AMS, Providence RI (1993)

20. Schauenburg, P., Schneider, H.-J., *On generalized Hopf Galois extensions*, arXiv:math.QA/0405184 (2004)

21. Schneider, H.-J., *Principal homogeneous spaces for arbitrary Hopf algebras*, Israel J. Math. 72, 167–195 (1990)

22. Schneider, H.-J., *Representation theory of Hopf Galois extensions*, Israel J. Math. 72, 196–231 (1990)

23. Schneider, H.-J., *Normal basis and transitivity of crossed products for Hopf algebras*, J. Algebra 152, 289–312 (1992)

24. Wisbauer, R., *Module and Comodule Categories - a Survey*, Proceedings of the Mathematics Conference, Birzeit (1998), Elyadi e.a. (ed.), World Scientific, 277-304 (2000)

25. Wisbauer, R., *Weak corings*, J. Algebra 245, 123–160 (2001)

26. Wisbauer, R., *On Galois corings*, Hopf algebras in non-commutative geometry and physics, S. Caenepeel and F. Van Oystaeyen (eds), LNPAM, Marcel Dekker, in press 2003

27. Wisbauer, R., *On Galois comodules*, arXiv math.RA/0408251 (2004)

ON THE FINITISTIC DIMENSION CONJECTURE *

CHANGCHANG XI

School of Mathematical Sciences,
Beijing Normal University,
100875 Beijing, P.R.China
E-mail: xicc@bnu.edu.cn

The famous finitistic dimension conjecture says that the supremum of the finite
projective dimensions of finitely generated modules over a given artin algebra is
always finite. This conjecture is over 40 year old. It has a close relationship with
the many other homological conjecture. The main purpose of the present note is
to survey some of the recent developments on the finitistic dimension conjecture.

1. A brief recall of the history

As we know, homological algebra was widely used in mathematics and other
fields. The homological invariants play certainly an important role in the
investigation of representations of groups, algebras and rings. One such
example can go back to Hilbert's famous syzygy theorem in 1890, which
gives precisely the global dimension of the polynomial algebra over a field
k.

Hilbert's syzygy theorem: gl.dim k$[x_1, ..., x_n] = n$.

Thus any module over k$[x_1, ..., x_n]$ can be resolved as an long exact
sequence of length at most n of free modules.

In 1940's, homological algebra stemmed from algebraic topology became
popular and was widely applied to the study of rings and algebras. Let us
just mention a few very famous names in this area: H.Cartan, S.Eilenberg,
S.MacLane, M.Auslander, D.Buchsbaum, M.Nagata, T.Nakayama, ...

The ring k$[x_1, ..., x_n]$ and its factor rings are the basic elements in the
algebraic geometry. The nice relationship between geometry and homological algebra can be seen from one beautiful result of Auslander-Buchsbaum-
Serre in 1955.

---

*This work is supported by the " 985 program" of the beijing normal university.

Auslander-Buchsbaum-Serre theorem: Let V be an algebraic variety over an algebraically closed field k, and R be the coordinate ring of V. Then V is smooth $\iff$ gl.dim$(R) < \infty$.

To investigate the algebras and modules with infinity dimension, the finitistic dimension was introduced:

Suppose R is an arbitrary ring, the finitistic dimension, denoted by fin.dim(A), is defined as follows:

fin.dim$(R) := sup\{pd(M) \mid M : \text{f.g. module and } pd(M) < \infty\}$

Fin.dim$(R) := sup\{pd(M) \mid M : \text{module and } pd(M) < \infty\}$

The first two elementary questions concerning the finitistic dimensions were the following proposed in 1960 in [**?**]:

(1) Are fin.dim(R) and Fin.dim(R) finite ?

(2) fin.dim(R)=Fin.dim(R) ?

The answers to the two questions was negative even for commutative noetherian rings. However, in 1960, H.Bass studied the two questions for non-commutative artin rings in [5], where he mentioned the following two conjectures of Rosenberg and Zelinsky (on finite dimensional algebras):

Finitistic dimension conjecture I:

For any an artin algebra A, its finitistic dimension is finite.

Finitistic dimension conjecture II:

For any artin algebra A, fin.dim$(A)=$ Fin.dim(A)

Conjecture II fails, this was settled by B.Zimmermann-Huisgen in 1992 by providing a counterexample in [28]. Conjecture I is still open. So, in this note, when we speak of finititsic dimension conjecture, we always mean the **finitistic dimension conjecture I**.

2. Connection with other conjectures

The finitistic dimension conjecture have been studied by many people, and its relationship with other famous homological conjectures was discovered. In particular, the following four conjectures are closely related to the finitistic dimension conjecture.

In 1958, Nakayama studied generalized Frobenius algebras in [**?**] and proposed the following conjecture.

Nakayama conjecture: If all I_j in a minimal injective resolution of an artin algebra A, say $0 \to {}_A A \to I_0 \to I_1 \to ...$, are projective, then A is self-injective.

As a generalization of Nakayama conjecture, Auslander-Reiten proposed the following conjecture in 1975 in [3].

Generalized Nakayama conjecture: If $0 \to {}_AA \to I_0 \to I_1 \to \dots$ is a minimal injective resolution of an artin algebra A, then any indecomposable injective is a direct summand of some I_j. Equivalently, if M is a finitely generated A-generator with $\mathrm{Ext}_A^i(M, M) = 0$ for all $i \geq 1$, then M is projective.

Later, in 1990, Colby-Fuller proposed the following conjecture in their paper [7].

Strong Nakayama Conjecture: If M is a non-zero f.g. module over an artin algebra A, then there is an $n \geq 0$ such that $\mathrm{Ext}_A^n(M, A) \neq 0$.

There is also the following Gorenstein symmetry conjecture:

Gorenstein symmetry conjecture: For any artin algebra A, if the injective dimension of ${}_AA$ is finite, then so is the injective dimension of A_A.

Note that *All* conjectures above are open. However, the following result reveals some relationship between these conjectures

Theorem 2.1.

- *If the finitistic dimension conjecture holds true, then the strong Nakayama conjecture holds true.*
- *If the strong Nakayama conjecture holds true, then the generalized Nakayama conjecture holds true.*
- *If the generalized Nakayama conjecture holds true, then the Nakayama conjecture holds true.*
- *If finitistic dimension conjecture holds true, then the Gorenstein symmetry conjecture holds true.*

Thus, the finitistic dimension possesses a strong homological property and can be far more revealing measures of homological complexity of an algebra at hand, while infinite global dimension often does not reveal much about that complexity.

For the proof of these statements one may refer to K.Yamagata's article [27] in Handbook of Algebra, vol.I, or [3].

3. Some known results

Here I shall recall some of the results on the finitistic dimension conjecture, which are known before 2002, and in the next part, I will report some new developments after 2002. (I apologize that the list of results below may not be complete.)

Let A be an artin algebra (or a finite dimensional algebra over a field k). We denote by A-mod the category of all finitely generated left A-modules. Given an A-module in A-mod, we denote by pro.dim(M) the projective dimension of M. Let $\mathcal{P}^\infty(A)$ be the full subcategory of A-mod consisting of all A-modules of finite projective dimension.

(1) H.Mochizuki proved in 1965 the following:

If the square of the radical vanishes, then the finitistic dimension conjecture for A is true.

(2) E.Green, E.Kirkman and J.Kuzmanovich showed in 1991 the following:

If A is a monomial algebra, then the finitistic dimension conjecture for A is true.

Recall that a finite dimensional algebra A, given by a quiver with relations, is called a monomial algebra if the relations consists only of paths of length at least two.

(3) E.Green and B.Zimmermann-Huisgen proved in 1991 the following result:

If the cube of the radical of A is zero, then the finitistic dimension conjecture for A is true.

(4) In the same year, Auslander and Reiten gave the following homological condition for the finitistic dimension conjecture to be true [4].

If $\mathcal{P}^\infty(A)$ is contravariantly finite in A-mod, then the finitistic dimension conjecture for A is true.

Recall that a subcategory $\mathcal{C}$ of A-mod is called *contravariantly finite* in A-mod if for any module M in A-mod there is a morphism $f : C \longrightarrow M$ such that $\mathrm{Hom}_A(C', f)$ is surjective for all C' in $\mathcal{C}$.

Note that in general, $\mathcal{P}^\infty(A) = \{M \mid pd(M) < \infty\}$ might not be contravariantly finite in A-mod.

(5) Y.Wang proved in 1994 the following in [20]

If an artin algebra A satisfies that $\mathrm{rad}^{2l+1}(A) = 0$ and A/rad^l is representation-finite, then the finitistic dimension conjecture for A is true.

(6) in 2000, I.Agoston,D.Happel,E.Lukas and L.Unger in [1] showed the following:

286

If A is a standardly stratified algebra, then the finitistic dimension conjecture is true.

This result follows also from the recollement argument in [11].

(7) K.Igusa and G.Todorov proved in 2002 the following:

If an artin algebra A is of the form eBe with B an algebra of global dimension at most three, and e an idempotent in B, then the finitistic dimension conjecture for A is true.

In particular, if the representation dimension of A is at most three, then the finitistic dimension conjecture for A is true. Here the representation dimension, introduced by Auslander in [?], is defined as follows:

$$\mathrm{rep.dim}(A) = \inf\{\mathrm{gl.dim}\,(\mathrm{End}_A(M)) \mid A \oplus D(A) \in \mathrm{add}(M)\}.$$

By using the description of the relationship between a projective resolution and an exact sequence of the modules in $\mathrm{add}(M)$, one can give a more direct alternative proof of the above fact: $\mathrm{rep.dim}(A) \leq 3$ implies that $\mathrm{fin.dim}(A) < \infty$. For a details see [25].

(8) C.C.Xi proved in 2002 in [23] the following:

If A is stably hereditary, then $\mathrm{rep.dim}(A) \leq 3$.

Recall that an artin algebra is called *stably hereditary* [23] if (1) each indecomposable submodule of an indecomposable projective module is either projective or simple, and (2) each indecomposable factor module of an indecomposable injective module is either injective or simple.

Note that the notion of stably hereditary algebra is a proper generalization of the notion of " stably equivalent to hereditary algebra".

(9) In 2002, K.Erdmann, Th.Holm, O.Iyama and J.Schroeer showed the following:

If B is a subalgebra of A with $\mathrm{rad}(B) = \mathrm{rad}(A)$ and if A is representation-finite, then $\mathrm{rep.dim}\,A \leq 3$.

In particular, the finitistic dimension conjecture is true for special biserial algebras and string algebras.

(10) For monomial algebras, Hongbo Shi has provided in [19] a graphic algorithm to calculate the finitistic dimension.

Let me also mention that a geometric approach to the finitistic dimension conjecture was discussed by Membrillo-Hernandez and Salmeron in

[17], and that the relationship between the finitistic dimension and the Ziegler spectrum is investigated by Krause in [15].

4. Some new results

In this section I shall first point out our idea to deal with the finitistic dimension of artin algebras, and then report some new developments on the finitistic dimension conjecture in the last a few years, and give example to illustrate the method we used. Finally, I shall mention some results on representation dimension.

4.1. *General question*

The most investigations on the finitstic dimension conjecture before 2002 are mainly concentrated on one single algebra. Our philosophy is: to approach the finitistic dimension by a series of "well-understood" of algebras. So, our general question may be formulated as follows:

Assumption: Let $A_0 \subseteq A_1 \subseteq \ldots \subseteq A_s$ be a finite chain of algebras with the same identity such that $\mathrm{rad}(A_i)$ is an (or a left) ideal in A_{i+1} for all i.

Question: If some of the bigger algebras in the chain have **finite** finitistic dimensions, what could we say about the finiteness of the finitistic dimension of the smallest algebra A_0 ?

Dually, we may use a family of quotient algebras of A to control the finitistic dimension of A.

The following fact is useful (see [24]). Given an algebra B over a field k, we can always embedded B into a full n by n matrix algebra over k. Thus, by idealizer method, there is a finite chain of subalgebras of $M_n(k)$, the full $n \times n$ matrix algebra over k:

$$B = A_0 \subseteq A_1 \subseteq \ldots \subseteq A_s \subset M_n(k)$$

such that $\mathrm{rad}(A_i)$ is a left ideal in A_{i+1} and A_s is representation-finite.

4.2. *New results*

In this subsection, I shall survey the results from [24] and [25]. For the proofs we refer the reader to the original papers.

The following result shows that our general question gives a new characterization of the finitistic dimension conjecture.

Theorem 4.1. *Let k be a field. Then the following are equivalent:*

(1) The finitistic dimension conjecture is true for all finite dimensional k-algebras.

(2) If $B \subseteq A$ is a pair of k-algebras with the same identity such that $rad(B)$ is a left ideal in A and if $fin.dim(A) < \infty$, then $fin.dim(A) < \infty$.

The next result is a partial answer to our general question.

Theorem 4.2. *Suppose $C \subseteq B \subseteq A$ is a chain of algebras with the same identity such that $rad(C)$ is an ideal in B and $rad(B)$ is a left ideal in A. If A is representation-finite, then $fin.dim(C) < \infty$.*

Note that this result extends the main result in [8]. For a chain with two terms, we may apply global dimension to bound the finititistic dimension of subalgebras. The following was proved in [25].

Theorem 4.3. *Suppose $B \subseteq A$ is a subalgebra of an artin algebra A with $rad(A) = rad(B)$. If $gl.dim(A) \leq 4$, then $fin.dim(B) < \infty$.*

So, we have controlled the finitistic dimension of a subalgebra by that of over-algebra. Next, we shall provide a result which show us that the finitistic dimension of an algebra can be controlled also by its factor algebras.

Theorem 4.4. *Let A be an artin algebra and let $I_j, 1 \leq j \leq n$ be a family of ideals in A with $I_1 I_2 \cdots I_n = 0$, such that $proj.dim(_A I_j) < \infty$ and $proj.dim(I_j)_A = 0$ for all $j \geq 3$. If A/I_1 and A/I_2 are representation-finite and if A/I_j has finite finitistic dimension for $j \geq 3$, then $fin.dim(A) < \infty$.*

As a direct consequence, we have a large class of algebras for which the finititsic dimension conjecture is true.

(1) Dual extensions of representation-finite algebras have finite finitistic dimensions.

(2) Trivially twisted extensions of representation-finite algebras have finite finitistic dimensions.

item[(3)] Hochschild extensions of representation-finite algebras have finite finitistic dimensions.

(4) For any two surjective algebra homomorphisms $f_i : A_1 \longrightarrow \bar{A}$ between algebras, if A_1 and A_2 are representation-finite, then the pullback algebra of the f_i has finite finitistic dimension.

For the proofs of (1),(2) and (3), we refer to [24], where the unexplained notion is precisely defined. For (4), we note that the kernel of f_i is an ideal of the pullback algebra A, and their product is zero. Since A modulo the kernel of f_i is representation-finite, the statement (4) follows immediately from the above theorem.

4.3. *Remarks*

(1) The recept to construct algebras $B \subseteq A$ wit read(B)=rad(A): Given an algebra A, we fix a decomposition of 1 into orthogonal primitive idempotents, say $1 = \sum_{j=1}^{n} e_j$. To define B, we just fix a partition of the set $I := \{1, 2, ..., n\}$, say $I = \cup_{i=1}^{m} I_i$, and put $f_i = \sum_{j \in I_i} e_j$. Now the algebra B is generated by $f_i, 1 \leq i \leq m$ together with rad(A). Clearly, A and B have the same identity and the same radical.

(2) The typical examples of algebras satisfying the conditions of Theorem 4.4 are the so-called trivially extensions of two representation-finite algebras. For non-trivially extensions we refer to [26] and the references therein.

(3) The ingredients in our proofs are the use of the function of Igusa-Todorov in [12], and some constructions to produce a suitable short exact sequences. For the convenience to the reader, we include here the lemma of Igusa and Todorov.

Let A be an artin algebra, that is, A is a finitely generated module over its center which is assumed to be a commutative artin ring. We denote by A-mod the category of all finitely generated left A-modules and by rad(A) the Jacobson radical of A. Given an A-module M, we denote by proj.dim(M) the projective dimension of M.

Let $K(A)$ be the quotient of the free abelian group generated by the isomorphism classes $[M]$ of modules M in A-mod modulo the relations:

(1) $[Y] = [X] + [Z]$ if $Y \simeq X \oplus Z$; and

(2) $[P] = 0$ if P is projective.

Thus $K(A)$ is a free abelian group with the basis of non-isomorphism classes of non-projective indecomposable A-modules in A-mod. Igusa and Todorov in [12] use the noetherian property of the ring of integers and define a function Ψ on this abelian group, which depends on the algebra A and takes values of non-negative integers.

Lemma 4.1. *(Igusa-Todorov) For any artin algebra A there is a function Ψ defined on the objects of A-mod such that*

(1) $\Psi(M) = proj.dim(M)$ if M has finite projective dimension.

(2) *If $add(M) \subseteq add(N)$, then $\Psi(M) \le \Psi(N)$. Moreover, if $add(M)= add(N)$, then $\Psi(M) = \Psi(N)$.*

(4) *If $0 \to X \to Y \to Z \to 0$ is an exact sequence in A-mod with $proj.dim(Z) < \infty$, then $proj.dim(Z) \le \Psi(X \oplus Y) + 1$.*

4.4. *Examples*

Now let us display two simple examples to illustrate the methods of our results.

Example 1. Let B be the algebra given by the following quiver

$$
1\circ \;\; \overset{\overset{\beta}{\longleftarrow}}{\underset{\underset{\alpha}{\longleftarrow}}{\overset{\gamma}{\longrightarrow}}} \;\; \circ 2
$$

with relations $\alpha\gamma = \gamma\alpha = \gamma\beta = 0$.

Note that this algebra, due to Igusa-Smalo-Todorov was given to show that $\mathcal{P}^\infty(A)$ is not contravariantly finite in A-mod.

By the recept that we have described, the algebra is a subalgebra of the following algebra A of global dimension 2:

$$
\circ \overset{}{\longleftarrow} \circ \underset{\gamma}{} \quad \overset{\overset{\beta}{\longleftarrow}}{\underset{\alpha}{\longleftarrow}} \quad \circ
$$

with relations $\alpha\gamma = 0$

We should note that the module category of a subalgebra B could be much more complicated than that of A.

Clearly, the algebra B is also a trivially twisted extension of two representation-finite algebras. Thus our Theorems can be applied to see the finiteness of finitistic dimension of B. (Of course, we have already known this.)

Example 2. Let A be an algebra (over a field) given by the following quiver with relations:

$$
\eta\xi = \gamma\delta, \quad \alpha^3 = \beta\delta = \alpha\delta = 0.
$$

Let B and C be the algebras given by the following quiver with relations, respectively:

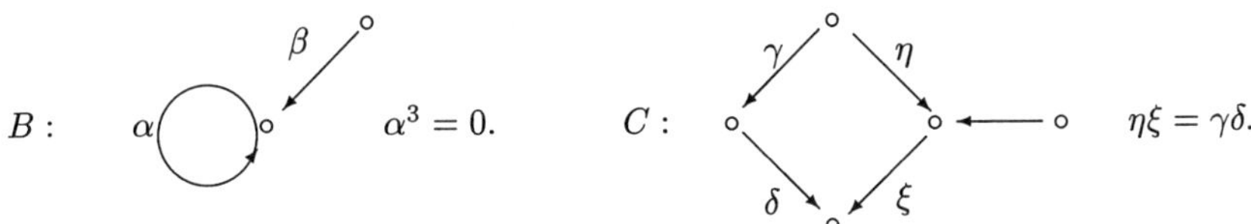

Suppose both γ and β have the same starting vertex 1 and the same ending vertex 2. Then A is the trivially twisted extension of B and C at the vertex $S = \{1, 2\}$. Since B and C are representation-finite, the algebra A has finite finitistic dimension.

4.5. *Some results on representation dimension*

Because of the close relationship between the representation dimension and the finitistic dimension conjecture, many people make efforts to calculate the representation dimension. Here I would like to report some new results on the subject.

In [25], the following result was proved.

Theorem 4.1. *Suppose $B \subseteq A$ is a subalgebra of an artin algebra A such that $rad(B)$ is an ideal in A. If A is stably hereditary, then $rep.dim(B) \leq 3$. In particular, the fin.dim($B < \infty$.*

Recently, F.Cohleo and I.M.Platzeck prove the following result.

Theorem 4.2. *(1) An artin algebras A such that the functor $Hom_A(D(A),)$ has finite length (or dually, $Hom_A(,A)$ has finite length) has representation dimension at most 3.*

(2) Trivial extensions of iterated tilted algebras has representation dimension at most 3.

The first result (1) is extended to the so-called laura algebras by Assem, Platzeck and Trepode more recently.

Also, I should mention that Th. Holm have calculated the representation dimension for many tame blocks of group algebras, and it turns out that those algebras have also representation dimension upper bounded by three.

More recently, R.Rouquier shows that there is an algebra of representation dimension n for any given n.

5. Some Open questions

In this section we mention some open questions related to the results in the note.

Question 1. Let C and B be two representation-finite algebras over a field. Does the trivially twisted extension of C and B at S has the representation dimension at most 3 ?

Question 2. Let A be an artin algebra and J an ideal in A with $J^3 = 0$. If A/J is rep-finite, is fin.dim$(A) < \infty$?

Note that if A/J^2 is representation-finite then the finitistic dimension of A is finite. This follows easily from Theorem 4. It is also well-known that if J is the Jacobson radical of A then the finitistic dimension conjecture for A is true.

Question 3. Let A and B be two artin algebras, and let $f : B \longrightarrow A$ be a surjective homomorphism of algebras such that the square of $\ker(f)$ vanishes. If the representation dimension of A is at most 3, is fin.dim$B < \infty$?

Question 4. Let A be an artin algebra and I an ideal in A with $I^2 = 0$. If A/I is representation-finite, is rep.dim$(A) \leq 3$?

This question has the positive answer in the case $I = \mathrm{rad}(A)$ or $I = \mathrm{rad}^n(A)$ with $n + 1$ the nilpotency index of $\mathrm{rad}(A)$.

Question 5. Let B be a subalgebra of an algebra A such that $\mathrm{rad}(B) = \mathrm{rad}(A)$.

Is fin.dim$(B) < \infty$ if gl.dim$(A) \leq 5$? (or more generally, if gl.dim$(A) < \infty$?)

Question 6. Suppose $A_4 \subseteq A_3 \subseteq A_2 \subseteq A_1$ is a chain of algebras with the same identity such that $\mathrm{rad}(A_i)$ is a left ideal in A_{i-1} for all i. If A_1 is representation-finite, is fin.dim$(A_4) < \infty$? (consider also the more general question.)

Acknowledgments

This research work is supported exclusively by the "985 Program" of BNU. The contents of the present note are an enlargement of my talk at the Fourth China-Japan-Korea International Symposium on Ring Theory, Nanjing, China. I would like to give my hearty thanks to the organizers Prof. Dr. Nanqing Ding and Jianlong Chen for their invitation, hospitality and excellent organization work. Also, I would like to thank my colleague Xiaosheng Zhu, Zhaoyong Huang and the students there for their help.

References

1. I. AGOSTON, D. HAPPEL, E.LUKACS AND L.UNGER, Finitistic dimension of standardly stratified algebras. *Comm. Algebra***28** (2000), no. 6, 2745–2752.
2. M. AUSLANDER, *Representation dimension of artin algebras.* Queen Mary College Mathematics Notes, Queen Mary College, London, 1971.
3. M. AUSLANDER AND I. REITEN, On a generalized version of the Nakayama conjecture. *Proc. Amer. Math. Soc.* **52** (1975), 69-74.
4. M. AUSLANDER AND I. REITEN, Applications of contravariantly finite subcategories. *Adv. in Math.* **85** (1990), 111-152.
5. H. BASS, Finititsic dimension and a homological generalization of semiprimary rings. *Trans. Amer. Math. Soc.* **95**(1960), 466-488.
6. F. U. COELHO AND M. I. PLATZECK, On the representation dimension of some classes of algebras. *J. Algebra* **275** (2004), no. 2, 615–628.
7. R. R. COLBY AND K. R. FULLER, A note on the Nakayama conjectures. *Tsukuba J. Math.* 14(1990), 343-352.
8. K. ERDMANN, T. HOLM, O. IYAMA AND J. SCHRÖER, Radical embedding and representation dimension. *Adv. Math.* **185** (2004), no. 1, 159–177.
9. E. L. GREEN, E. KIRKMAN AND J. KUZMANOVICH, Finitistic dimensions of finite-dimensional monomial algebras. *J. Algebra* **136** (1991), no. 1, 37–50.
10. E. L. GREEN AND B.ZIMMERMANN-HUISGEN, Finitistic dimension of artin rings with vanishing radical cube. *Math. Z.* **206** (1991), 505-526.
11. D. HAPPEL, Reduction techniques for homological conjectures. *Tsukuba J. Math.***17** (1993), no. 1, 115–130.
12. K. IGUSA AND G. TODOROV, On the finitistic global dimension conjecture for artin algebras. Preprint, (2002), 1-4.
13. K. IGUSA AND D. ZACHARIA, Syzygy pairs in a monomial algebra. *Proc. Amer. Math. Soc.* **108** (1990), 601-604.
14. O. IYAMA, Finiteness of representation dimension. *Proc. Amer. Math. Soc.* **131** (2003), no.4, 1011-1014.
15. H.KRAUSE, Finitistic dimension and Ziegler spectrum. *Proc. Amer. Math. Soc.***126**(1998), no. 4, 983–987.
16. Y. M. LIU AND C. C. XI, Constructions of stable equivalences of Morita type for finite dimensional algebras I., to appear in Trans. A.M.S. Preprint is available at http://math.bnu.edu.cn/~ccxi/Papers/Articles/mstable.pdf/
17. F.H.MEMBRILLO-HERNANDEZ AND L.SALMERON, A geometric approach to the finitistic dimension conjecture. *Arch. Math.***67**(1996), 448-456.
18. T. NAKAYAMA, On algebras with complete homology.*Abh. Math. Sem. Univ. Hamburg* **22** (1958), 300–307.
19. H.B.SHI, Finitistic dimension of monomial algebras. *J. Algebra* **264** (2003), no. 2, 397–407.

294

20. Y. WANG, A note on the finitistic dimension conjecture. *Comm. in Algebra* **22** (1994), no. 7, 2525–2528.
21. A. WIEDEMANN, Integral versions of Nakayama and finitistic dimension conjectures. *J. Algebra* **170** (1994), no.2, 388-399.
22. C. C. XI, On the representation dimension of finite dimensional algebras. *J. Algebra* **226** (2000), 332-346.
23. C. C. XI, Representation dimension and quasi-hereditary algebras. *Adv in Math.* **168**, 193 (2002).
24. C. C. XI, On the finitistic dimension conjecture I: related to representation-finite algebras. *J.Pure Appl. Alg.* **193**, 287 (2004). Erratum to " On the finitistic dimension conjecture I. related to representation-finite algebras [J.P.A.A. 193 (2004)287-305]". Preprint is avialable at: http://math.bnu.edu.cn/ ~ccxi/ Papers/Articles/correctum.pdf
25. C. C. XI, On the finitistic dimension conjecture II: related to finite global dimension. Preprint is avialable at: http://math.bnu.edu.cn/~ccxi /Papers/ Articles/correctum.pdf
26. C. C. XI, Twisted doubles of algebras. I. Deformations of algebras and the Jones index. Algebras and modules, II (Geiranger, 1996), 513–523, CMS Conf. Proc., 24, Amer. Math. Soc., Providence, RI, 1998.
27. K. YAMAGATA, Frobenius Algebras. In: Handbook of Algebra. Vol.1 (1996), 841-887.
28. B. ZIMMERMANN-HUISGEN, Homological domino effects and the first finitistic dimension conjecture. *Invent. Math.* **108**(1992), no. 2, 369–383.
29. B. ZIMMERMANN-HUISGEN, The finitistic dimension conjectures—a tale of 3.5 decades. Abelian groups and modules (Padova, 1994), 501–517, *Math. Appl.*, **343**, Kluwer Acad. Publ., Dordrecht, 1995.

GALOIS COVERINGS OF SELF-INJECTIVE ALGEBRAS BY TWISTED REPETITIVE ALGEBRAS

KUNIO YAMAGATA*

*Tokyo University of Agriculture and Technology,
Nakacho 2-24-16, Koganei, Tokyo 184-8488, Japan
E-mail addresses: yamagata@cc.tuat.ac.jp*

This is a survey of the results on self-injective algebras with Galois coverings, mainly obtained by A. Skowroński and the author. The aim is to introduce from Ref.26 some criterion theorems for self-injective algebras to have Galois coverings by repetitive algebras.

1. Introduction

This paper is a survey of some results on self-injective algebras with Galois coverings from joint works with A. Skowroński and Y. Ohnuki - K. Takeda: Refs. 23-26, and 17. The aim is, however, not to show all such results in an expository work, but to introduce from Ref.26 some criterion theorems for selfinjcetive algebras to have Galois coverings with specific admissible groups by repetitive algebras. Throughout this paper, all algebras are associative algebras over a fixed field K, and assumed to be finite dimensional and basic with identity unless otherwise stated.

Covering techniques introduced by P. Gabriel and K. Bongartz in Ref. 7, and further basic properties were developed by Gabriel and P. Dowbor - A. Skowroński in Refs. 13, 9. Representation-finite self-injective algebras were classified by C. Riedtmann Ref. 19 where a prototype of coverings was appeared. The coverings by repetitive algebras was introduced by D. Hughes-J. Waschbüsch Ref. 16 to study representations of trivial extension algebras of representation-finite type. On the other hand, since coverings by repetitive algebras were applied to representation-infinite self-injective algebras by Skowroński in Ref. 22, they have played an important

---

*Work supported by the Japan Society for the Promotion of Science, Grant in Aid for Science Research (c) (1) no. 155 400 12

role in the representation theory of representation-infinite self-injective algebras. See Ref. 10, 11, 12 for applications to representations of groups. Thus it is an important problem to characterize the self-injective algebras which have coverings by repetitive algebras of algebras with easier properties (e.g. without oriented cycles in the ordinary quivers). In this paper, we consider the coverings by repetitive algebras with infinite cyclic admissible groups generated by Nakayama-positive automorphisms (see Subsec. 3.2), and show a ring theoretical characterization for self-injective algebras to have those coverings. See Ref. 1 for representations of algebras and Ref. 28 for Frobenius algebras.

The author wishes to express his thanks to Y. Ohnuki for several helpful comments.

2. Coverings

2.1. *Locally bounded categories*

We denote by D the standard duality $\mathrm{Hom}_K(-,K)$. For a finite dimensional basic algebra Λ, $D\Lambda$ is called the *standard duality module*, and Λ is self-injective if and only if $\Lambda \cong D\Lambda$ as left (or right) Λ-modules. A K-category is a category whose Hom-sets are K-vector spaces and composition of morphisms is K-bilinear.

A K-category R is said to be *locally bounded* (Ref. 7) if R is small and the following conditions are satisfied:

    (a) distinct objects of R are non-isomorphic;
    (b) the algebra $R(x,x)$ is local for any object x of R;
    (c) $\bigoplus_{y\in R}\dim R(x,y)$ and $\bigoplus_{y\in R}\dim R(y,x)$ are finite for any object x of R.

A locally bounded K-category R is identified with an algebra Λ over K which is not necessarily finite dimensional but with a complete set of orthogonal primitive idempotents $\{e_i\}$, i.e., $\Lambda = \bigoplus_i \Lambda e_i = \bigoplus_i e_i\Lambda$. The correspondence is given by the relation; $\Lambda = \bigoplus_{i,j} R(i,j), e_i = 1_i$ the identity morphism on i, and $e_j\Lambda e_i = R(i,j)$. In particular, a bounded K-category, i.e., the object set is finite and each Hom-set is finite dimensional, corresponds to a finite dimensional K-algebra. We freely identify a locally bounded K-category and the corresponding K-algebra.

2.2. *Covering functors*

Let R, Λ be locally bounded K-categories.

A functor $F : R \to \Lambda$ is called a *covering functor* if, for any $x \in R$ and $a \in \Lambda$, the induced K-homomorphisms

$$\bigoplus_{F(y)=a} R(x,y) \to \Lambda(F(x),a), \qquad \bigoplus_{F(y)=a} R(y,x) \to \Lambda(a,F(x))$$

are isomorphic (K. Bongartz-P. Gabriel; Ref. 7).

Let G be a subgroup of the group $\mathrm{Aut}(R)$ of K-automorphisms which acts *freely* on the object set of R (i.e., $gx \neq x$ for each object x of R and $1 \neq g \in G$). Then the category R/G is defined as follows: the objects of R/G are the G-orbits $\bar{x}$ of objects x of R. A morphism $f : a \to b$ in R/G is a family of morphisms $f = (_y f_x) \in \prod_{\substack{\bar{x}=a \\ \bar{y}=b}} R(x,y)$ which satisfy $g(_y f_x) = {}_{gy} f_{gx}$ for all $g \in G$ and $x,y \in R$. Composition $h := f'f$ of $f : a \to b$ and $f' : b \to c$ is defined by $_z h_x = \sum_{\bar{y}=b} {}_z f'_y \cdot {}_y f_x$ for any x, z with $\bar{x} = a, \bar{z} = c$. This sum makes sense because R is locally bounded.

There is a canonical functor $F : R \to R/G$. In fact, let $F(x) = \bar{x}$ for any object x of R. For a morphism $u : x \to y$ in R, let $F(u) = (_t f_s) : \bar{x} \to \bar{y}$ where $_t f_s = g(u)$ if $s = g(x)$ and $t = g(y)$ for some $g \in G$, and $_t f_s = 0$ otherwise. Then F is a covering functor what we call a *Galois covering* with the *admissible group* G (P. Gabriel; Ref. 13). It should be noted that R/G is a finite dimensional algebra if and only if R has finitely many G-orbits. If an algebra Λ is isomorphic to R/G, we say that Λ has a Galois covering $R \to \Lambda$ with the admissible group G and we denote it by $R \xrightarrow{G} \Lambda$ simply.

Two Galois coverings $F_1 : \mathcal{C}_1 \to A$ and $F_2 : \mathcal{C}_2 \to A$ are said to be *isomorphic* over A if there is an equivalence $\eta : \mathcal{C}_1 \to \mathcal{C}_2$ with $F_1 \cong F_2\eta$.

2.3. *Repetitive categories*

Let B be a K-algebra with a fixed complete set of orthogonal primitive idempotents $\{e_1,\ldots,e_m\}$, and σ an automorphism of the bounded K-category B. Let $B_n = B, DB_n = DB$ be copies of B and DB respectively $(n \in \mathbb{Z})$. The *twisted repetitive category* $\widehat{B}_\sigma$ of B is by definition the direct sum of K-modules

$$\widehat{B}_\sigma = \bigoplus_{n \in \mathbb{Z}} B_n \oplus (DB_n)_\sigma,$$

and with multiplication given by

$$\left(\sum_{i \in \mathbb{Z}}(b_i, f_i)\right) \cdot \left(\sum_{i \in \mathbb{Z}}(c_i, g_i)\right) = \sum_{i \in \mathbb{Z}}(b_i c_i, b_i g_i + f_i c_{i+1})$$

for $b_i, c_i \in B_i$ and $f_i, g_i \in (DB_i)_\sigma$. Here, M_σ for a right B-module M stands for a right B-module with right operation $x \cdot b = x\sigma(b)$ for $x \in M, b \in B$. $\widehat{B}_\sigma$

is an infinite dimensional K-algebra without identity, but with a complete set of orthogonal primitive idempotents $\{e_{n,i} \mid n \in \mathbb{Z}, i = 1, \ldots, m\}$ where $e_{n,i}$ is a copy of e_i in B_n. Moreover, $\widehat{B}_\sigma$ is self-injective. The (standard) *repetitive category* $\widehat{B}$ is $\widehat{B}_{id}$ (Ref. 16). The *Nakayama automorphism* $\nu_{\widehat{B}_\sigma}$ of $\widehat{B}_\sigma$ is the automorphism of $\widehat{B}_\sigma$ whose restriction to each $B_n \oplus (DB_n)_\sigma$ is the identity to $B_{n+1} \oplus (DB_{n+1})_\sigma$. An automorphism φ of $\widehat{B}_\sigma$ is said to be *positive* if $\varphi(e_{n,i})$ belongs to $\sum_{r \geq n} B_r$ for any (n, i). $\widehat{B}_\sigma$ is described as the doubly infinite matrix algebra, with all B_n's on the diagonal,

$$
\widehat{B}_\sigma = \begin{pmatrix}
\ddots & \ddots & & & & 0 \\
 & B_{n-1} & (DB_{n-1})_\sigma & & & \\
 & & B_n & (DB_n)_\sigma & & \\
 & & & B_{n+1} & \ddots & \\
0 & & & & & \ddots
\end{pmatrix}
$$

which consists of all matrices having only finitely many entries different from zero, and addition and multiplication are naturally defined as those of matrices by using zero map $DB \otimes DB \to 0$.

For an algebra B and an automorphism σ, $\widehat{B}_\sigma / \langle \varphi \rangle$ is a finite dimensional self-injective algebra for a positive automorphism φ of $\widehat{B}_\sigma$, and $\widehat{B}_\sigma / \langle \nu_{\widehat{B}_\sigma} \rangle$ is isomorphic to $B \ltimes (DB)_\sigma$ as an algebra. The following fact is proved in Refs. 27,17.

Proposition 2.1. *For an algebra B and an automorphism σ, the following assertions hold;*

(1) K-categories $\widehat{B}_\sigma$ and $\widehat{B}$ are isomorphic.

(2) $B \ltimes DB_\sigma$ is isomorphic to an algebra $\widehat{B}/\langle \varphi \nu_{\widehat{B}} \rangle$ with a positive automorphism φ.

(3) $B \ltimes DB_\sigma \cong B \ltimes DB$ if and only if σ is inner.

It follows from the proposition that all twisted categories of an algebra B are isomorphic, but it should be noted that all coverings $\widehat{B}_\sigma \to A$ are not necessarily isomorphic over A. See Example 4.1 in the section 4.

3. Positive Galois coverings

3.1. *Periodic categories*

Let $\mathcal{C}$ be a K-category $\mathcal{C}$ with a fixed automorphism $\nu_{\mathcal{C}}$ satisfying the following two conditions:

    (a) $\nu_{\mathcal{C}}$ acts freely on the object class of $\mathcal{C}$,

    (b) $\mathcal{C}(\nu_{\mathcal{C}}(x), x) \neq 0$ for all objects $x \in \mathcal{C}$.

Such a category is called a *periodic* category, and denoted by $(\mathcal{C}, \nu_{\mathcal{C}})$ when the automorphism $\nu_{\mathcal{C}}$ is specified. A typical example of the periodic category is the category $(\widehat{B}_\sigma, \nu_{\widehat{B}_\sigma})$. Our aim is to characterize categorically Galois coverings by twisted repetitive categories, by making use of periodic categories.

A full convex subcategory C of $\mathcal{C}$ is said to be a *quasi-core* if the object class of C is a complete set of representatives of the $\nu_{\mathcal{C}}$-orbits in the object class of $\mathcal{C}$, where a convex subcategory is by definition a path closed subcategory, i.e., if $x \to x_1 \to \cdots \to x_m \to y$ is a path in $\mathcal{C}$ with $x, y \in C$, then all x_i's belong to C. By $\mathrm{Aut}(\mathcal{C}, \nu_{\mathcal{C}})$ we understand the set of automorphisms of $\mathcal{C}$ commuting with $\nu_{\mathcal{C}}$.

Let $\mathcal{C}$ be a periodic K-category with a quasi-core C. Then, an automorphism $h \in \mathrm{Aut}(\mathcal{C}, \nu_{\mathcal{C}})$ is said to be *positive* if $h(Obj(C)) \subseteq \bigcup_{i \geq 0} \nu_{\mathcal{C}}^i(Obj(C))$ (c.f.,2.3), and $\nu_{\mathcal{C}}$-*positive* (or *Nakayama-positive*) if $h = \varphi \nu_{\mathcal{C}}$ for a positive automorphism φ.

Example 3.1. . For a bounded K-category B and an automorphism σ, $(\widehat{B}_\sigma, \nu_{\widehat{B}_\sigma}^n)$ $(n \geq 1)$ has a quasi-core $B_0 \vee \cdots \vee B_{n-1}$.

For an algebra B, there is a fully faithful functor from the bounded derived category of B to the stable module category of B, $F : D^b(B) \to \underline{\mathrm{mod}}\,\widehat{B}$; and moreover, F is dense if and only if $gl.dim\,B$ is finite (D. Happel, Refs. 14, 15). This may show some importance of the repetitive category.

The following well-known theorem by C. Riedtmann and D. Hughes-J. Waschbüsch (Refs. 19, 8, 20, 21 and Ref. 16) shows that a representation-finite self-injective algebra over an algebraically closed field is described by a Galois covering with admissible group generated by an automorphism. Two self-injective algebras A and Λ are said to be *socle-equivalent* if the factor algebras $A/\operatorname{soc} A$ and $\Lambda/\operatorname{soc} \Lambda$ are isomorphic.

Theorem 3.1. *Let A be a representation-finite self-injective algebra over an algebraically closed field K. Then there is a tilted algebra B of Dynkin* *type Δ and a positive automorphism φ of B such that A is socle-equivalent to $\widehat{B}/\langle\varphi\rangle$. Moreover, A is isomorphic to $\widehat{B}/\langle\varphi\rangle$, unless char $K = 2$ and $\Delta = D_{3m}$.* (See Refs. 20, 21 for more details.)

3.2. *Positive Galois coverings of self-injective algebras*

Definition 3.1. Let A be a self-injective algebra. A Galois covering $F : \mathcal{C} \xrightarrow{G} A$ is said to be *positive* if $\mathcal{C}$ is a periodic K-category, say $(\mathcal{C}, \nu_{\mathcal{C}})$, with a quasi-core, and G is an infinite cyclic group generated by a $\nu_{\mathcal{C}}$-positive automorphism h of $\mathcal{C}$, and the following conditions are satisfied:

(G0) $F\nu_{\mathcal{C}} = \nu_A F$;

(G1) If $F(x) = a$ for objects $x \in \mathcal{C}$ and $a \in A$, then F canonically induces the isomorphisms of K-modules

$$F : \mathcal{C}(x, x) \oplus \mathcal{C}(h(x), x) \xrightarrow{\sim} A(a, F(x)) = A(a, a)$$

$$F : \mathcal{C}(x, x) \oplus \mathcal{C}(x, h^{-1}(x)) \xrightarrow{\sim} A(F(x), a) = A(a, a),$$

(G2) For any objects $x \in \mathcal{C}$ and $a \in A$ with $F(x) \neq a$, there are objects $y, z \in \mathcal{C}$ with $F(y) = a = F(z)$ such that F canonically induces the isomorphisms of K-modules

$$F : \mathcal{C}(y, x) \xrightarrow{\sim} A(a, F(x)), \quad F : \mathcal{C}(x, z) \xrightarrow{\sim} A(F(x), a).$$

In the above definition, $\nu_{\mathcal{C}}$ is also called the *Nakayama automorphism* of $\mathcal{C}$, which generalizes the Nakayama automorphism $\nu_{\widehat{B}_\sigma}$ of a repetitive category $\widehat{B}_\sigma$.

Definition 3.2. A self-injective algebra A is said to be *positive* if there is a positive Galois covering $\mathcal{C} \xrightarrow{G} A$ by a periodic K-category $\mathcal{C}$, and we denote the covering by $\mathcal{C} \xrightarrow{g} A$ simply when G is generated by a positive automorphism g.

For an algebra B and an automorphism σ of B, as shown in Proposition 2.1, $\widehat{B}_\sigma$ and $\widehat{B}$ are isomorphic. However, positive Galois coverings of a self-injective algebra A by $\widehat{B}_\sigma$ and $\widehat{B}$ are not necessarily isomorphic over A. In fact, coverings $F_{\widehat{B}_\sigma} : \widehat{B}_\sigma \xrightarrow{g} A$ and $F_{\widehat{B}} : \widehat{B} \xrightarrow{h} A$ are isomorphic over A if and only if σ is an inner automorphism (Ref. 26). See Ref. 26 [Sec. 5] for a uniqueness problem of positive Galois coverings.

Example 3.2. Let $K \ni \lambda \neq 0, \pm 1$, and let A be the self-injective algebra over K defined by the following quiver and relations

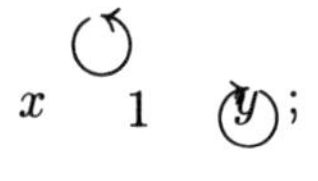

$$x \quad\overset{\circlearrowleft}{1}\quad \overset{\circlearrowright}{\textcircled{y}}; \qquad yx = \lambda xy, \; x^2 = 0, \; y^2 = 0,$$

and $B = K + Kx$ the subalgebra of A and σ_λ the automorphism of B with $\sigma_\lambda(1) = 1$ and $\sigma_\lambda(x) = \lambda x$. Then $A \cong B \ltimes (DB)_{\sigma_\lambda}$, but $A \not\cong B \ltimes (DB)$, as K-categories. Let $\mathcal{C}_\lambda$ be the locally bounded K-category given by the following quiver and relations:

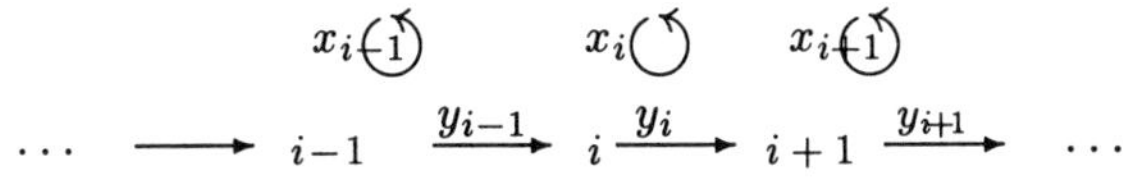

$$\cdots \longrightarrow \underset{i-1}{\bullet} \overset{y_{i-1}}{\longrightarrow} \underset{i}{\bullet} \overset{y_i}{\longrightarrow} \underset{i+1}{\bullet} \overset{y_{i+1}}{\longrightarrow} \cdots$$

$$y_i x_i = \lambda x_{i+1} y_i, \; x_i^2 = 0, \; y_{i+1} y_i = 0,$$

then it is a K-category with Nakayama automorphism $\nu_{\mathcal{C}_\lambda}$ such that $\nu_{\mathcal{C}_\lambda}(i) = i + 1$, $\nu_{\mathcal{C}_\lambda}(x_i) = x_{i+1}$, $\nu_{\mathcal{C}_\lambda}(y_i) = y_{i+1}$ $(i \in \mathbb{Z})$. It is easy to see that the positive Galois covering $F_{\mathcal{C}_\lambda} : \mathcal{C}_\lambda \overset{\nu_{\mathcal{C}_\lambda}}{\longrightarrow} A$ is isomorphic to the positive Galois covering $F_{\widehat{B}_{\sigma_\lambda}} : \widehat{B}_{\sigma_\lambda} \overset{g}{\longrightarrow} A$ where $g = \nu_{\widehat{B}_{\sigma_\lambda}}$. Moreover, $F_{\widehat{B}_{\sigma_\lambda}}$ is not isomorphic to the positive Galois covering $\widehat{B} \overset{\nu_{\widehat{B}}}{\longrightarrow} A$, while $\widehat{B}_{\sigma_\lambda}$ is isomorphic to $\widehat{B}$ as a K-category. See Refs. 26, 17 for more details.

Recall that an algebra Λ is said to be *symmetric* if there is a symmetric non-degenerate bilinear form $\Lambda \times \Lambda \to K$. This is equivalent to say that $\Lambda \cong D\Lambda$ as Λ-bimodules, and hence a symmetric algebra is self-injective, obviously.

Theorem 3.2. *A basic and connected symmetric algebra A is positive by a repetitive category if and only if A is a trivial extension of a basic and connected algebra.*

Proof. If A is a trivial extension of an algebra B, then there is a positive Galois covering $F_{\widehat{B}} : \widehat{B} \overset{\nu_{\widehat{B}}}{\longrightarrow} A$. Conversely, assume that there is a covering $F : \widehat{B} \overset{g}{\longrightarrow} A$, where $g = \varphi\nu_{\widehat{B}}$, that is, $A \cong \widehat{B}/\langle g \rangle$. It then follows from Ref. 17 that A is isomorphic to the trivial extension of B by DB. $\qquad\square$

This theorem shows that a positive symmetric algebra is a splittable Hochschild extension algebra of B by DB, because so is $B \ltimes DB$. The following example shows that not all non-symmetric Hochschild extension algebras are positive if the base field K is not algebraically closed. See Ref. 17 for details.

Example 3.3. Let $K = \mathbb{Z}_2(a,b,c)$ be the rational function field with three invariants a,b,c over the prime field $\mathbb{Z}_2$. Let $L = K[X,Y,Z]/(X^2 - a, Y^2 - b, Z^2 - c)$ be the factor ring of the polynomial ring $K[X,Y,Z]$ with three variables, and let $x = \overline{X}$, $y = \overline{Y}$ and $z = \overline{Z}$ where $\bar{f}$ denotes the residue class of $f \in K[X,Y,Z]$ in L. A 2-cocycle $\alpha : L \times L \to L$ of the K-algebra L is defined by the equality

$$\alpha(x^l y^m z^n, x^{l'} y^{m'} z^{n'}) = x^{l+l'-1} y^{m+m'-1} z^{n+n'-1}(lm'z + mn'xy),$$

where the numbers l,m,n,l',m',n' are 0 or 1. Then it is shown that the extension algebra A of L by the 2-cocycle α is a *non-symmetric* self-injective, and A satisfies the required property by Proposition 2.1, because L is a simple K-algebra.

Problem 1. Are there non-positive, non-symmetric Hochschild extension algebras by the standard duality modules over an algebraically closed field?

4. Criterion Theorems

A quasi-core C of a periodic K-category $\mathcal{C}$ is said to be a *core* of $\mathcal{C}$ if $\mathcal{C}(\nu_{\mathcal{C}}^i(x), \nu_{\mathcal{C}}^j(y)) = 0$ for all $x,y \in C$ and $i,j \in \mathbb{Z}$ with $i - j \geq 2$. For example, $(\widehat{B}_\sigma, \nu_{\widehat{B}_\sigma})$ has a core B_0.

Theorem 4.1. *For a basic and connected self-injective K-algebra A, the following assertions are equivalent.*

   (1) *A is positive.*
   (2) *A is positive by a twisted repetitive category of an algebra.*
   (3) *A is positive by a repetitive category of an algebra.*
   (4) *There is an ideal I of A such that, for some $e = e^2 \in A$,*
      (a) *the right annihilator $r_A(I)$ of I in A is eI,*
      (b) *the canonical algebra homomorphism $eAe \xrightarrow{nat} eAe/eIe$ splits.*

Moreover, in these cases, the factor algebra A/I may be taken as the algebra in (2) or (3).

An idempotent e is called a *residual identity* of A/I if e is a sum of possibly minimal number of orthogonal primitive idempotents with the property that the residue class of e in A/I is identity and $1 - e \in I$. A residual identity always exists for an ideal I, and it is uniquely determined within inner automorphisms. The following lemma is proved in Ref. 26.

Lemma 4.1. *Let A be a self-injective algebra, e a non-zero idempotent of A and I an ideal of A. Then e is a residual identity of the factor algebra A/I if $r_A(I) = eI$.*

Proof. It is easily shown that there are no nonzero idempotent e' of A such that $e'e = e' = ee'$ and $e' \in I$. Moreover, $1 - e$ belongs to I, because $l_A r_A(I) = I$ by a theorem of Nakayama (see Ref. 28 [Theorem 2.2.3]). $\square$

See Ref. 23 for other properties of the ideal I.

Theorem 4.2. *Let A be a basic and connected self-injective K-algebra having an ideal I with $r_A(I) = eI$ for a non-zero idempotent e in A. Then A is socle-equivalent to a positive self-injective algebra. Moreover, in the case when K is algebraically closed and A/I has no oriented cycles in the quiver, A is isomorphic to a positive self-injective algebra determined by A/I.*

Proof. This is essentially same as Ref. 24 [Theorem 4.1]. In fact, A is socle-deformed to a self-injective algebra satisfying (4) in Theorem 4.1. See Ref. 23 [Sec. 4]. $\square$

In the above theorem the positive self-injective algebra socle-equivalent to A is constructed concretely from A and I by *socle deformation.*

It should be noted that the annihilator condition $r_A(I) = eI$ does not imply the split-condition $(4)(b)$ in Theorem 4.1, so that the socle-equivalence in the above theorem is not replaced by isomorphism, in general. A counterexample is given in Refs. 24 [Example 4.2], 25. If the repetitive algebra of A/I is however given by an algebra without oriented cycles in the ordinary quiver, then the Hochshild module $H^2(eAe/eIe, eIe)$ is zero (23), and hence the criterion condition is described by the annihilator condition $(4)(a)$ only.

Theorem 4.3. *Let K be an algebraically closed field. Then there exists a positive Galois covering $F : \widehat{B} \to A$ by the repetitive category of an algebra B without oriented cycles in the quiver if and only if there is an ideal I of A such that $r_A(I) = eI$ for some $e = e^2 \in A$ and the quiver of A/I has no oriented cycles.*

There are socle-equivalent positive self-injective algebras whose Galois coverings are non-isomorphic repetitive algebras of connected algebras.

Example 4.1. Let A_1 and A_2 be the algebras defined by the following quiver Q and relations:

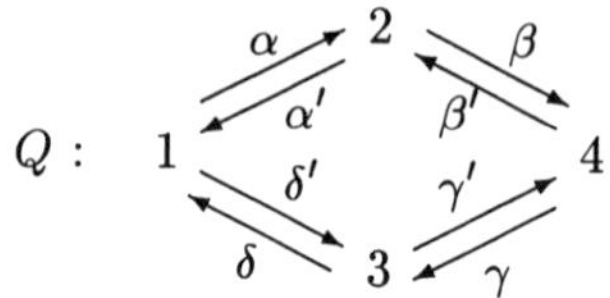

$$A_1 : \alpha\alpha' = \beta'\beta, \ \beta\beta' = \gamma'\gamma, \ \gamma\gamma' = \delta'\delta, \ \delta\delta' = \alpha'\alpha,$$
$$0 = \alpha\delta = \delta\gamma = \gamma\beta = \beta\alpha = \alpha'\beta' = \beta'\gamma' = \gamma'\delta' = \delta'\alpha',$$

$$A_2 : \beta\alpha = \gamma'\delta', \ \gamma\beta = \delta'\alpha', \ \delta\gamma = \alpha'\beta', \ \alpha\delta = \beta'\gamma',$$
$$0 = \alpha'\alpha = \beta'\beta = \gamma'\gamma = \delta'\delta = \alpha\alpha' = \beta\beta' = \gamma\gamma' = \delta\delta'.$$

Then A_1 and A_2 are socle-equivalent self-injective algebras with cubed zero radical. Their Nakayama permutations are $\begin{pmatrix} 1\ 2\ 3\ 4 \\ 1\ 2\ 3\ 4 \end{pmatrix}$ and $\begin{pmatrix} 1\ 2\ 3\ 4 \\ 4\ 3\ 2\ 1 \end{pmatrix}$, respectively. Let $I_i\,(i = 1, 2)$ be the ideals of A_i generated by $\alpha', \beta', \gamma', \delta'$ in A_i, respectively. The factor algebras $A_i/I_i\,(i = 1, 2)$ are isomorphic to the connected algebra B_1 whose quiver is the subquiver of Q with arrows $\alpha, \beta, \gamma, \delta$ and such relations that any composition of two arrows are zero. Moreover, it is easy to see that I_1 satisfies the condition (4) in the criterion theorem 4.1, but does not I_2 because $I_2{}^2 \neq 0$. Thus A_1 is a positive self-injective algebra by $\widehat{B_1}$, but A_2 has no positive Galois covering by $\widehat{B_1}$. On the other hand, let J_2 be the ideal of A_2 generated by $\delta, \delta', \beta, \beta'$. Then $J_2{}^2 = 0$ and J_2 satisfies the condition (4), so that A_2 is a positive self-injective algebra by $\widehat{B_2}$, where B_2 is the direct product of two copies of the algebra $B^{(0)}$ whose quiver is the subquiver of Q with arrows α, α' and relations $\alpha'\alpha = \alpha\alpha' = 0$.

A more general construction of algebras with decomposable covering spaces like A_2 in the above is given in the following.

Proposition 4.1. *Let C be a connected algebra, and let $B^{(0)}, \ldots, B^{(r)}$ be copies of C, and B the direct product of $B^{(i)}\,(i = 0, \ldots, r)$. Let σ be an automorphism of B whose restriction to each $B^{(i)}$ is an algebra isomorphism to $B^{(i+1)}$ for $i = 0, \ldots, r-1$ and to $B^{(0)}$ for $i = r$. Then the trivial extension algebra A of B by DB_σ is connected and has a Galois covering by $\widehat{B}$ with an admissible group generated by a single element.*

Proof. It is clear that the trivial extension algebra A of B by DB_σ is connected and has the Galois covering $\widehat{B_\sigma} \to A$. Hence, by Theorem 4.1, A

has a Galois covering by $\widehat{B}$ with an admissible group generated by a single element. $\qquad\qquad\square$

In the above example, it is not difficult to see that A_2 has a positive Galois covering by $\widehat{B^{(0)}}$ with admissible group generated by $\nu^2_{\widehat{B^{(0)}}}$. In fact, in Theorem 4.1, take $e = e_1 + e_2$ and I the ideal generated by β, δ', e_3 and e_4, which satisfy the condition (4) in Theorem 4.1.

For self-injective algebras we considered the positive Galois coverings with admissible groups generated by Nakayama-positive automorphisms. But there are important self-injective algebras with positive Galois coverings whose admissible groups are generated by an automorphism which is not necessarily Nakayama-positive. See Theorem 5.1 and Refs. 22, 5, for example.

Problem 2. Find a criterion theorem for a self-injective algebra to have a Galois covering by a repetitive algebra with infinite cyclic admissible group generated by an automorphism which is not necessarily Nakayama-positive.

References

1. M. Auslander, I. Reiten, S. O. Smalo, *Representation Theory of Artin Algebras*, Cambridge Studies in Advanced Math. 36, Cambridge University Press, 1995.
2. I. Assem, A. Skowroński, *On tame repetitive algebras*, Fund. Math. 142 (1993) 59–84.
3. J. Białkowski, A. Skowroński, *Selfinjective algebras of tubular type*, Colloq. Math. 94 (2002) 175–194.
4. J. Białkowski, A. Skowroński, *Socle deformations of selfinjective algebras of tubular type*, J. Math. Soc. Japan 56 (2004) 687–716.
5. R. Bocian, A. Skowroński, *Weakly symmetric algebras of Euclidean type*, J. reine angew. Math. (2005), in press.
6. R. Bocian, A. Skowroński, *Socle deformations of selfinjective algebras of Euclidean type*, Preprint (Toruń 2003).
7. K. Bongartz, P. Gabriel, *Covering spaces in representation theory*, Invent. Math. 65 (1982) 331–378.
8. O. Bretscher, C. Läser, C. Riedtmann, *Self-injective algebras and simply connected algebras*, Manuscripta Math. 36 (1981) 253–307.
9. P. Dowbor, A. Skowroński, *Galois coverings of representation-infinite algebras*, Comment. Math. Helv. 62 (1987) 311–337.
10. K. Erdmann, A. Skowroński, *On Auslander-Reiten components of blocks and self-injective biserial algebras*, Trans. Amer. Math. Soc. 330 (1992) 165–189.
11. R. Farnsteiner, A. Skowroński, *Classification of restricted Lie algebras with tame principal block*, J. reine angew. Math. 546 (2002) 1–45.

12. R. Farnsteiner, A. Skowroński, *The tame infinitesimal groups of odd characteristic*, Preprint (Toruń 2003).

13. P. Gabriel, *The universal cover of a representation-finite algebra*, In: Representations of Algebras, Lecture Notes in Math. 903, pp. 68–105, Springer Verlag, 1981.

14. D. Happel, *On the derived category of a finite-dimensional algebra*, Comment. Math. Helv. 62 (1987) 339–388.

15. D. Happel, *Auslander-Reiten triangles in derived categories of finite-dimensional algebras*, Proc. Amer. Math. Soc. 112 (1991) 641–648.

16. D. Hughes, J. Waschbüsch, *Trivial extensions of tilted algebras*, Proc. London Math. Soc. 47 (1983) 347–364.

17. Y. Ohnuki, K. Takeda, K. Yamagata, *Automorphisms of repetitive algebras*, J. Algebra 232 (2000) 708–724.

18. Z. Pogorzały, A. Skowroński, *Selfinjective biserial standard algebras*, J. Algebra 138 (1991) 491–504.

19. C. Riedtmann, Algebren, Darstellungsköcher, Überlagerungen und zurück, Comment. Math. Helv. 55 (1980), 199-224.

20. C. Riedtmann, *Representation-finite self-injective algebras of class A_n*, In: Representation theory, II (Proc. Second Internat. Conf., Carleton Univ., Ottawa, Ont., 1979), pp. 449–520, Lecture Notes in Math., **832**, Springer, Berlin, 1980.

21. C. Riedtmann, *Representation-finite self-injective algebras of class D_n*, Compositio Math. 49 (1983) 231-281 .

22. A. Skowroński, *Selfinjective algebras of polynomial growth*, Math. Ann. 285 (1989) 177–199.

23. A. Skowroński, K. Yamagata, *Socle deformations of self-injective algebras*, Proc. London Math. Soc. 72 (1996) 545–566.

24. A. Skowroński, K. Yamagata, *Galois coverings of selfinjective algebras by repetitive algebras*, Trans. Amer. Math. Soc. 351 (1999) 715–734.

25. A. Skowroński, K. Yamagata, *On selfinjective artin algebras having nonperiodic generalized standard Auslander-Reiten components*, Colloq. Math. 96 (2003) 235–244.

26. A. Skowroński, K. Yamagata, *Positive Galois coverings of selfinjective algebras*, Adv in Math., in press.

27. K. Yamagata, *Representations of non-splittable extension algebras*, J. Algebra 115 (1988) 32–45.

28. K. Yamagata, *Frobenius algebras*, In: Handbook of Algebra 1, Vol.1, Elsevier, 1996, 841–887.

THE CLEBSCH-GORDAN DECOMPOSITION FOR QUANTUM ALGEBRA $w\mathfrak{sl}_q(2)$

SHILIN YANG[*]

College of Applied Sciences
Beijing University of Technology, 100022, Beijing, P.R. China
E-mail: slyang@bjut.edu.cn

HONG WANG

China Civil Affairs College, 065201, P. R. China

The aim of this paper is to study $w\mathfrak{sl}_q(2)$-representations. It is classified that all finite dimensional integrable highest weight modules. The problem of decomposition of $V \otimes_{\mathbb{C}} W$ for two finite dimensional integrable modules V and W is also considered.

Introduction

Throughout, we assume that the basic field is the complex number field $\mathbb{C}$. All algebras, modules and vector spaces are over $\mathbb{C}$ unless otherwise specified. $\mathbb{N}$ denotes the set of non-negative numbers. Let q be a parameter with q being not a root of unity.

F. Li and S. Duplij [6] constructed a quantum algebra $w\mathfrak{sl}_q(2)$, which is generated by the four variables $E, F, K, \overline{K}$ with the relations:

$$K\overline{K} = \overline{K}K = J, \; JK = K, \overline{K}J = \overline{K} \tag{1}$$

$$KE = q^2 EK, \overline{K}E = q^{-2}E\overline{K} \tag{2}$$

$$KF = q^{-2}FK, \overline{K}F = q^2 F\overline{K} \tag{3}$$

$$EF - FE = \frac{K - \overline{K}}{q - q^{-1}}. \tag{4}$$

This is an interesting example of weak Hopf algebras in the sense of [4]. In the paper [6], the authors gave a description of the structure theory of

---

[*]Partially supported by the National Science Foundation of China (grant No. 10271014) and the Fund of Elitist Development of Beijing City (grant No. 20042D0501518)

$\mathfrak{wsl}_q(2)$ in detail, such as its basis, group-like elements, regular quasi-R-matrix and so on. However, the representation theory of $\mathfrak{wsl}_q(2)$ is not concerned.

The present paper is to study finite dimensional integrable highest weight modules of $\mathfrak{wsl}_q(2)$, then to consider the problem of decomposition of $V \otimes_\mathbb{C} W$ for two finite-dimensional integrable highest weight $\mathfrak{wsl}_q(2)$-modules V and W. The main results are Theorem 1.1 and Theorem 2.1. Theorem 1.1 is to classify all finite dimensional integrable highest weight modules of $\mathfrak{wsl}_q(2)$. The proof is similar to the classic one for the quantized enveloping algebra $U_q(\mathfrak{sl}_2)$ of the three complex semisimple Lie algebra. It is mentioned that not all finite dimensional indecomposable modules of $\mathfrak{wsl}_q(2)$ are irreducible. The result is different from that all finite dimensional indecomposable modules of $U_q(\mathfrak{sl}_2)$ are irreducible highest weight modules. Theorem 2.1 is devote to the Clebsch-Gordan decomposition for finite dimensional integrable highest weight modules of $\mathfrak{wsl}_q(2)$. From the result, we can conclude that the modules $V \otimes_\mathbb{C} W$ and $W \otimes_\mathbb{C} V$ are not isomorphic as $\mathfrak{wsl}_q(2)$-modules in general. This is different from the one of $U_q(\mathfrak{sl}_2)$. The proof of Theorem 2.1 is more difficult.

1. Finite dimensional integrable highest weight modules

As a generalization of Hopf algebra, the concept of weak Hopf algebra was introduced and studied in [4, 5]. In this sense, a weak Hopf algebra $(H, \mu, \eta, \Delta, \varepsilon)$ is both bialgebra and there exists a weak antipode $T \in \hom_k(H, H)$ of H such that $T * I * T = T$ and $I * T * I = I$, where I is an identity map of H and $*$ is the convolution product.

According to [6], the quantum algebra $\mathfrak{wsl}_q(2)$ is a weak Hopf algebra. The comultiplication Δ, the counit ε and the weak antipode T are given by the following formulas

$$\Delta(E) = 1 \otimes E + E \otimes K, \Delta(F) = F \otimes 1 + \overline{K} \otimes F, \tag{5}$$

$$\Delta(K) = K \otimes K, \ \Delta(\overline{K}) = \overline{K} \otimes \overline{K}, \tag{6}$$

$$\varepsilon(E) = \varepsilon(F) = 0, \varepsilon(K) = \varepsilon(\overline{K}) = 1, \tag{7}$$

$$T(E) = -E\overline{K}, T(F) = -KF, T(K) = \overline{K}, T(\overline{K}) = K. \tag{8}$$

It is noticed that $J \neq 0$. If $J = 1$, $\mathfrak{wsl}_q(2)$ is isomorphic to $U_q(\mathfrak{sl}_2)$. In the present paper, we always assume that $J \neq 0$ and $J \neq 1$. This means that K and $\overline{K}$ are both not invertible and $\mathfrak{wsl}_q(2)$ is not a Hopf algebra.

The following notations will be used in the sequel.

$$[m] = \frac{q^m - q^{-m}}{q - q^{-1}} \text{ for } m \geq 0, \quad [m]! = [1][2] \cdots [m],$$

$$[0]! = 1, \quad \begin{bmatrix} n \\ t \end{bmatrix} = \frac{[n]!}{[t]![n-t]!}.$$

The formula

$$EF^m = F^m E + [m] F^{m-1} \frac{q^{-(m-1)} K - q^{m-1}\overline{K}}{q - q^{-1}}$$

holds in $\mathfrak{wsl}_q(2)$.

Lemma 1.1. *Let V be a $\mathfrak{wsl}_q(2)$-module and $0 \neq v \in V$. If $Kv = \lambda v$ for some $\lambda \in \mathbb{C}$, then there exists a unique element $\bar\lambda \in \mathbb{C}$ such that $\overline{K}v = \bar\lambda v$. Precisely, if $\lambda \neq 0$, $\bar\lambda = \lambda^{-1}$; if $\lambda = 0$, then $\bar\lambda = 0$.*

Proof. Assume that $Kv = \lambda v$, we have

$$K\overline{K}Kv = \overline{K}\lambda^2 v = Kv = \lambda v.$$

If $\lambda \neq 0$, we get $\overline{K}v = \lambda^{-1}v$. If $\lambda = 0$, since $\overline{K}K\overline{K} = \overline{K}$, we have $\overline{K}K\overline{K}v = \overline{K}v$. Hence $\overline{K}v = 0$. The claim has been proved. $\square$

Let V be a $\mathfrak{wsl}_q(2)$-module and λ be a scalar, we denote V^λ the subspace of all vectors v in V such that $Kv = \lambda v$. The scalar λ is called a weight of V if $V^\lambda \neq \{0\}$. It is easy to see that

$$EV^\lambda \subseteq V^{q^2\lambda}, \ FV^\lambda \subseteq V^{q^{-2}\lambda}. \tag{9}$$

An element $v \neq 0$ of V is said to be a highest weight vector of weight λ if $Ev = 0$ and $Kv = \lambda v$. A $\mathfrak{wsl}_q(2)$-module is said to be a highest weight module of highest weight λ if it is generated by a highest weight vector of weight λ. We say that V is an *integrable* module if for any $0 \neq v \in V$, there exists a positive integer r_0 such that for all $r \geq r_0$, $E^r v = F^r v = 0$.

Let v be a highest weight vector of weight λ, set $v_0 = v$ and $v_p = \frac{1}{[p]!}F^p v$ for $p > 0$. We denote by V the vector space spanned by $\{v_i | i \geq 0\}$. It is straightforward to see that

$$Kv_p = \lambda q^{-2p}v_p, \overline{K}v_p = \bar\lambda q^{2p}v_p, \tag{10}$$

$$Ev_p = \frac{q^{-(p-1)}\lambda - q^{p-1}\bar\lambda}{q - q^{-1}}v_{p-1}, \tag{11}$$

$$Fv_{p-1} = [p]v_p \tag{12}$$

and V is a $\mathfrak{wsl}_q(2)$-module. The following is similar to [3, Proposition V.I.3.2].

Lemma 1.2. *Any nonzero finite-dimensional integrable $\mathfrak{wsl}_q(2)$-module V contains a highest weight vector. Furthermore, E and F are nilpotent as operators on V.*

Proof. Since V is finite-dimensional, there exists a nonzero vector w and a scalar λ such that $Kw = \lambda w$. If $Ew = 0$, the vector w is a highest weight vector and we are done. If not, let us consider the sequence of vectors $E^n w$ where n runs over the non-negative integers.

If $\lambda \neq 0$, According to (9), it is a sequence of eigenvectors with distinct eigenvalues. Consequently, there exists an integer n such that $E^n w \neq 0$ and $E^{n+1} w = 0$. The vector $E^n w$ is a highest weight vector. If $\lambda = 0$, we have to use the condition that V is integrable. In this case, there exists an integer n such that $E^n w \neq 0$ and $E^{n+1} w = 0$. The vector $E^n w$ is a highest weight vector.

In order to show that the action of E on V is nilpotent, let $v_1, \cdots, v_n$ be the basis of V over $\mathbb{C}$. By the assumption, there exist $r_1, \cdots, r_n$ such that $E^{r_i} v_i \neq 0$ but $E^{r_i+1} v_i = 0$. Let $r = \max\{r_1, \cdots, r_n\}$, then for all i, $E^{r+1} v_i = 0$ and it follows that $E^{r+1} V = 0$. Hence E acting on V is nilpotent. The same argument works for F. $\qquad\square$

The following lemma is similar to the case of $U_q(\mathfrak{sl}_2)$ (see [3, Theorem VI.3.5]).

Lemma 1.3. *(1) Let V be a finite-dimensional $\mathfrak{wsl}_q(2)$-modules generated by a highest weight vector v of weight $\lambda \neq 0$, then:*

(1) The scalar λ is of the form $\lambda = \varepsilon q^n$, where $\varepsilon = \pm 1$ and n is the integer defined by $\dim(V) = n + 1$.

(2) Setting $v_p = \frac{1}{[p]!} F^p v$. We have $v_p = 0$ for $p > n$ and, in addition, the set $\{v = v_0, v_1, \cdots, v_n\}$ is a basis of V.

(3) The operator K acting on V is diagonalizable with the $(n+1)$ distinct eigenvalues $\{\varepsilon q^n, \varepsilon q^{n-2}, \cdots, \varepsilon q^{-n+2}, \varepsilon q^{-n}\}$.

(4) Any other highest weight vector in V is a scalar multiple of v and of weight λ.

(5) The module V is simple.

(2) Two finite-dimensional $\mathfrak{wsl}_q(2)$-modules generated by highest weight vectors of the same weight are isomorphic.

Lemma 1.3 implies that, up to isomorphism, there exists a unique simple $\mathfrak{wsl}_q(2)$-module of dimension $n+1$ and generated by a highest weight vector of weight εq^n. We denote this module by $V_{\varepsilon,n}$. In this case, the formulas (10)-(12) can be written as follows for $V_{\varepsilon,n}$

$$Kv_p = \varepsilon q^{n-2p} v_p, \ \overline{K}v_p = \varepsilon q^{-n+2p} v_p,$$
$$Ev_p = \varepsilon\,[n-p+1]\,v_{p-1},$$
$$Fv_p = [p]\,v_{p+1}.$$

We have found that all finite-dimensional simple highest weight $\mathfrak{wsl}_q(2)$-modules under the assumption that $\lambda \neq 0$. In general, for any $\mathfrak{wsl}_q(2)$-module V, we denote by V_0 the subspace of all vectors in V such that $Kv = 0$. By Lemma 1.1, we have $\overline{K}v = 0$. It is easy to see that V_0 is in fact a $\mathfrak{wsl}_q(2)$-module. If $0 \neq v \in V_0$, then $Kv = \bar{K}v = 0$. Also,

$$EF^j v = \left(F^j E + [j]\,F^{j-1}\frac{q^{-(j-1)}K - q^{j-1}\overline{K}}{q - q^{-1}} \right) v = F^j Ev.$$

This means that for each pair (i,j), $E^i F^j v = F^j E^i v$.

Let $W(n)$ be the vector space spanned by the basis $\{v_i | 0 \leq i \leq n\}$. There is a $\mathfrak{wsl}_q(2)$-module structure on $W(n)$ defined by

$$Ev_i = 0,$$
$$Kv_i = 0,$$
$$Fv_i = v_{i+1} \text{ for } 0 \leq i \leq n - 1, \text{ and}$$
$$Fv_n = 0.$$

It is obvious that $W(n)$ is an indecomposable $\mathfrak{wsl}_q(2)$-module of dimension $n + 1$.

Theorem 1.1. *Assume that M is a finite dimensional integrable highest weight module of dimension $n + 1$. Then $M \cong W(n)$ or $M \cong V_{\varepsilon,n}$.*

Proof. It is easy to see that there exists an highest weight vector v_0 such that $M = \mathfrak{wsl}_q(2)v_0$ and

$$Ev_0 = 0, \quad Kv_0 = \lambda v_0.$$

If $\lambda \neq 0$, we have shown that $M \cong V_{\varepsilon,n}$ by Lemma 1.3. If $\lambda = 0$, M has to be spanned by $v_0, Fv_0, \cdots, F^n v_0, \cdots$. Moreover, we have $F^n v_0 \neq 0$ but $F^{n+1}v_0 = 0$ since M is integrable and $\dim M = n+1$. It is easy to see that $\{v_0, Fv_0, \cdots, F^n v_0\}$ is linear independent. Let $v_i = F^i v_0$. Then

$$Ev_i = EF^i v_0 = F^i(Ev_0) = 0, \quad Kv_i = q^{-2i}F^i Kv_0 = 0,$$

and $Fv_n = 0$, $Fv_i = v_{i+1}$ for $0 \leq i \leq n - 1$. It follows that $M \cong W(n)$. $\square$

Corollary 1.1. *Any finite dimensional integrable simple $\mathfrak{wsl}_q(2)$-module M is either $V_{\varepsilon,n}$ or $W(0)$.*

Proof. Indeed, since V is integrable, there exists a highest weight vector v_0 such that

$$Ev_0 = 0, \ Kv_0 = \lambda v_0.$$

Let V be the submodule generated by v_0. Since M is simple, we have $V = M$. It is obvious that if $\lambda \neq 0$, then $M \cong V_{\varepsilon,n}$. If $\lambda = 0$, then M is generated by $\{F^i v_0 | 0 \leq i \leq n \text{ for some } n\}$. If $n \geq 1$, then M has a non-trivial submodule. Therefore, $n = 0$ and it follows that $M \cong W(0)$. $\square$

Let V be the vector space with a $\mathbb{C}$-basis $\{X^i Y^j | 0 \leq i \leq m, 0 \leq j \leq n\}$. There is a $\mathfrak{wsl}_q(2)$-module structure on V defined as follows

$$K \cdot (X^i Y^j) = \bar{K} \cdot (X^i Y^j) = 0,$$
$$E \cdot (X^i Y^j) = X^{i+1} Y^j, \text{ for } 0 \leq i < m, \ E(X^m Y^j) = 0,$$
$$F \cdot (X^i Y^j) = X^i Y^{j+1}, \text{ for } 0 \leq j < n, \ F(X^i Y^n) = 0.$$

We denote this module by $M(m,n)$. It is noticed that $M(0,n) \cong W(n)$.

It is well known that there is a Graded Lex Order on $\mathbb{N} \times \mathbb{N}$. Precisely, given two pairs $(i_1, j_1), (i_2, j_2) \in \mathbb{N} \times \mathbb{N}$, we say that $(i_1, j_1) > (i_2, j_2)$ if and only if $i_1 + j_1 > i_2 + j_2$, or $i_1 + j_1 = i_2 + j_2$ and in the vector $(i_1 - i_2, j_1 - j_2)$, the left-most non-zero entry is positive.

Proposition 1.1. *The $\mathfrak{wsl}_q(2)$-module $M(m,n)$ is an indecomposable module. It is not simple if $(m,n) > (0,0)$.*

Proof. If $M(m,n)$ is decomposable. For example, $M = M_1 \oplus M_2$. Let $0 \neq v_1 \in M_1$,

$$v_1 = \sum_{(i,j)} a_{i,j} X^i Y^j.$$

We assume that (i_0, j_0) is the minimal in the index set. We get that

$$E^{m-i_0} F^{n-j_0} \cdot v_1 = a_{i_0,j_0} X^m Y^n \in M_1.$$

Hence $X^m Y^n \in M_1$. Similar argument shows that $X^m Y^n$ also belongs to M_2. This concludes a contradiction. Therefore, $M(m,n)$ is indecomposable.

If $(m,n) > (0,0)$, the non-irreducibility of $M(m,n)$ is obvious. $\square$

2. The Clebsch-Gordan decomposition

Let V and W be $\mathfrak{wsl}_q(2)$-modules. The tensor product $V \otimes W$ over $\mathbb{C}$ is again a left $\mathfrak{wsl}_q(2)$-module defined by

$$E \cdot (v \otimes w) = v \otimes Ew + Ev \otimes Kw,$$
$$F \cdot (v \otimes w) = Fv \otimes w + \overline{K}v \otimes Fw,$$
$$K \cdot (v \otimes w) = Kv \otimes Kw,$$
$$\overline{K} \cdot (v \otimes w) = \overline{K}\, v \otimes \overline{K}\, w.$$

for all $v, w \in W$.

We now prove the quantum version of Clebsch-Gordan formula for finite-dimensional integrable highest weight $\mathfrak{wsl}_q (2)$-modules. We denote by $V_n = V_{1,n}$ and

$$(n + 1)W(m) = \underbrace{W(m) \oplus \cdots \oplus W(m)}_{n+1 \text{ copies}}.$$

The main theorem of this section is as follows.

Theorem 2.1. *Let m, n be two nonnegative integers. Then there exist isomorphisms of $\mathfrak{wsl}_q (2)$-modules*

$$(1) \quad V_n \otimes V_m \cong \oplus_{\ell=0}^{\min(m,n)} V_{n+m-2\ell};$$
$$(2) \quad V_m \otimes W(n) \cong \oplus_{\ell=0}^{\min(m,n)} W(n + m - 2\ell);$$
$$(3) \quad W(n) \otimes V_m \cong M(m, n);$$
$$(4) \quad W(m) \otimes W(n) \cong (n + 1)W(m).$$

Proof. (1). For the proof of (1), it is more or less the same as [3, Theorem VII.7.1].

(2). It is noticed that

$$E(v_i \otimes w_j) = v_i \otimes Ew_j + Ev_i \otimes Kw_j = 0.$$
$$K(v_i \otimes w_j) = 0.$$

Recall that

$$\Delta(F^r) = \sum_{i=0}^{r} q^{i(r-i)} \begin{bmatrix} r \\ i \end{bmatrix} F^{r-i}\overline{K}^i \otimes F^i.$$

314

Therefore,

$$F^{m+n}(v_0 \otimes w_0) = \sum_{i=0}^{m+n} q^{i(m+n-i)} \begin{bmatrix} m+n \\ i \end{bmatrix} F^{m+n-i}\overline{K}^i v_0 \otimes F^i w_0$$

$$= [m-1]! \begin{bmatrix} m+n \\ n \end{bmatrix} v_m \otimes w_n \neq 0,$$

and

$$F^{m+n+1}(v_0 \otimes w_0) = 0.$$

In particular, we choose $m = 1$. Let $v = v_0 \otimes w_0$. We have $F^{n+1}v = [n+1]v_1 \otimes w_n \neq 0$ and $F^{n+2}v_0 = 0$. On the other hand, we have

$$F^n v = F^n(v_0 \otimes w_0) = \sum_{i=0}^{n} q^{i(n-i)} \begin{bmatrix} n \\ i \end{bmatrix} F^{n-i}\overline{K}^i v_0 \otimes F^i w_0$$

$$= q^{-n}v_0 \otimes w_n + q^{-(n-1)}[n]v_1 \otimes w_{n-1}.$$

Now we choose two $a, b \in \mathbb{C}$ and consider the element

$$w = av_1 \otimes w_0 + bv_0 \otimes w_1.$$

It is easy to see that

$$F^{n-1} \cdot w = a \sum_{u+v=n-1} \begin{bmatrix} n-1 \\ u \end{bmatrix} q^{uv+v} F^u \cdot v_1 \otimes F^v \cdot w_0$$

$$+b \sum_{u+v=n-1} \begin{bmatrix} n-1 \\ u \end{bmatrix} q^{uv-v} F^u v_0 \otimes F^v \cdot w_1$$

$$= bq^{-n+1}v_0 \otimes w_n + \left(aq^{n-1} + b\frac{q^{n-1} - q^{-n+1}}{q - q^{-1}} \right) v_1 \otimes w_{n-1}$$

and

$$F^n \cdot w = a \sum_{u+v=n} \begin{bmatrix} n \\ u \end{bmatrix} q^{uv+v} F^u \cdot v_1 \otimes F^v \cdot w_0$$

$$+b \sum_{u+v=n} \begin{bmatrix} n \\ u \end{bmatrix} q^{uv-v} F^u v_0 \otimes F^v \cdot w_1$$

$$= \left(aq^n + b\frac{q^n - q^{-n}}{q - q^{-1}} \right) v_1 \otimes w_n.$$

Now we take $w = av_1 \otimes w_0 + bv_0 \otimes w_1$ where $a = -(q^n - q^{-n})$ and $b = q^n(q - q^{-1})$. Then $F^{n-1} \cdot w \neq 0$ and $F^n \cdot w = 0$. Finally, we have to show that

$$\{v, Fv, \cdots, F^{n+1}v, w, Fw, \cdots, F^{n-1}w\}$$

is linearly independent. It suffices to show that $F^{n+1}v$, $F^{n-1}w$ are linearly independent. This is a obvious fact. Therefore, there are two sub-modules which are isomorphic to $W(n+1)$ and $W(n-1)$ and their sum is direct. For dimension reasons we get that

$$V_1 \otimes W(n) \cong W(n+1) \oplus W(n-1) \text{ for } n \geq 1.$$

It is obvious that $V_1 \otimes W(0) \cong W(1)$. Now we can apply the induction on m. We assume that the assertion is proved for $\leq m$. Consider the tensor product $V_1 \otimes V_m \otimes W(n)$. By the claim (1) and assumption, we have

$$\begin{aligned}
V_1 \otimes V_m \otimes W(n) &\cong [V_{m+1} \oplus V_{m-1}] \otimes W(n) \\
&\cong V_{m+1} \otimes W(n) \oplus V_{m-1} \otimes W(n) \\
&\cong V_{m+1} \otimes W(n) \oplus \bigoplus_{\ell=0}^{\min(m-1,n)} W(m-1+n-2\ell).
\end{aligned}$$

On the other hand,

$$\begin{aligned}
V_m \otimes V_1 \otimes W(n) &\cong V_m \otimes (W(n+1) \oplus W(n-1)) \\
&\cong V_m \otimes W(n+1) \oplus V_m \otimes W(n-1) \\
&\cong \bigoplus_{\ell=0}^{\min(m,n+1)} W(n+m+1-2\ell) \\
&\quad \oplus \bigoplus_{\ell=0}^{\min(m,n-1)} W(n+m-1-2\ell).
\end{aligned}$$

Noting that

$$V_1 \otimes V_m \otimes W(n) \cong V_m \otimes V_1 \otimes W(n)$$

and comparing the above identities we get that

$$V_{m+1} \otimes W(n) \cong \bigoplus_{\ell=0}^{\min(m+1,n)} W(n+m+1-2\ell).$$

Therefore,

$$V_m \otimes W(n) \cong \bigoplus_{\ell=0}^{\min(m,n)} W(n+m-2\ell)$$

for all non-negative integers m, n.

(3). Recall that $\Delta(E) = 1 \otimes E + E \otimes K$, $\Delta(F) = F \otimes 1 + \overline{K} \otimes F$, and $\Delta(K) = K \otimes K$. We consider the case $W(n) \otimes V(m)$. It is obvious that

$$E(w_i \otimes v_j) = w_i \otimes Ev_j + Ew_i \otimes Kv_j = [m - j + 1]w_i \otimes v_{j-1},$$
$$F(w_i \otimes v_j) = Fw_i \otimes v_j + \overline{K}w_i \otimes Fv_j = w_{i+1} \otimes v_j,$$
$$K(w_i \otimes v_j) = 0.$$

Consider the element $w_0 \otimes v_m \in W(n) \otimes V(m)$. It is noticed that

$$F^i E^{m-j}(w_0 \otimes v_m) = [j+1][j] \cdots [m-j]F^i w_0 \otimes v_j = [j+1][j] \cdots [m-j]w_i \otimes v_j.$$

This means that

$$W(n) \otimes V_m \cong M(m, n).$$

(4). Finally, let $\omega_i(j) = v_i \otimes w_j$ and W_j be the submodule generated by $\omega_0(j)$ for $0 \le j \le n$. It is easy to see that

$$E\omega_0(j) = E(v_0 \otimes w_j) = v_0 \otimes Ew_j + Ev_0 \otimes Kw_j = 0$$

and

$$K\omega_0(j) = K(v_0 \otimes w_j) = Kv_0 \otimes Kw_j = 0.$$

We also see that

$$F\omega_i(j) = F(v_i \otimes w_j) = Fv_i \otimes w_j + \overline{K}v_i \otimes Fw_j = v_{i+1} \otimes w_j = \omega_{i+1}(j).$$

Therefore, $W_j \cong W(m)$ for all $0 \le j \le n$. It is noticed that $\{\omega_i(j)|i = 0, \cdots, m; j = 0, \cdots, n\}$ is a basis of $W(m) \otimes W(n)$. Comparing the dimension of $W(m) \otimes W(n)$ with $(n + 1)W(m)$, we get that

$$W(m) \otimes W(n) \cong (n + 1)W(m).$$

The proof of the theorem is finished. $\qquad\square$

References

1. G. Böhm, F. Nill, and K. Szlachányi, Weak Hopf algebras I. Integral theory and C^*-structure, J. Algebra **221**, 385-438(1999).
2. T. Hayashi, An algebra related to the fusion rules of Wess-Zumino-Witten models, *Lett. Math Phys.* **22** (1991), 291-296.
3. C. Kassel, Quantum Groups, GTM 155, Springer-Verlag, 1995.
4. F. Li, Weak Hopf algebra and some new solutions of Yang-Baxter equation. *J. Algebra* **208**, 72-100(1998).
5. F. Li, Solutions of Yang-Baxter equation in endomorphism semigroup and quasi-(co)braided almost bialgebras. *Comm. in Algebra* **28**, 2253-2270(2000).
6. F. Li, S. Duplij, Weak Hopf Algebra and Singular Solutions of Quantum Yang-Baxter Equation, *Commun. Math. Phys.* **225**, 191-217(2002).

COMPUTATION OF THE PROJECTIVE DIMENSION OF FINITELY GENERATED MODULES OVER POLYNOMIAL RINGS

ZHONG YI*

Department of Mathematics, Guangxi Normal University,
Guilin, Guangxi, 541004, P. R. China
zyi@mailbox.gxnu.edu.cn

Let k be a field. Let $A = k[x_1, x_2, ..., x_n]$ be the polynomial ring over k of n indeterminators. Let M be a finitely generated A-module. By using the Groebner bases of modules, in this paper, we give an algorithm to compute the projective dimension of M.

1. introduction

Let k be a field. Let $A = k[x_1, x_2, ..., x_n]$ be the polynomial ring over k of n indeterminators. In [1] the theory of groebner bases of modules over A was developed. By using this theory, the homomorphism modules of any two finitely generated A-modules M and N was explicitely calculated. For a module M, the projective dimension of M is an important index of M in ring theory and module theory; but its definition and the description of its properties are all non-constructive. Usually, it is very difficult to know the concret value of the projective dimension of a giving module. The purpose of this paper is to give an algorithm, by which we can calculate the explicit value of any giving finitely generated module over the polynomial ring $A = k[x_1, x_2, ..., x_n]$. The main ideal is by using the theory of groebner bases of modules. We fulfill our task by solving the following questions. Also the following questions are interesting ones for their own right and may be used in the discuss of other questions related with module properties.

(i) (membership problem for arbitrary finitely generated modules) Let M be a finitely generated A-module. Let N be a submodule of M (denoted

*Supported by NSF of China (10271021), NSF of Guangxi (0135005), EYTP of MOE of China (2002-40)

318

by $N \preceq M$). For an arbitrary element $m \in M$, determine whether $m \in N$ or not.

(ii) (direct summand criterion) Let M be a finitely generated A-module and let $N \preceq M$, determine whether N is a direct summand of M or not.

(iii) (projectivity criterion) Determine whether M is a projective A-module or not.

(iv) (projective resolution and projective dimension) Calculate a shortest projective resolution of M and thus find out the projective dimension of M.

2. Main Results

In this section, we begin to study the four questions proposed in the above section one by one. We use [1] as the reference of groebner bases of modules, use [2] as a reference about results of modules and rings, and use [3] as a reference about homological concepts and properties. In this section we always suppose M is a finitely generated A-module and N is a submodule of M, if not otherwise stated. It is well-known that every finitely generated module is isomorphic to a factor module of a free module of finite rank. As it was done in [1], when we say that we have an explicitly given f.g. A-module M, we mean that we are giving $L = < a_1, ..., a_m >$ for explicit $a_1, ..., a_m \in A^s$ such that $M \cong A^s/L$ for some explicit isomorphism. When M is a submodule of A^s, M is explicitly given means we have explicitly given elements $m_1, ... m_t \in A^s$ such that $M = < m_1, ..., m_t >$, or more generaly, if we have an explicitely given submodule L of A^s, the submodule $M = < m_1 + L, ..., m_t + L >$ of A^s/L is explicitly given.

(i) Membership problem. Let s be a positive integer, let $M = < f_1, ..., f_m >$, where $f_1, ..., f_m \in A^s$, be a submodule of A^s. If $f \in A^s$, by using a Groebner bases of M, we can determine algorithmically whether $f \in M$ or not, see $[1, \S 3.6]$ for details. Now we consider the similar problems for arbitrary f.g. modules, not just for submodules of A^s. Let M be a f.g. A-module. As pointed out above, we may suppose that $M \cong A^s/L$, s is a positive integer, $L = < f_1, ..., f_m > \preceq A^s$. Suppose N is an explicitly given submodule of M, such that $N = < g_1 + L, ..., g_n + L >$, where $g_i \in A^s$. Let m be an explicitly given element of M, we may suppose that $m = f + L$, $f \in A^s$. We hope to determine whether $m \in N$ or not.

Let $K = < f_1, ..., f_m, g_1, ..., g_n >$. Then $L \preceq K \preceq A^s$ and $N = K/L$. By [2, 3.8 Corollary], we know that $m = f + L \in N$ if and only if $f \in K$. Since

$L \preceq K \preceq A^s$, by the discusion of $[1, \xi 3.6]$, we can determine algorithmically whether $f \in K$ or not. If $f \in K$, we can find elements $a_1, ..., a_m, b_1, ..., b_n \in A$, such that

$$f = a_1 f_1 + ... + a_m f_m + b_1 g_1 + ... + b_n g_n \qquad (1)$$

thus

$$m = f + L = b_1(g_1 + L) + ... + b_n(g_n + L) \qquad (2)$$

So we solved problem (i) algorithmically. It is just a simple generalization of $[1, \xi 3.6 \text{ (i)}]$.

(ii) Let M be a f.g. A-module and let $N \preceq M$. Now we determine algorithmically whether N is a direct summand of M or not. We recall that N is a direct summand of M, by definition, means that there exists a submodule N' of M such that each element m of M can be uniquely decomposed as a sum $m = n + n'$, where $n \in N$ and $n' \in N'$. There are also many equivalent descriptions about direct summands of modules, but they are all in the existence way, which are not constructive. We don't have an explicitly given method to judge whether a submodule is a direct summand or not yet. By using (i), which is based on the calculation of the groebner bases of modules, we can obtain such a criterion.

Lemma 2.1. *Let M be an A-module and let N be a submodule of M. Let $i : N \to M$ be the embedding map. Then $Hom_A(M, N)i$ is a submodule of $Hom_A(N, N)$ and N is a direct summand of M if and only if $1_N \in Hom_A(M, N)i$.*

Proof. It is obvious that $Hom_A(M, N)i$ is a submodule of $Hom_A(N, N)$. Since N is a direct summand of M if and only if the exact sequence

$$0 \longrightarrow N \overset{i}{\longrightarrow} M \longrightarrow M/N \longrightarrow 0$$

splits. This sequence splits if and only if there exists a homomorphism $\alpha : M \longrightarrow N$ such that $1_N = \alpha i$, this is equivalent to $1_N \in Hom_A(M, N)i$.

Theorem 2.2. *Let M be a f.g. A-module and let N be a submodule of M. We can determine whether N is a direct summand of M by the following procedure:*

(a) using the algorithm given in [1] *to calculate a representation of* $Hom_A(N,N) \cong A^s/L$;

(b) using the method givening in [1] *to calculate a representation of* $Hom_A(M,N)i$ *as a submodule of* $Hom_A(N,N) \cong A^s/L$, *that is, determine a submodule K of A^s such that $Hom_A(M,N)i \cong K/L$, where $L \preceq K \preceq A^s$;*

(c) using the method described in (i) to determine whether $1_N \in Hom_A(M,N)i$ or not;

(d) if $1N \in Hom_A(M,N)i$ is true, then N is a direct summand of M, otherwise N is not a direct summand of M.

The above theorem gives an explicit algorithm to determine whether a submodule of a f.g. module is a direct summand or not.

(iii) Now we determine whether a f.g. module is a projective module or not. At first, let's recall the definition and some equivalent characterizations of projective modules.

Definition 2.3. Let M be an R-module. If for each diagram

$$
\begin{array}{c}
M \\
\downarrow f \\
H \xrightarrow{\ \pi\ } K \longrightarrow 0
\end{array}
$$

of R-modules and $R-$homomorphisms, where π is an epimorphism, there exists an R-homomorphism $g : M \longrightarrow H$ such that $f = \pi g$, then M is called a projective R-module. The following theorem summarizes some characterizations of projective modules. It's a combination of [4, Theorem 2 on p.65], [4, Theorem 3 on p.66] and [4, Proposition 3 on p.135].

Proposition 2.4. *Let M be an R-module. The following are equivalent:*
(a) M is a projective module;
(b) M is a direct summand of a free R-module;
(c) $Hom_R(M,-)$ is an exact functor;
(d) $Ext^1_R(M,N) = 0$ for each R-module N.

Now we can given an algorithmically method to determine the projectivity of a finitely generated A-module M.

Theorem 2.5. *Let $A = k[x_1, x_2, ..., x_n]$. Let $M \cong A^s/L$ be a f.g. A-module, where $L \preceq A^s$. Then M is projective if and only if L is a direct* *summand of A^s. Thus giving a f.g. A-module $M \cong A^s/L$, using Theorem 2.2 to determine whether L is a direct summand of A^s, we can determine whether M is projective or not.*

Proof. It is clear that

$$0 \longrightarrow L \xrightarrow{i} A^s \xrightarrow{\pi} A^s/L \longrightarrow 0 \tag{3}$$

is an exact sequence. If $M \cong A^s/L$ is projective, then the above sequence is split. Thus L is a direct summand of A^s. Conversely, if L is a direct summand of A^s, then the above sequence splits. Thus $M \cong A^s/L$ is isomorphic to a direct summand of A^s, so M is projective.

(iv) Let $M \cong A^{s_0}/L_0$ be a f.g. A-module. We hope to fine a shortest projective resolution of M and thus determine the projective dimension of M. We may suppose $M \neq 0$, that is $L_0 \not\lneq A^{s_0}$. At first we have a short exact sequence

$$0 \longrightarrow L_0 \xrightarrow{i_0} A^{s_0} \xrightarrow{\pi_0} A^{s_0}/L_0 \longrightarrow 0 \tag{4}$$

Using Theorem 2.5 to determine whether M is projective or not. If M is projective, then

$$0 \longrightarrow M \xrightarrow{1_M} M \longrightarrow 0 \tag{5}$$

is a projective resolution of M.

Suppose M is not projective. Since $L_0 \not\lneq A^{s_0}$ is f.g., using the method given in [1, ξ3.7 pp.161-168] to compute the syzygy module of a finite base of L_0, denote it as L_1, then $L_0 \cong A^{s_1}/L_1$, where s_1 is an integer. So we have a short exact sequence

$$0 \longrightarrow L_1 \xrightarrow{i_1} A^{s_1} \xrightarrow{\pi_1} A^{s_1}/L_1 \longrightarrow 0 \tag{6}$$

Using Theorem 2.5 to determine whether $L_0 \cong A^{s_1}/L_1$ is projective or not. If L_0 is projective, then

$$0 \longrightarrow L_0 \xrightarrow{i_0} A^{s_0} \xrightarrow{\pi_0} M \longrightarrow 0 \tag{7}$$

is a projective resolution of M. If L_0 is not projective, since $L_1 \not\stackrel{<}{=} A^{s_1}$ is finitely generated, again using the method given in [1, §3.7 pp.161-168] to compute the syzygy module of a finite base of L_1, denote the syzygy module by L_2, then $L_1 \cong A^{s_2}/L_2$. So we have a short exact sequence

$$0 \longrightarrow L_2 \stackrel{i_2}{\longrightarrow} A^{s_2} \stackrel{\pi_2}{\longrightarrow} L_1 \longrightarrow 0 \tag{8}$$

Using Theorem 2.5 to determine whether L_1 is projective or not. If L_1 is projective then

$$0 \longrightarrow L_1 \stackrel{i_1}{\longrightarrow} A^{s_1} \stackrel{\pi_1}{\longrightarrow} A^{s_0} \stackrel{\pi_0}{\longrightarrow} M \longrightarrow 0 \tag{9}$$

is a projective resolution of M. If L_1 is not projective, then repeat the above procedure. Thus for each positive integer j, we have exact sequences

$$0 \longrightarrow L_j \stackrel{i_j}{\longrightarrow} A^{s_j} \stackrel{\pi_j}{\longrightarrow} L_{j-1} \longrightarrow 0 \tag{10}$$

and

$$0 \longrightarrow L_j \stackrel{i_j}{\longrightarrow} A^{s_j} \longrightarrow A^{s_{j-1}} \longrightarrow ... \longrightarrow A^{s_0} \longrightarrow M \longrightarrow 0 \tag{11}$$

By [4, Theorem 7 on p.182] we know that $gl.\dim.k[x_1, x_2, ..., x_n] = n$. Thus for each f.g. A-module M, $pr.\dim.M \preceq n$. Suppose that $pr.\dim.M = k \preceq n$. Using [4, §7.5 Theorem 11], we know that $L_0, ..., L_{k-2}$ can not be projective and L_{k-1} must be projective. Thus the smallest k such that L_{k-1} is projective determines the projective dimension of M and

$$0 \longrightarrow L_{k-1} \longrightarrow A^{s_{k-1}} \longrightarrow A^{s_{k-2}} \longrightarrow ... \longrightarrow A^{s_0} \longrightarrow M \longrightarrow 0 \tag{12}$$

gives a shortest projective resolution of M.

Summarizing the above analysis, we have

Theorem 2.6. *Let* $M \cong A^{s_0}/L_0$ *be a f.g. A-module. Using the procedure described above to determine exact sequences*

$$0 \longrightarrow L_j \xrightarrow{i_j} A^{s_j} \xrightarrow{\pi_j} L_{j-1} \longrightarrow 0 \qquad (13)$$

and

$$0 \longrightarrow L_j \xrightarrow{i_j} A^{s_j} \longrightarrow A^{s_{j-1}} \longrightarrow \ldots \longrightarrow A^{s_0} \longrightarrow M \longrightarrow 0 \qquad (14)$$

Then there exists a positive integer k ($\leq n$) such that L_{k-1} is projective.

Let k be the smallest k such that L_{k-1} is projective. Using Theorem 2.5 to determine whether M is projective. If M is projective, then

$$0 \longrightarrow M \xrightarrow{1_M} M \longrightarrow 0 \qquad (15)$$

is a projective resolution of M and $\mathrm{pr.\,dim\,}.M = 0$. If M is not projective, then

$$0 \longrightarrow L_{k-1} \longrightarrow A^{s_{k-1}} \longrightarrow A^{s_{k-2}} \longrightarrow \ldots \longrightarrow A^{s_0} \longrightarrow M \longrightarrow 0 \qquad (16)$$

is a shortest projective resolution of M and $\mathrm{pr.\,dim\,}.M = k$.

Remark 2.7. (a) In the special case of $M = A = k[x_1, x_2, ..., x_n]$. (i) is the ideal membership problem discussed in [5] and [1, §2.1(i)]. (ii) can only true when $N = 0$ or $N = M = A$, because A is an indecomposible A-module. (iii) and (iv) are trivial since A is projective.

(b) It is a conjecture of Serre that any f.g. projective module P over the polynomial ring $A = k[x_1, x_2, ..., x_n]$ in n commutative variables over a field k is, in fact, a free module. This conjecture was proved by Quillen and Suslin independtly in 1976, see [6, 3.25, p.67] for details. Thus the shortest projective resolution of M given in Theorem 2.6 is, in fact, a free resolution of M.

(c) In [1, Theorem 3.10.4], a totally constructive process was given to show that a f.g. module over $A = k[x_1, x_2, ..., x_n]$ has a free resolution of length less than or equal to n. But we don't know whether the free resolution given by the algorithm of [1, 3.10] is a shortest one or not. In Theorem 2.6 we give a shortest free resolution of M.

Let $A = k[x_1, x_2, ..., x_n]$. Let M be a f.g. A-module. We would like to mention the following questions, which are closely related with the questions we considered above.

324

(ii)' decide whether M is indecomposible or not;

(ii)" since M is f.g. and A is Noetherian, M has finite indecomposible decomposition. Find an algorithm to calculate such a finite indecomposible decomposition of M.

(iii)' determine whether M is injective or not.

(iv)' determine a shortest injective resolution of M and thus obtain the injective dimension of M.

Remark 2.8. (a) It seems that (iii)' and (iv)' are difficult, because usually injective modules are not f.g..

(b) Since in Noetherian rings f.g. flat modules are just f.g. projective modules, so the corresponding flat property is the same as projective property.

Acknowledgments

This work is completed while the author visited Institute of Mathematics and System Sciences, Chinese Academy of Sciences. The author is very grateful to Prof. Gao Xiaoshan and Prof. Li Ziming for their kindness and to the Mathematics Mechanization Research centre, the Academy of Mathematics and System Sciences for their support and hospitality.

References

1. W. W. Adams and P. Loustaunau, An Introduction to Groebner Bases, Graduate studies in Mathematics, Vol. 3, American Mathematical Society, 1994.
2. F. W. Anderson, and K. R. Fuller, Rings and Categories of Modules, New York Springer-Verlag Inc, New York, 1973.
3. J. J. Rotman, An Introduction to Homological Algebra, Academic Press, (New York), 1979.
4. D. G. Northcott, An Introduction to Homological Algebra, Cambridge University Press, 1960.
5. T. Becker and V. Weispfenning, Groebner Bases: A Computational Approach to Commutative Algebra, Springer Verlag, Berlin and New York, 1993.
6. C. Faith, Rings and Things and a Fine Array of Twentieth Century Associative Algebra, American Mathematical Society, 1999.